FORGING HANDBOOK

FORGING HANDBOOK

Thomas G. Byrer, Editor

S.L. Semiatin, Associate Editor

Donald C. Vollmer, Associate Editor

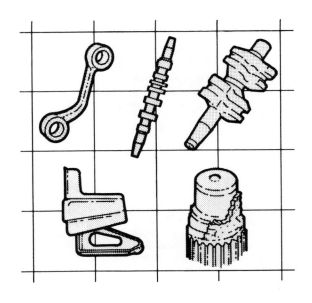

FORGING INDUSTRY ASSOCIATION

AMERICAN SOCIETY FOR METALS

**Library of Congress Catalog Number: 85-071789
ISBN: 0-87170-194-4
SAN: 204-7586**

Forging Industry Association is an Ohio Corporation not for profit composed of United States and Canadian producers of forgings and producers of raw materials, equipment, or major equipment components commonly used in the forging industry. The Association, with its predecessor organizations, has served the Forging Industry since 1913.
55 Public Square, Cleveland, Ohio 44113
216/781-6260

Founded in 1913, American Society for Metals is one of the world's largest technical societies, with over 52,000 members in 98 countries. A not-for-profit organization, ASM is dedicated to advancement of technical knowledge through the exchange of ideas and information on metallurgical and materials engineering.
Metals Park, Ohio 44073
216/338-5151

PREFACE

The preparation of any book of this nature involves the efforts of many people. My thanks first to the Handbook Committee* of the Forging Industry Association for their confidence and support throughout this endeavor. In particular, my acknowledgments to Bob McCreery, Chairman, and committee members John Kromberg and Al Sabroff for their special assistance and advice. A special thanks goes to Alex Clarke of FIA, who was a constant supporter throughout.

Obviously, a note of appreciation goes to all contributing authors who donated their time and talents to this effort. They are acknowledged in each section of the Handbook. Those who contributed photographs and slides for use in the Handbook are also to be thanked for their cooperation.

My associate editors bear special mention. Lee Semiatin was willing and obviously able to step in and add or revise portions of the Handbook when needed in addition to preparation of major subsections. Don Vollmer provided yeoman service from beginning to end, including the organizing of all photos and figures—a formidable task!

My thanks to Nancy Hill McClary of Battelle's Report and Library Services Department for her editorial and organizational efforts once the pieces began to come together to make the whole. As always, Brenda Block Koone provided invaluable assistance to the editor throughout the project. Her support is especially appreciated. Finally, a special thanks goes to Kathleen Mills at the American Society for Metals for helping us with many publication details.

Our objective was to compile a document that would reflect the dynamic nature of the U.S. forging industry and the importance of its products to our U.S. industrial base. This document was written to serve as a resource for expert and newcomer alike seeking knowledge and answers. Time will tell if we have succeeded.

T.G. Byrer
Columbus, Ohio

*FIA Handbook Committee: Robert H. McCreery (Chairman), Ervin Beatty, Norman F. Gustafson, J.F. Kromberg, William McElroy, Raymond W. Pollard, Alvin M. Sabroff, and Frank K. Smith

CONTENTS

ix

1 INTRODUCTION AND APPLICATIONS

Condensed Glossary of Forging Terms

A complete glossary can be found following Section 4. Illustrations courtesy of Forging Industry Association Handbook Committee.

Aircraft quality. Stock and forgings for aircraft and other critical applications produced under closely controlled melting and fabricating practices to minimize nonmetallic inclusions, segregation, and surface and internal flaws.

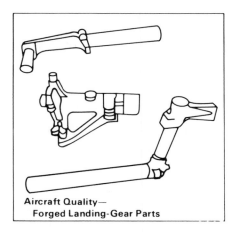

**Aircraft Quality—
Forged Landing-Gear Parts**

Air-lift hammer. A gravity drop hammer whose ram is raised for each stroke by an air cylinder.

Alloy. A metal containing additions of other metallic or nonmetallic elements to enhance its metallic properties. Commonly used to denote materials with relatively high amounts of alloying elements—for example, "alloy steels" as differentiated from carbon steels. Nearly all nonferrous applications involve al-

loys, sometimes with only small amounts of alloying elements.

Annealing. A heat treatment at a controlled temperature and cooling rate that reduces hardness and strength of a metal, improves formability, or develops a desired microstructure. Machinability may be improved or degraded, depending on the material involved.

Anvil. A large, heavy metal block that supports the frame structure and holds the stationary die of a forging hammer. Also, the metal block on which blacksmith forgings are made.

Austenite. A solid solution of iron and one or more alloying elements that is characterized by a face-centered cubic crystal structure. In the common engineering steels, the high-temperature austenite phase transforms into pearlite, bainite, or martensite, depending on the cooling rate. Certain grades of stainless steel are austenitic at room temperature by virtue of alloying and heat treatment.

Backward extrusion. Forcing metal to flow in a direction opposite to the motion of a punch or die.

Banded structure. A chemically segregated or aligned structure that parallels the direction of metalworking.

Batch furnace. A furnace for heating materials where all loading and unloading is done through a single door or slot.

Bed. Stationary platen of a press to which the lower die assembly is attached.

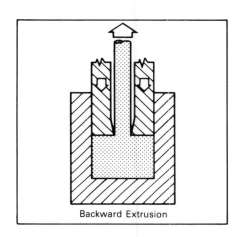

Backward Extrusion

Blank. A piece of stock (also called a "slug" or "multiple") from which a forging is made.

Blast cleaning. A process for cleaning or finishing metal objects with a shower of high-velocity abrasive particles (grit, sand, or shot).

Blocker-type forging. A forging that approximates the general shape of the final part with relatively generous finish allowance and radii. Such forgings are sometimes specified to reduce die costs where only a small number of forgings are desired and the cost of machining each part to its final shape is not excessive.

Blocker (blocking impression). The impression in the die (often one of a series of die impressions) that imparts the approximate shape to the part preparatory to forging to the final shape in the finisher dies.

Blow. The impact or force delivered by one work stroke of the forging equipment.

Board hammer. A gravity drop hammer where the ram is raised by attached wood boards. The boards are driven upward by action of contra-rotating rolls, then released. Energy for forging is obtained by the mass and velocity of the freely falling ram and the attached upper die.

Boss. A relatively short protrusion or projection on the surface of a forging, often cylindrical in shape.

Burnt. Permanently damaged metal caused by heating conditions that produce incipient melting or intergranular oxidation.

Carbon steel. Steel that derives its properties mainly from the addition of carbon, without substantial amounts of other alloying elements.

Case hardening. A process in which the surface layer or "case" of a ferrous alloy is made substantially harder than the interior or "core." Typical methods are carburizing, cyaniding, carbonitriding, nitriding, induction hardening, and flame hardening. Commonly used to produce forged parts with a combination of good wear resistance and good toughness.

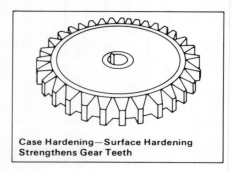

Case Hardening—Surface Hardening Strengthens Gear Teeth

Close-tolerance forging. A forging held to unusually close dimensional tolerances so that little or no machining is required after forging.

Coining. The process of lightly deforming all or some portion of a forged part to obtain closer tolerances or smoother surfaces or to eliminate draft.

Cold shut. A defect such as a lap that forms whenever metal folds over itself during forging. This can occur where vertical and horizontal surfaces intersect.

Cold working. Permanent plastic deformation of a metal at a temperature below its recrystallization point—low enough to produce strain hardening.

Compressive strength. The maximum stress that a material under compression can withstand without permanent deformation (or fracture, in brittle materials).

Conventional forging design. A forging characterized by design complexity and tolerances that fall within the broad range of general forging practice. These de-

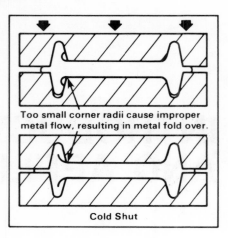

Too small corner radii cause improper metal flow, resulting in metal fold over.

Cold Shut

signs often require little or no machining, except where precision detail is required.

Counterblow forging equipment. Forging equipment with two opposed rams that simultaneously strike the workpiece from opposite sides.

Creep. Flow or plastic deformation of a metal subjected to long-term stresses below its normal yield strength.

Critical points. Temperatures at which metals undergo phase changes or transformations.

Dendrites. "Pine tree-shaped" crystals produced during freezing of an ingot.

Descaling. The process of removing oxide scale from heated stock prior to or during forging operations, using such means as light blows, wire brushes, scraping devices, or water spray.

Die blocks. The metal blocks, usually of heat-treated tool steel, containing the impressions that impart the desired shape during forging.

Die life. The productive life of a die impression, usually measured in terms of the number of forgings produced before the impression has worn beyond acceptable tolerances.

Die lubricant. A material sprayed, swabbed, or otherwise applied during forging to reduce friction and/or provide thermal insulation between the workpiece and the dies.

Die match. The alignment of the upper (moving) and lower (stationary) dies in a hammer or press. An allowance for misalignment (or mismatch) is included in forging tolerances.

Die set. A die holder with built-in guides to ensure alignment of mating dies and tools during operation.

Die shift. Misalignment of the top and bottom dies in the plane of the parting line that can occur during forging. This

condition must be corrected to maintain forging tolerances.

Die sinking. The process of machining impressions in die blocks.

Diffusion. The movement of atoms within a material, usually from regions of high concentration to regions of low concentration, to achieve greater homogeneity of the material.

Directional properties. Mechanical properties that vary with the direction when a material or part is tested, resulting from structural fibering.

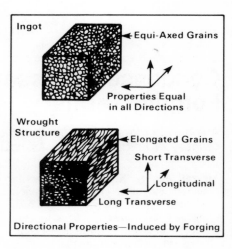

Directional Properties—Induced by Forging

Draft. Taper on the sides of a forging (and the forging die impression) that is necessary for removal of the workpiece from the dies—commonly between 5 and 7°.

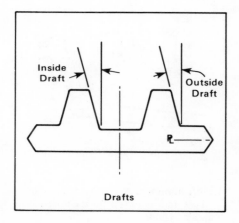

Drafts

Drawing out. A forging operation in which the cross section of stock is reduced and the stock lengthened between flat or simple contour dies.

Drop forging. The process of shaping metal between dies in a drop hammer, or the parts thus made.

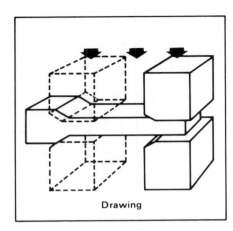

Drawing

Drop hammer. A general term applied to forging hammers where the energy for forging is provided by gravity, steam, or compressed air.

Drop Hammer

Ductility. The relative ability of metals to deform under stress without fracture.

Dye penetrant testing. Inspection methods for detecting surface flaws with penetrating liquids containing dyes or fluorescent materials.

Elastic limit. The maximum stress a metal can withstand without permanent deformation.

Electrical conductivity. The capacity of a material to conduct electrical current, sometimes used to measure the degree of aging in aluminum alloys.

Electroslag remelting (ESR). A metal refining process in which the metal in the form of a consumable electrode is

remelted through a layer of slag. ESR alloys have fewer inclusions, greater uniformity, increased soundness, improved forgeability, and superior properties.

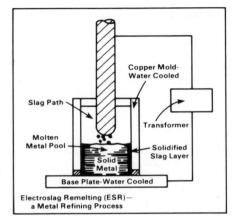

Electroslag Remelting (ESR)—
a Metal Refining Process

Elongation. A measure of the ability of a material to deform plastically in a uniform manner. Usually expressed as the percentage of permanent stretch before rupture in a tensile test.

Embrittlement. A loss of ductility that can occur in metals as a result of mechanical or chemical defects introduced during processing or environmental exposure.

Endurance (or fatigue) limit. The maximum stress a metal can withstand without failure under prolonged cyclic loading.

Equiaxed structure. Grains that are essentially spherical in shape.

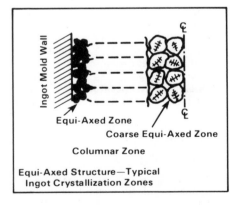

Equi-Axed Structure—Typical
Ingot Crystallization Zones

Fatigue. The tendency of metals to crack and break when subjected to cyclic stresses well below the ultimate tensile strength. Failure by fatigue is the most frequent cause of mechanical failure of structural and machine components.

Ferrite. A solid solution of iron and one or more alloying elements that is characterized by a body-centered cubic crystal structure. Ferrite is the principal constituent of cast irons and steels with very small amounts of carbon and other alloying elements.

Fiber stress. Localized stress at a point or line on a section over which stress is not uniform, such as the cross section of a beam under a bending load.

Fillet. The radius at the concave intersection of two surfaces.

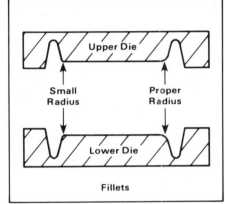

Fillets

Finish allowance. The amount of stock left on the surface of the forging for machining. Also called "machining allowance" or "forging envelope."

Finisher (finishing impression). The die impression that imparts the final shape to a forged part.

Flakes. Short, discontinuous internal fissures in ferrous metals caused by localized internal stresses during cooling after hot working. In some nonferrous metals, such as aluminum, flakes are small voids caused by hydrogen in the metal.

Flash. Excess metal that is extruded in a thin layer between the dies at the parting line and later removed by trimming. Flash helps control the flow of metal into the die cavities.

Flash extension. That portion of flash remaining on the part after trimming— usually included in the normal forging tolerances.

Flat die forging (open die forging). Forging between flat or simple contour dies by repeated strokes and manipulation of the workpiece. Also known as "hand" or "smith" forging.

Flow stress. The stress required to cause plastic deformation of metals.

Forgeability. The relative ability of metals and alloys to deform without rupture during forging.

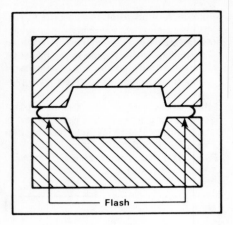

Flash

Forging stresses. Elastic stresses caused by forging, or cooling from the forging temperature. These stresses can be removed by subsequent heat treatment.

Fracture stress. The maximum load at fracture divided by the actual fracture area.

Free machining. A term used to describe metals that have alloying additions which reduce the tool force required in machining operations. Sulfur or lead in small amounts is used in "free machining" steels.

Grain. An individual crystal in a metal or alloy. See *Equiaxed structure*.

Grain flow. Fiberlike grain pattern developed by the shaping of metal in dies during forging. Grain flow imparted by good forging techniques enhances metal properties.

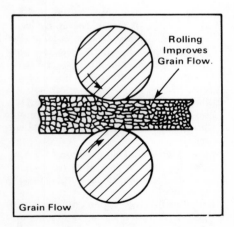

Rolling Improves Grain Flow.

Grain Flow

Grain size. The average area or volume of grains in polycrystalline metals; usually expressed as average diameter or number of grains per unit of area or volume.

Hardenability. The depth and degree of hardness that can be produced in a metal,

usually steel, by heating above the transformation temperature and quenching.

Hardening. Any process or treatment that increases the hardness and strength of a metal. The two most common methods are heat treatment and cold working.

Hardness. The resistance of a metal to indentation; most commonly expressed as numerical values derived from measurements on Brinell and Rockwell hardness testing devices.

Heading. An upsetting process used to form heads on the ends of rods or wire, as in bolt or rivet making.

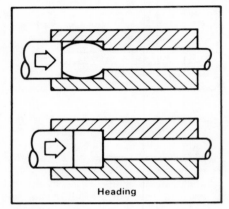

Heading

Heat. A term used to identify the material produced from a melting operation. Different heats of the same material can vary in chemical composition within prescribed limits. Stock from a single heat will have a consistent analysis and more uniform properties.

Heat treatment. A sequence of controlled heating and cooling operations applied to a solid metal to impart desired properties.

Homogenizing. A heat treatment at high temperature to reduce chemical segregation by diffusion of alloying elements.

Homologous temperature. The ratio of the metalworking temperature (T) to the melting point of the metal (T_m), both expressed in Kelvin (°C + 273). This ratio can be used to classify forging processes: cold, $T/T_m < 0.3$; warm, $T/T_m = 0.3$ to 0.5; and hot, $T/T_m > 0.5$.

Hot inspection. An in-process visual examination of forgings, using gauges, templates, or other nondestructive inspection methods to ensure quality.

Hot shortness. Brittleness in hot metal.

Hot working. The mechanical working of metal at a temperature above its recrystallization point—a temperature high enough to prevent strain hardening.

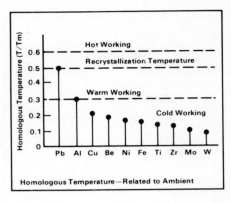

Homologous Temperature—Related to Ambient

Impact strength. A measure of the ability of a material to sustain high-velocity loading in the presence of a notch. The notched-bar impact strength of an alloy is the best single indicator of its engineering serviceability. Fiber structures developed in forging significantly influence the impact strength of the metal.

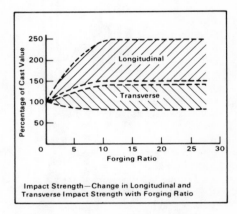

Impact Strength—Change in Longitudinal and Transverse Impact Strength with Forging Ratio

Impression die forging. Shaping of hot metal by compression completely within the cavities of two dies that enclose the workpiece on all sides.

Inclusions. Particles of nonmetallic compounds of metals and impurity elements that are present in ingots and are carried over in wrought products. The shape and distribution of inclusions are changed by plastic deformation and contribute to directionality in metals.

Ingot. A casting intended for subsequent rolling, forging, or extrusion.

Ironing. A press operation used to obtain a more exact alignment of the various parts of a forging or to obtain a better surface condition.

Isothermal. Constant and uniform temperature.

Isothermal forging. Forging process in which a constant and uniform temperature is maintained in the workpiece during forging by heating the dies to the same temperature as the workpiece.

Killed steel. Steel treated with additions of silicon or aluminum to the melt to minimize the oxygen content so that no reaction occurs between carbon and oxygen during solidification.

Lap. A surface defect appearing as a seam, caused by the folding over of fins or sharp corners and their subsequent rolling or forging into the surface.

Longitudinal. The direction in a wrought metal product parallel to the principal direction of working. In forgings, this is usually the direction of the grain flow or fiber structure. See *Directional properties*.

Machinability. The relative ease with which materials can be shaped by cutting, drilling, or other chip-forming processes.

Macrostructure. The structure and internal condition of a polished and etched metal surface observed with the naked eye or under low magnification up to about 10×.

Magnetic particle testing. A nondestructive method of surface and subsurface inspection of ferromagnetic materials. The metal is magnetized and coated with iron powder, which adheres to lines of magnetic flux leakage caused by defects. Commonly used to inspect forgings for seams, cold shuts, cracks, etc.

Malleable. The ability of a metal to be deformed by rolling or forging.

Mandrel forging. The process of rolling and forging a hollow blank over a mandrel in order to produce a weldless ring or heavy-wall tube.

Martensite. An unstable structure in quench-hardened steel that has the maximum hardness of any of the transformation products of austenite. Tempering stabilizes the martensitic structure and produces useful engineering properties.

Matching draft. Adjustment of draft angles (usually increased) on parts with unsymmetrical ribs and side walls to make the surfaces of the forging meet at the parting line.

Mechanical properties. Measurements that describe the reaction of a material to applied force, or the relationship between stress and strain—for example, the modulus of elasticity, tensile strength, and impact strength.

Mechanical working. Subjecting metal to pressure exerted by rolls, hammers, or presses to change its shape or physical properties.

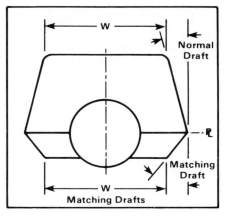

Microstructure. The structure and internal condition of metals observed by microscopic examination of a polished and etched surface at high magnifications.

Modulus of elasticity. The ratio of stress to strain within the elastic range; also known as Young's modulus. It is a measure of the stiffness of a material and its ability to resist deflection when loaded.

No-draft forging. A forged shape with extremely close tolerances and little or no draft, requiring a minimum of machining to produce the final part. Mechanical properties can be enhanced by this closer control of grain flow and retention of surface material.

Normalizing. A heat treatment for steels that involves heating above the transformation temperature and cooling in air to refine grain size for better response to hardening heat treatments, to improve machinability, or to provide desired mechanical properties.

Notch sensitivity. The loss of impact, endurance, or static strength of a metal that results from surface defects such as deep

Mechanical Press

scratches or notches that act as stress raisers.

Notch toughness. The resistance of a metal to crack propagation under applied stress. In fracture mechanics, notch toughness measurements help predict the type and size of flaw that will cause fracture in service.

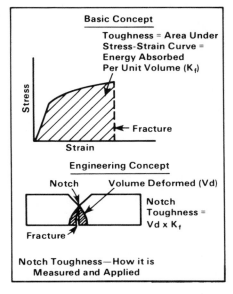

Basic Concept
Toughness = Area Under Stress-Strain Curve = Energy Absorbed Per Unit Volume (K_f)

Engineering Concept

Notch Toughness = $Vd \times K_f$

Notch Toughness—How it is Measured and Applied

Overheated. Metal with an undesirable coarse grain structure due to exposure to an excessively high temperature. Unlike a "burnt" structure, the metal is not permanently damaged but can be corrected by heat treatment and/or mechanical working.

Parting line. The line along the surface of a forging where the dies meet, usually at the largest cross section of the part. Flash is formed at the parting line.

Pearlite. An iron-carbon alloy (about 0.80% carbon in a plain carbon steel) with a structure of alternate layers of ferrite and iron carbide.

Phase. A physically homogeneous form of a metal or alloy.

Physical properties. Characteristics of materials that are of a basic nature, such as density and electrical conductivity.

Plastic deformation. The permanent change in the shape of material by the action of applied forces. The ability of metals to flow in a plastic manner without fracture is the fundamental basis for all metal-forming processes.

Plasticity. The ability of a metal to undergo permanent deformation without rupture.

Platter. The entire workpiece upon which the forging equipment performs work, including the flash, tonghold, and as many parts as are made at one time.

Powder metals. Metals and alloys in the form of fine particles (usually in the range of 1 to 1000 μm). Shaped objects can be produced from powders by compaction and bonding of the particles under high pressures and temperatures.

Precipitation hardening. A heat treatment that develops an increase in strength and hardness by precipitation of a constituent from a supersaturated solid solution. This is a common form of heat treatment for aluminum alloys.

Press forging. Shaping of metal between dies by mechanical or hydraulic presses. Usually this is accomplished with a single work stroke of the press at each die station.

Proof load. The load at which a product will just begin to permanently deform. Rated working loads of a part are commonly based on a percentage of the proof load. Since proof-load tests are nondestructive, they can be used to establish product integrity.

Proportional limit. The maximum stress that a material can sustain without deviating from a straight-line relationship between stress and strain.

Punch. The movable tool in a trimming press or upsetter.

Quenching. Rapid cooling from a high temperature by contact with liquids, gases, or solids. The cooling rate during quenching is important in heat treatment because it controls the degree of hardening of most alloys.

Ram. The dropping weight of a hammer or the power-actuated platen of a press to which the moving die is attached.

Recovery. Removal of residual stresses in strain-hardened metals without substantial change in grain structure.

Recrystallization. The formation of a new grain structure either by annealing a cold-worked metal or by heating a metal above its transformation temperature.

Reduction in area. A tensile ductility parameter that measures the ability of a material to deform plastically in a localized fashion. Generally stated as a percentage of original area of section.

Refractory metals. The group of metals with melting points above 3400 °F that offer the highest elevated-temperature engineering properties of commercially available metals; most commonly, columbium, tantalum, molybdenum, tungsten, and their alloys.

Residual alloys. Elements present in a metal in minute quantities but not added intentionally.

Residual stress. Stresses that are present in a free metal body, usually as a result of nonuniform deformation or severe

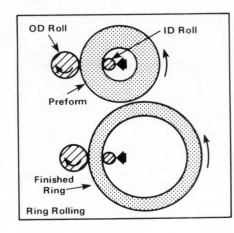

Ring Rolling

temperature gradients during quenching.

Ring rolling. A process of rolling a preform between two rotating rolls. This forging process reduces the ring wall section while increasing the diameter.

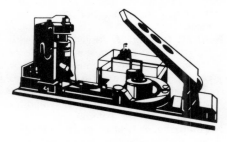

Ring Mill

Roll forging. A process of shaping stock between two driven rolls that rotate in opposite directions and have one or more matching sets of grooves in the rolls. Used to produce finished parts or preforms for subsequent forging operations.

Scale. The oxide surface layer that is formed by heating metals in an oxygen-rich atmosphere. Steel forging scale is loosely adherent and easily removed.

Screw press. A high-speed press in which the ram is activated by a large screw assembly that is powered by various types of drive mechanisms.

Segregation. A nonuniform distribution of alloying elements in a metal that occurs during solidification of an ingot or shape casting.

Shear. Deformation in which parallel planes within the metal are displaced by sliding but retain their parallel relation to each other.

Shearing. A process of mechanically cutting metal bars to the proper stock length for forging a part of a given weight.

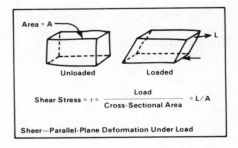

Sheer—Parallel-Plane Deformation Under Load

Shrinkage. The thermal contraction of metal during cooling after hot forging. Die impressions are made oversize according to precise shrinkage scales to meet design dimensions and tolerances.

Slot furnace. A common batch-type forge furnace where stock is charged and removed through a slot opening.

Solid solution. A solid metal in which an alloying element is dissolved without forming a second phase.

Solution heat treatment. Heating an alloy to a temperature at which a second element will form a solid solution with the base metal, and then rapidly cooling to retain the constituent in solution.

Spheroidizing. Prolonged annealing of steel that produces a rounded or globular form of the carbides present. A softening treatment commonly used prior to severe cold-working operations.

Steam hammer. A drop hammer where the ram is raised for each stroke by a double-acting steam cylinder. The energy delivered to the workpiece is supplied by the ram and attached upper die driven downward by steam pressure. Energy delivered during each stroke may be varied.

Stock. Raw material used to make forgings.

Strain. Deformation expressed as a pure (dimensionless) number—for example, the change in length divided by the original length.

Strain hardening. An increase in hardness and strength caused by plastic deformation below the recrystallization temperature (cold or warm working).

Strain rate. The rate of deformation—for example, the percent elongation or reduction per second.

Stress. Internal force reactions set up on a body when it is subjected to a load. Calculated by dividing the load by the cross-sectional area over which it acts.

Stress raisers. Design features such as sharp corners or mechanical defects such as notches that act to intensify the stress at these locations.

Stress relieving. Reducing residual stresses (recovery) in a metal object by con-

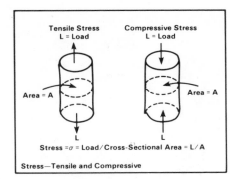

Stress—Tensile and Compressive

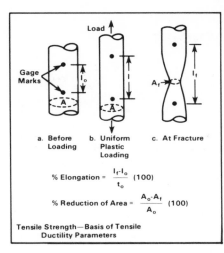

% Elongation = $\dfrac{l_f - l_o}{l_o}$ (100)

% Reduction of Area = $\dfrac{A_o - A_f}{A_o}$ (100)

Tensile Strength—Basis of Tensile Ductility Parameters

trolled heating and cooling without substantial changes in grain structure. Can sometimes be done mechanically by cold compression or stretching.

Superalloys. A term broadly applied to iron-base, nickel-base, and cobalt-base alloys, often quite complex, that exhibit high elevated-temperature mechanical properties and oxidation resistance.

Swage (swedge). Operation of reducing or changing cross-sectional area by revolving the stock under rapid impact blows.

Temper. The degree of hardening produced in a metal by heat treatment or cold working.

Tempering. Reheating a quench-hardened or normalized steel to a subcritical temperature to develop desired mechanical properties.

Tensile properties. Mechanical properties of a metal when loaded in tension, including tensile strength, yield strength, proportional limit, elongation, and reduction of area.

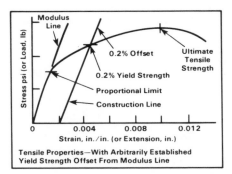

Tensile Properties—With Arbitrarily Established Yield Strength Offset From Modulus Line

Tensile (or ultimate) strength. The maximum stress a metal will withstand in tension. Calculated as the maximum load divided by the original cross-sectional area of a specimen pulled to failure in a tensile test.

Tolerance. The permissible deviation from a specification for any design characteristic.

Toughness. Ability of a metal to absorb energy without failure when a load is applied.

Transformation. A phase change that occurs in pure metals at a specific temperature and in alloys over a range of temperatures.

Transverse. The direction in a wrought metal product perpendicular to the principal direction of working. See *Directional properties*.

Trimming. The process of removing flash or excess metal from a forging. The part can be either hot or at room temperature.

Tumbling. A process for removing scale from forgings in a rotating container by impacting them with each other in a medium of abrasive particles.

Ultrasonic testing. A nondestructive inspection method for locating segregation or structural defects within a metal by means of sound waves.

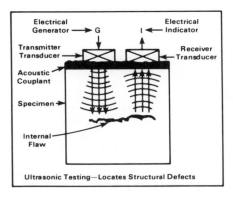

Ultrasonic Testing—Locates Structural Defects

Upsetter. A horizontal forging machine where the workpiece is gripped between two grooved dies and deformed by a punch that exerts force on the end of the stock.

Upsetter

Upsetting. Working metal to increase the cross-sectional area of a portion or all of the stock.

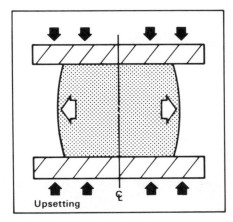

Upsetting

Vacuum refining. Melting and/or casting in vacuum to remove gaseous contaminants from a metal.

Warm working. Deformation at elevated temperatures below the recrystallization temperature. The flow stress and rate of strain hardening are reduced with increasing temperature; thus, lower forces are required than in cold working.

Yield point. The stress at which a marked increase in deformation occurs without increasing the load; observed in low- and medium-carbon steels.

Yield strength. The stress at which a material exhibits a specific amount of permanent deformation. In tensile tests, usually measured as the stress at 0.2% elongation.

Basic Forging Advantages

The process of hot working metals has long been used to ensure strength, toughness, reliability, and the highest quality in a wide variety of products. Today, these characteristics assume even greater importance as operating temperatures, loads, and stresses increase, and as reliability and toughness become more critical.

Products are being designed with forged components that can accommodate the highest possible loads and stresses. Recent advances in forging technology, such as the forging of previously "unforgeable" materials, have greatly increased the range of properties available in forgings. Economically, forged products are becoming even more attractive because of their inherent superior reliability, improved tolerance capabilities, and the higher efficiency with which forgings may be machined and further processed by automated methods.

Directional Strength

Controlling deformation during the forging process (usually at elevated temperatures) results in greater metallurgical soundness and improved mechanical properties of the material. In most cases, forging stock has been preworked to refine the dendritic structure of the ingot and to remove defects or porosity from the casting process. This produces directional alignment (or "grain flow") for important directional properties in strength, ductility, and resistance to impact and fatigue. As shown in Figure 1-1, these properties are deliberately oriented in directions requiring maximum strength. Working the material between impression dies also achieves recrystallization and grain refinement that yield the maximum strength potential of the material with the minimum property variation from piece to piece.

Properly developed grain flow in forgings closely follows the outline of the component (Figure 1-2). In contrast, bar stock and plate have unidirectional grain flow; any changes in contour will cut the flow lines, exposing grain ends, and render the material more liable to fatigue and more sensitive to stress corrosion. A casting has no grain flow or directional strength.

Because forgings are designed to approximate final part shape, they make better use of material than parts machined from bar stock or plate. How close a forging approximates the desired part configuration depends on a compromise between the cost for more elaborate dies versus the cost for subsequent machining.

Compared to welded fabrications, forgings are stronger because weld efficiencies are rarely 100% and welds are seldom

FIGURE 1-3. Forged fan hub of C1141 steel weighs 40 lb and has a $13^1/_2$-in. O.D.

totally free of porosity. Also, since a weld is a metallurgical notch, it requires elaborate inspection procedures in highly stressed components; a forging does not.

Structural Integrity

The degree of structural reliability achieved in a forging is unexcelled by any other metalworking process. There are no internal gas pockets or voids that could cause unexpected failure under stress or impact. Forgings achieve superior chemical uniformity by dispersing any massive segregation of alloys or nonmetallics that might otherwise cause an unpredictable response to heat treatment or jeopardize component performance under load.

For example, concern for reliability prompted the manufacturer of a reversible fan used on engines for off-highway equipment to convert to a forged fan hub, as shown in Figure 1-3. The previous hub (not a forging) performed according to design specifications, but part discontinuities or undetected internal flaws could have caused it to shatter were it struck by a foreign object during its 180 rpm rotation. At no increase in overall cost, the forged fan hub provides an extraordinary degree of safety due to its structural integrity.

To the designer, the structural integrity of forgings means realistic safety factors based on material that will respond predictably to its environment without costly special processing to correct for internal defects. To the production man, the inherent structural reliability of forgings means reduced inspection requirements, uniform response to heat treatment, and consistent machinability. All these factors contribute to faster production rates and lower costs.

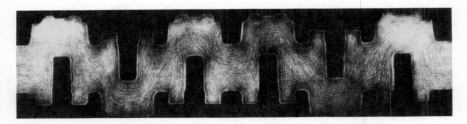

FIGURE 1-1. Polished and etched surface of sectioned crankshaft.

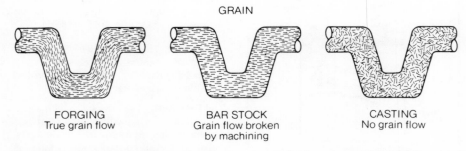

GRAIN

FORGING
True grain flow

BAR STOCK
Grain flow broken
by machining

CASTING
No grain flow

FIGURE 1-2. Schematic representation of grain structure in forging, bar stock, and casting.

Impact Strength

Designers and materials engineers are recognizing the increasing importance of resistance to impact and fatigue as a portion of total component reliability. Forged products have the toughness and strength to resist nominal loading, plus the ductility to resist failure under unexpected shock loading that may exceed design criteria. Through proper grain flow orientation, the forging process can develop the maximum impact strength and fatigue resistance possible in a material, with greater values than any other metalworking process and, thus, greater life expectancy.

The resulting higher strength-to-weight ratio can be used to reduce section thickness in part designs without jeopardizing performance characteristics or safety (Figure 1-4). Weight reduction, even in parts produced from less expensive materials, can amount to a considerable cost savings over the life of a product run.

Uniformity

The consistency of material from one forging to the next, and between separate quantities of forgings produced months or years apart, is extremely high. Forged parts are made through a carefully controlled sequence of production steps, rather than random flow of material into the desired shape. Uniformity in composition and structure, piece to piece and lot to lot, ensures reproducible response to heat treatment, minimum variation in machinability, and consistent property levels of finished parts.

Dimensional characteristics are also remarkably stable. In the truest sense, successive forgings are produced from the same die impression. Because die impressions exert positive control over all contours of the forged part, the possibility of transfer distortion is eliminated. For relatively long production runs, where grad-

ual wearing of the die impression can become a factor, corrective die maintenance procedures are established to hold any variations within tolerance limits.

Metallurgical Spectrum

Because virtually all metals can be forged, the range of physical and mechanical properties available in forged products spans the entire spectrum of ferrous and nonferrous metals.

Aluminum alloys are readily forged—primarily for structural and engine applications in the aircraft and transportation industries where temperatures do not exceed about 400 °F (200 °C). Forged aluminum combines low density with good strength-to-weight ratio.

Magnesium forgings are usually employed at service temperatures lower than 500 °F (260 °C) (certain alloys provide short-term service to 700 °F, or 370 °C). Magnesium forgings offer the lowest density of any commercial metal.

Copper, brass, and bronze are well suited to forging and are important for applications requiring corrosion resistance, and for their electrical and thermal conductivity.

Low-carbon and low-alloy steels comprise the greatest volume of forgings produced for service applications up to 900 °F (480 °C). Advantages are relatively low material cost, ease of processing, and good mechanical properties. The varied response of these steels to heat treatment offers the designer a wide choice of properties in the finished forging.

Special-alloy steels permit forgings with more than 300 ksi (or 300,000 psi) yield strength at room temperature. These steel forgings are important in transportation, mining, industrial and agricultural equipment, as well as high-stress applications in missiles and aircraft.

Stainless steels are widely used (particularly where corrosion is a problem) in

pressure vessels and steam turbines and many other applications in the chemical, food processing, petroleum, and hospital services industries. Stainless steel forgings are used for high-stress service at temperatures up to 1250 °F (675 °C) and low-stress service to 1800 °F (980 °C) and higher.

Nickel-base superalloy forgings are preeminent for service in the 1200 to 1800 °F (650 to 980 °C) range. They are particularly important for creep-rupture strength and oxidation resistance. Structural shapes, turbine components, and fittings and valves are common applications.

Titanium forgings are used primarily at temperatures to 1000 °F (540 °C) and are important for high strength, low density, and excellent corrosion resistance. Alloys of titanium offer yield strengths in the 120 to 180 ksi range at room temperature. Configurations nearly identical to steel parts are forgeable and 40% lighter in weight. Applications include aircraft engine components and structurals, ship components, and valves and fittings in transportation and chemical industries.

Refractory metal forgings have become important for high-temperature applications involving advanced chemical, electrical, and nuclear propulsion systems and flight vehicles. Columbium, molybdenum, tantalum, and tungsten and their alloys are now forged for applications requiring enhanced resistance to creep in high thermal environments.

Beryllium forgings are used primarily in nuclear, structural, and heat sink applications. Special forging techniques have been developed to process beryllium in sintered, ingot, or powdered form.

Zirconium and hafnium forgings are produced in relatively limited quantities and are used almost exclusively in nuclear applications.

For cryogenic applications, forgings of certain aluminum, austenitic stainless steel, titanium, and nickel-base alloys have the necessary toughness, high strength-to-weight ratios, and freedom from ductile-brittle transition problems.

Physical Limits

Forgings are produced economically in an extremely broad range of sizes, as illustrated in Figure 1-5. With the increased use of special punching, piercing, shearing, trimming, and coining operations, there have been substantial increases in the range of economical forging shapes and the feasibility of improved precision. However, parts with small holes, internal passages, reentrant pockets, and severe

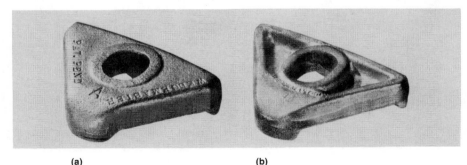

(a)   (b)

FIGURE 1-4. Forged truck wheel-rim clamp (a) of C1045 steel with reduced weight of 40% and increased strength of 2½ times more than previous clamp (b).

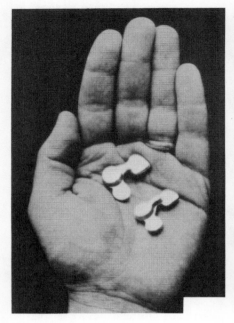

(a)

(b)

FIGURE 1-5. Forged parts ranging in size from (a) the less-than-an-inch desk calculator part to (b) the towering jumbo-jet engine support "banjo" weighing more than 5,000 lb.

draft limitations usually require more elaborate forging and tooling and more complex processing, and are therefore usually more economical in larger sizes.

Compatibility

Forgings are equal or superior to metal parts produced by other methods in their adaptability to other manufacturing processes.

Heat Treatment. The characteristically uniform refinement of crystalline structure in forged components ensures superior response to all forms of heat treatment, the maximum possible development of desired properties, and unequaled uniformity.

Welding. Forged components of weldable materials have a near absence of structural defects. (Note that the word porosity is not even associated with forgings.) Fine grain size is also inherent to the forging process. Therefore, the material at or near welding surfaces offers the best possible opportunity for strong, efficient welds by any welding technique.

Machining. The near absence of internal discontinuities or surface inclusions in forgings provides a dependable machining base for metal-cutting processes such as turning, milling, drilling, boring, broaching, and shear spinning and for shaping processes such as electrochemical ma-

chining, chemical milling, electric-discharge machining, and plasma-jet techniques.

Assembly. Forged parts are readily fabricated by welding, bolting, or riveting. More importantly, single-piece forgings can often be designed to eliminate the need for assemblies (Figure 1-6).

Surface Conditioning. In many applications, forgings are ready for use without surface conditioning or machining. How-

ever, forged surfaces are suited to plating, polishing, painting, or treatment with decorative or protective coatings.

Economic Advantages

The superior functional advantages of forgings invariably translate into economic benefits:

Source Flexibility. The forging process can be quickly responsive to variations in product demand once dies are available at the forge plant. A relatively large number of qualified forging sources also ensure best possible service.

Low Rejection Rates. Cost-conscious users of mechanical components are placing increasing importance on the low rejection rates they can obtain with forgings. It is common for an entire manufacturing run to be completed without a single rejected part. This is particularly significant when internal voids or flaws discovered during or after machining can mean a total loss of investment in machine time, labor, and tools. Also, the optimum responses of forgings to heat treatment mean reduced likelihood of rejections due to inconsistencies in property levels.

Machining Economy. Because dimensional relationships from one forging to another are not subject to wide or erratic variations, forgings lend themselves well to machining in fixtures with permanent settings. Setup and cutting efficiency can be optimized, and the possibility of operator error through slower target-locating methods is eliminated. Freedom from surface inclusions, particularly nonmetallics,

(a)

(b)

FIGURE 1-6. Gear and ring drive assembly converted from (a) a three-piece riveted and welded fabrication to (b) a two-piece weldment of forged parts.

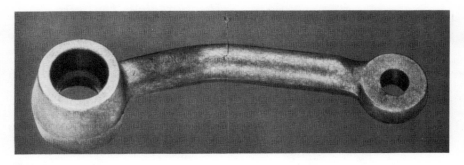

FIGURE 1-7. Pierced and coined bosses on this automobile idler arm forging cut drilling costs 20% and eliminated a broaching operation.

helps ensure decreased tool wear and minimum tool breakage, because sudden contact with subsurface voids or areas of differential hardness is avoided.

Weight Savings. Beyond the obvious weight savings from higher strength-to-weight ratios, other savings often result. A reduction in material can mean less surface area to machine. Piercing and trimming operations during forging can also reduce gross part weight while eliminating machining operations (Figure 1-7).

Applications: Past and Present

As Figure 1-8 shows, forging technology has progressed significantly in the 200-plus years since America was born. Blacksmiths were America's original forgemen and her earliest industry. As the nation moved westward and grew industrially after the War of Independence, enterprising blacksmiths opened the first true forge plants, equipping them with tilt hammers and hiring other blacksmiths to work for them. By 1860, drop forging was an established industry. Figure 1-9 illustrates a collection of early drop forgings.

The typical forge was equipped with a hearth fed by bellows, tongs, small anvils, and an elaborate hood and chimney. Typical products were needles, hunting knives, scissors, hearth fenders, nails, locks, spurs, stirrups, coats of mail, and candlesticks. Along with its use in weapons of war, wrought iron was used also in the decorative arts. Elaborate ornamental door hinges, gates, fences, balconies, lanterns, and torch holders, for example, were produced in great quantities. Important, too, were castings of bronze and brass, such as bells, guns, statuary, and doors.

Americans have depended on the forging industry for most of their basic tools since the days when the blacksmith made them by hand, one at a time, on his anvil. Tools were indispensible to westward-pushing pioneers who had to clear forests to plant crops and build houses, furniture, roads, and cities.

American Axes

In American forges the axe underwent a transformation from the Old World's ancient multipurpose tool to an entire range of specialized models: the huge broad axe for logging, the chisel-and-mortise axe, joiners' and shipbuilders' axes, and the axette (hatchet) (Figure 1-10). The American broad axe was the first to be made as a standardized, mass-produced, precision tool.

Hammers

The "king of tools," the hammer has always been used by its maker, the forgeman, to pound out hundreds of products, including other tools. Without the hammer's pioneering blows on anvils, and then on closed impression dies, American civilization could not have been fashioned (Figure 1-11).

Cutlery

Forged knives, shears, and scissors (Figure 1-12) have long held a competitive edge in quality. Many cutlery companies are really specialty forging companies. Until the early 20th century, production methods were similar to those used by blacksmiths, with suitable variations, such as hammering blades on indented surfaces called "bosses." One of today's leading cutleries, founded in 1848, began using power drop hammers in 1906 and revolutionized tailor's shears in 1912 by switching from malleable iron to forged steel handles.

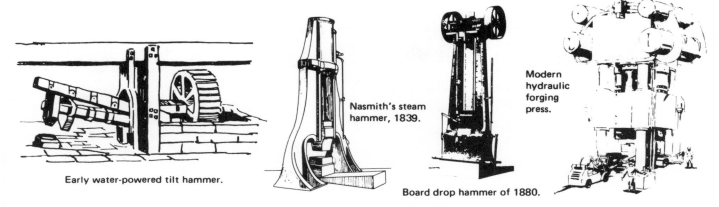

Early water-powered tilt hammer.

Nasmith's steam hammer, 1839.

Board drop hammer of 1880.

Modern hydraulic forging press.

FIGURE 1-8. 200 years of forging technology.

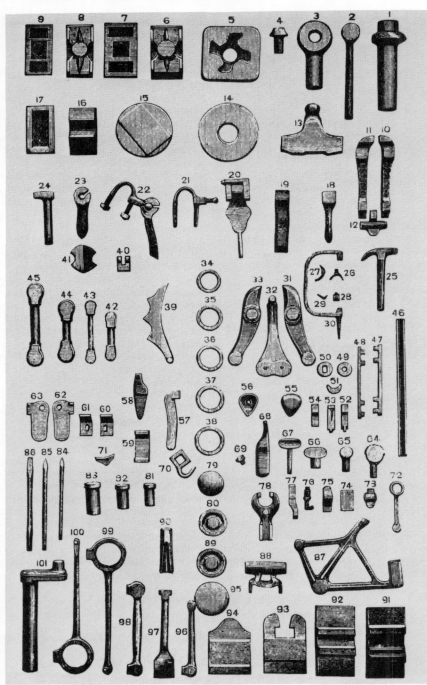

FIGURE 1-9. Collection of early drop forgings, which were small and of wrought iron. Many were harness and wagon forgings. Source: Ref. 1.

The Horseless Carriage

Many forging companies that made carriage hardware quickly adapted to the new "horseless carriages" and "gasoline buggies" of the 1880s and 1890s. The widespread acceptance of these newfangled contraptions depended heavily on strong, dependable forgings, such as the first heat-treated forged crankshaft (Figure 1-13) that appeared on the 1902 Locomobile and lasted the life of the engine—a phenomenon previously unheard of.

Low-cost, mass-produced cars became a reality in 1908 partly because the maker of the famous Model T decreed that all its stressed parts must be forged. Thus strong, tough, lightweight forgings helped put America on wheels.

Major Breakthroughs

Several breakthroughs occurred in the development of the forging industry. The first major technological breakthrough was precision forging dies, developed at the forge shop (later, arms factory) of Samuel Colt. These permitted the production of duplicate interchangeable parts at high speed.

The second breakthrough was the gravity drop hammer, developed almost simultaneously (1839) in England and America. Steam lifted the ram on the former; on the latter, rope twisted around a water-powered shaft lifted it.

In the 1860s came the board drop hammer, an American invention; and, in the 1880s, the hydraulic forging press. Both are still used today, technologically refined in numerous ways and, in some models, greatly increased in size.

With these improvements and refinements has come a vast array of forged products used today. Table 1-1 shows a listing of the ten major markets for forgings and typical forged parts. Figure 1-14 through 1-19 show the wide variety of products being made today.

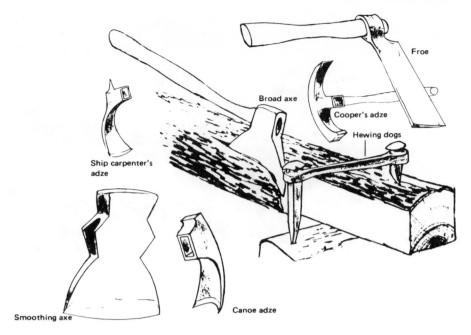

FIGURE 1-10. American axes.

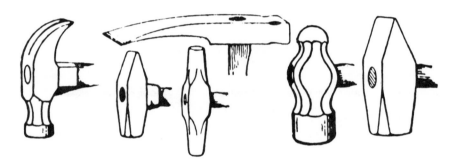

FIGURE 1-11. Types of hammers.

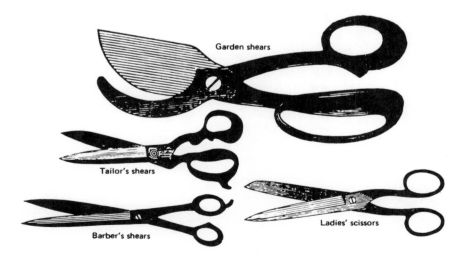

FIGURE 1-12. Types of cutlery.

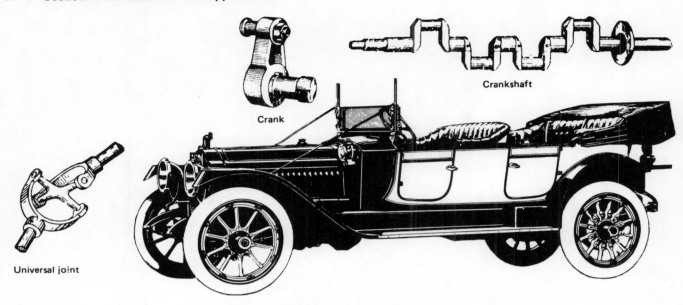

FIGURE 1-13. Early automobile and forged parts used in it.

TABLE 1-1. Major Markets for Forgings and Typical Forged Parts

Market	Typical forgings	Market	Typical Forgings
Aerospace (aircraft, aircraft engines, missiles)	Landing gear, airframe members, jet engine compressor blades, fuel nozzle supports, hot-gas manifolds	Railroad equipment	Rail joints and anchors, gear stops, wheels, axles, handbrake levers and gears
Automotive, truck, and trailer	Wheels, connecting rods, axle spindles, universal joints, truck door locks, trailer couplers, crankshafts	Oil-field machinery and equipment	Valve bodies, bonnets, and gates and retainers; drill bits and scraper blades; pipe tees, fittings, and threaded openings
Off-highway equipment	Rock-ripper teeth, wire-rope sockets, drilling-rig bit bases, dragline chains, crawler-tractor pivot (linkage) pins, trencher digging-wheel parts	Metalworking and special industry machinery	Mill-roll ends, gears, die cams, conveyor-line parts, covers, flanges, nozzles, mounting rackets, shafts, tire molds
Farm machinery and equipment	Sickle guards, cultivator shanks, tractor engine parts, plow-beam heads	Steam engines and turbines (except locomotives)	Pump casings, turbine discs and blades, heads, housings
Plumbing fixtures, valves, and fittings	Tees, elbows, valve bodies, fittings	Mechanical-power transmission equipment, including bearings	Ring forgings for bearing assemblies, rotor spiders and crowns
Ordnance (except missiles)	Artillery-shell canister and candle bases, tank-hull armorplate		

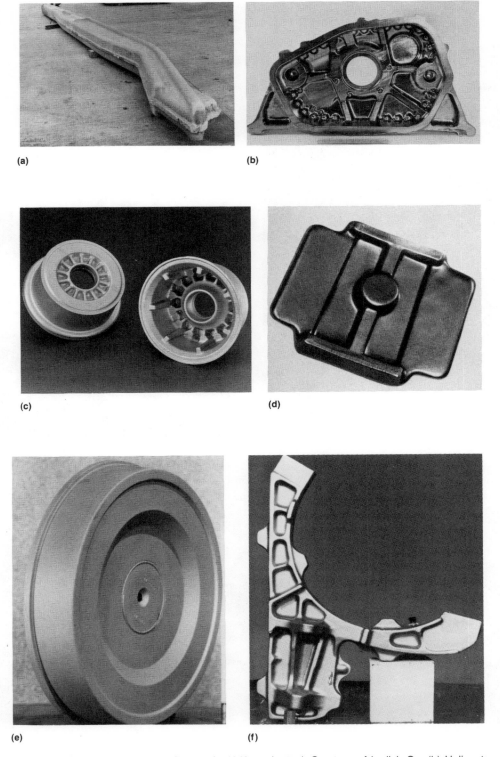

FIGURE 1-14. Aerospace applications. (a) Boeing 747 flap track, 4340 mod. steel. Courtesy of Ladish Co. (b) Helicopter gearbox housing, ZK60A-T5 magnesium alloy. Courtesy of Alcoa. (c) Aircraft wheels. Courtesy of Weber Metals. (d) Engine support, Waspaloy. (e) Isothermally forged JT80 second-stage fan disk, titanium alloy. Courtesy of Ladish Co. (f) F-15 airframe component, Ti-6Al-4V. Courtesy of Ladish Co.

(a)

(b)

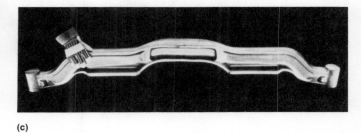

(c)

(d)

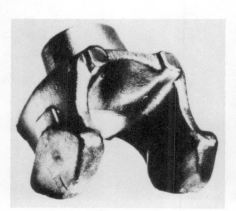

(e)

(f)

FIGURE 1-15. Automobile and truck applications. (a) Steering link, 4137 steel. Courtesy of Ladish Co. (b) Crankshaft. (c) Truck axle. Courtesy of T&W, Forging Division. (d) Truck bracket, 1030 steel. (e) Universal joint. Courtesy of Meadville Forging Co. (f) Forged gears. Courtesy of Eaton Corp.

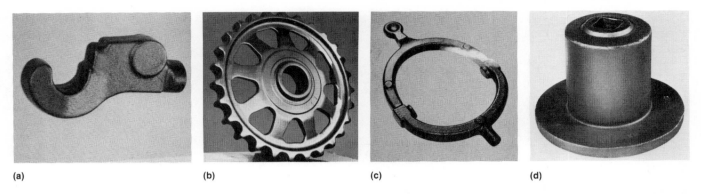

(a)   (b)   (c)   (d)

FIGURE 1-16. Off-highway equipment applications. (a) Tractor hitch link, A4145 steel. (b) Drive sprocket, magnesium-boron steel. Courtesy of Ladish Co. (c) Steering clutch release yoke, C1026 steel. (d) Steel truck assembly part. Courtesy of Canton Drop Forge Co.

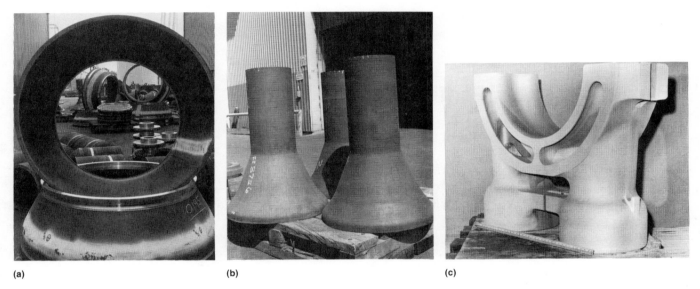

(a)   (b)   (c)

FIGURE 1-17. Steam turbine applications. (a) Turbine case. Courtesy of Cameron Iron Works. (b) Turbine shafts. Courtesy of Cameron Iron Works. (c) 410 CB nozzle box. Courtesy of Ladish Co.

(a)   (b)   (c)

FIGURE 1-18. Railroad and petrochemical applications. (a) Railroad coupler assembly. Courtesy of Meadville Forging Co. (b) A4130 steel oil field valve body. Courtesy of Interstate Southwest Forge Co. (c) Coupling fittings. Courtesy of Meadville Forging Co.

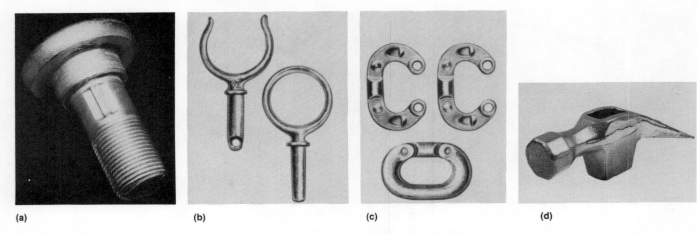

(a) (b) (c) (d)

FIGURE 1-19. Miscellaneous applications. (a) Cold forged lawn mower blade bolt. Courtesy of National Machinery Co. (b) Oar locks. (c) Chain links. (d) Hammer head. Courtesy of Meadville Forging Co.

Current Industry Profile

By Charles G. Scofield, Director of Marketing,
Forging Industry Association, Cleveland, Ohio

The State of the Industry in the 1980s

Today, the forging industry covers an extremely broad range of services and technologies available to the marketplace. It still meets the relatively "low tech/blacksmith" requirements of industry, but the focus of the '80s and '90s is on the high volume and/or high technology demands of users and potential users of forgings. The forged product is a part of just about everything used in daily life—albeit hidden, for the most part, from the untrained eye. Forgings are vital to washing machines, lawn mowers, hand tools, automobiles, trucks, buses, railroads, farm and off-highway equipment, gas and oil production, and aerospace and military hardware. The industry is small (somewhat more than $5 billion in 1983), but is vital to the well-being of the United States.

Forging Industry Concerns: A 20-Year Outlook

The industry has a number of concerns facing it in the next 10 to 20 years; it is important to note that in increasing numbers, forging executives are thinking and planning that far ahead. Such forward visibility is not an idle exercise, it is an absolute necessity. Survival is not the name of the game—prosperity is.

Some of the concerns and challenges are:

- Government regulations and controls
- Radical design changes in U.S./Canadian-produced passenger cars
- Insufficient new product development involving the use of forgings
- Viable alternatives to forgings—composites, ceramics, castings
- Foreign competition

Government Regulations and Controls. The United States Occupational Safety and Health Act of 1970 and its accompanying Environmental Protection Agency impact on U.S. industry was substantially accepted by industry for its intent: that is, to provide the workforce with a safe and healthy place of employment and to correct an acknowledged abuse of two natural resources—air and water. The problem has been not the philosophy of workplace safety and the protection of natural resources, but rather the cost of implementing the resulting laws and regulations. Few national and/or regional governments have imposed tighter controls in the areas of safety and natural resource protection than have the governments of the United States and Canada.

The price has been a heavy burden for the individual forging company to pay. The monies expended to meet regulation requirements for hearing protection, neigh-

borhood noise control, machine guards, and air and water cleanliness are on the order of hundreds of millions of dollars—money part of which, of course, would otherwise have been invested in plant expansion and modernization. Some metal-forming companies have simply gone out of business because the cost of bringing a plant up to regulation levels approached or exceeded the net worth of the company. For the most part, the industry has conformed to Occupational Safety and Health Administration as well as Environmental Protection Agency standards.

Radical Design Changes in Passenger Cars. Passenger car and truck applications have long been a major market for North American forgers. The typical eight-cylinder automobile produced by Chevrolet, Chrysler, or Ford in the 1970s contained some 70 forgings in the form of connecting rods, valves, axles, hubs, and tie rods. In a good production year of nine million vehicles, that represented significant tonnage of forged pieces (Figure 1-20).

To meet fuel efficiency standards, the auto industry was forced to reduce both the size and the weight of its vehicles. The effect on Detroit's suppliers, as Table 1-2 shows, was catastrophic. Over a 5-year period, Ford Motor Company cut the steel content of its average car by 31%, which included a permanent loss of forging business. Added to that was the fact that of the 5,750,000 passenger cars sold in the United States in 1982, almost 28% were imported. This resulted in further loss of business to North American forgers.

Insufficient New Product Development. The forging industry has been largely

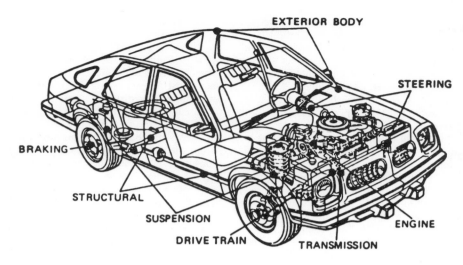

FIGURE 1-20. Applications for forgings in automobiles. Source: Ref. 2.

TABLE 1-2. Changes in Material Use in Automobiles (1977–1985)

Material	Material use by weight, lb 1977	1982	1985
Plastics	165	224	225
Aluminum	110	133	135
High-strength steel	105	252	270
Cold rolled steel	820	510	490
Hot rolled steel	1419	864	760
Cast iron	620	352	315
Other	521	369	338

Source: Ref. 2.

a reactive industry and not a proactive industry insofar as developing new uses for forgings. Despite the very best of intentions and some periods of forged product development from within, for the most part the forging industry waits for things to happen. It is imperative that the industry *begin* to *make* things happen, that new product design groups be penetrated even before new products get on the drafting board. The luxury of waiting to see what the design engineer has determined has disappeared. A strong influence on his process selection is a must. Sales training programs are increasingly emphasizing this fact.

Perhaps even more important is the theory of "controlling one's own destiny"— developing expanded services through marketing and engineering initiatives, such as machining, painting, assembling, packaging, distributing, and shipping. The proprietary product to which an individual company would have exclusive rights under worldwide patent laws is not a possibility to be overlooked.

Viable Alternatives to Forgings. Composites are an extremely exciting and viable threat to forgings—not only in aerospace applications, but in commercial forging applications as well. The United States is again considering the building of one or two hydraulic forging presses, each with a possible rated forging capacity of 200,000 tons. This is a controversial issue and has many facets, one being the use of composites as a substitute for large forged airframe parts. Neither the major potential users of forgings that could be produced from such large presses (airframe builders) nor the forgers themselves agree on the need, desirability, or practicality of

large presses; however, the potential of large composite airframe sections will be a significant factor in the final decision to proceed or to not proceed with the design and building of new presses at or near the 200,000-ton level.

The substitution of nonmetallics in important high-volume markets such as passenger cars and trucks is another threat to the forger. Interest is high in composite drive shafts and leaf springs for trucks. After these parts prove themselves, composites in glass fiber/resin matrices for structural truck parts have a tremendous future.

Yet another interesting development to watch is the "impossible plastic engine" (Figure 1-21). The two-liter, four-cylinder engine is fuel-injected, has a twin overhead cam cylinder head, and develops 318

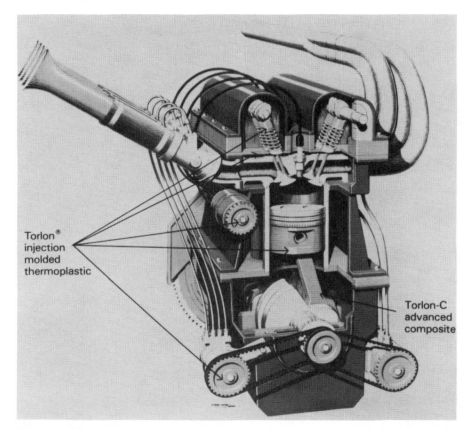

Torlon® injection molded thermoplastic

Torlon-C advanced composite

FIGURE 1-21. Schematic of the "impossible plastic engine." Reprinted with permission of Amoco Chemical Corp.

hp. It weighs 160 lb, about half the weight of its metal counterpart. Plastic is used for the timing gears, valve spring retainers, tappets, piston pins, intake valve stems, piston skirts, and connecting rods. The engine block is made of plastic reinforced with graphite fibers. A comparable, current production metal engine weighs 415 lb and develops a maximum of 188 hp.

Castings are a continuing threat to the forging industry. Foundry research and development efforts over the past 20 years have made tremendous strides in bringing high-technology casting potential to the level where it has gained the attention of design engineers responsible for even critical aerospace applications. The turbine blade business is testimony to that. For example, consider two case histories of conversions from forgings to castings (Figure 1-22). A large-volume steel forging was converted to a ductile iron casting at a part cost savings of 32%, and a high-technology nuclear application stainless steel forging was converted to a gray iron casting.

The same is true of ceramics. The jet engine industry has a great interest in material substitution because of the possibility of a shortage of critical materials needed to produce such sophisticated products. It is felt by many that there is a real and immediate need for American jet engine manufacturers to have alternatives to the traditionally used metal forgings. The U.S. Air Force and Navy assign a great deal of research and development money to this area of materials technology. If breakthroughs are achieved in these high-technology areas, commercial applications will follow.

Foreign Competition. There is a growing concern over the impact that imports are having in the U.S. and Canadian marketplace. A number of metalworking industries have joined together in what is known as the Metalworking Fair Trade Coalition. At present the Coalition is made up of 25 trade organizations representing more than $75 billion in sales revenue, some 25,000 plants, and normally employing almost 1.5 million persons.

The purpose of the Coalition is to attract the attention of the U.S. and Canadian governments to possible unfair trade practices on the part of some foreign companies and/or their governments. The Coalition now has the strength to gain government recognition, cooperation, and positive action to improve understanding of fair trade practices among world trading partners.

The Coalition is not protectionist; it supports free trade policies. However, to date, most foreign laws concerning trade are dramatically different from U.S. and Canadian laws. Our laws seemingly favor the importer. Some foreign governments not only permit but actually encourage and participate in dumping, subsidize export industries, grant ultraliberal credit terms, and manipulate exchange rates.

Foreign governments quite naturally want their companies to succeed, because they recognize that successful businesses mean jobs, taxes, and technical progress. They often play a coordinating role to systematically reach their goal: higher export volume of selected products. U.S. companies do not enjoy that sort of government support.

Fair trade simply has not existed. When a foreign metalworking company with its government's financial support has wanted to obtain a foothold in the North American market—the largest in the world—it has offered selling prices currently ranging from 20 to 40% below those of U.S. competitors. American purchasing agents, of course, have taken advantage of such drastic price reductions. Note that these are *price* reductions, not *cost* reductions. The costs are there; they simply cannot be avoided.

The Coalition represents many metalworking industries that were technological leaders in their respective fields. In some cases their brand-new automated equipment has literally been the best available. Yet, these efficient operations have commonly been underpriced by margins of 25% or more. This is not free trade, nor is it fair trade.

If this "mismatch" continues (i.e., the small U.S. or Canadian firm of 50 employees competing with a foreign firm backed by the economic power of its government), industry will rapidly deteriorate as a viable producer of durable goods. Such is the case with the U.S. fastener industry. Figure 1-23 shows import activity from 1960 to 1982. In that period, imported fasteners rose to a place of dominance in a critical industry. This came about by means of extremely clever marketing on the part of exporters and what has to be termed an indifferent attitude on the part of the U.S. government.

Conclusion

The forging industry is a maturing industry that must adjust to a variety of threats and opportunities. Like many basic indus-

```
Casting . . . . . . . . . . . . . Clutch cup
Application . . . . . . . . . Farm tractor power takeoff
Specification . . . . . . . . 100-70-03 ductile iron
Size . . . . . . . . . . . . . . 6 in. × 5 in.
Redesign . . . . . . . . . . From steel forging
Cost savings . . . . . . . . Part cost reduced 32%;
                              $112,000 annual savings
Design advantages . . . . . Substantial cost savings
                              while still retaining ten-
                              sile and yield strengths
Company . . . . . . . . . . International Harvester,
                              Rock Island, IL
Foundry . . . . . . . . . . . International Harvester,
                              Memphis, TN
```

```
Casting . . . . . . . . . . . Shielding plug
Application . . . . . . . . . Nuclear power reactor
Specification . . . . . . . . Class 40 gray iron
Size . . . . . . . . . . . . . . 27 in. × 4 in.
Redesign . . . . . . . . . . From stainless steel forg-
                              ing
Cost savings . . . . . . . . Part cost reduced 80%;
                              $200,000 annual savings
Design advantages . . . . Better friction properties
                              with no loss in shielding
                              characteristics
Company . . . . . . . . . . Canadian General Electric
                              Co., Peterborough, On-
                              tario
Foundry . . . . . . . . . . . Unknown
```

FIGURE 1-22. Two case histories of conversion from forgings to castings. Reprinted with permission of the Iron Castings Society.

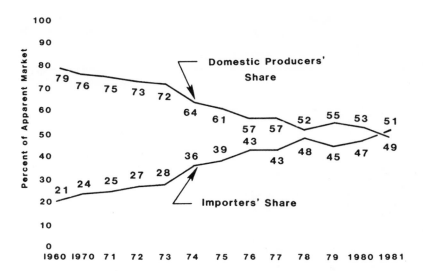

FIGURE 1-23. Domestic producers' and importers' shares of apparent U.S. market for bolts, large screws, and nuts by weight. Figures for 1981 are averages through the third quarter. Third quarter: importers, 56%; domestic producers, 44%. Source: Ref. 2.

tries of North America, the forging industry is going through a "change of life." There will be fatalities, there will be survivors, and there will be those who prosper. Those who prosper have their game plan in place today: vision, courage, financial credibility (whether small business or large), and leadership.

Future Trends

By Ian R. Williamson, Executive Vice President,
Forging Developments (International), Incorporated, Cleveland, Ohio

Contributing Author: G. W. Kuhlman,
Manager of Research and Advanced Technologies,
Alcoa Research Laboratories, New Kensington, Pennsylvania

The drop forging industry has been in existence since the end of the 19th century, when Nasmith invented and demonstrated his steam hammer. Since then, the industry has passed slowly through several phases of development, and recently has experienced a technological revolution. The effects, together with various economic pressures, are being felt by forging companies throughout the world. As the industry changes, some companies will cease to exist and others will grow in size, output, and stature.

These changes may be brought about by the adoption of one or more new processes. Thus, some forging companies realize that they are no longer safe, because the local market is expanding. The competition is coming not only from within the forging industry, but from other manufacturing companies as well. These com-

bined pressures will eventually force the most reluctant forging company to make its manufacturing techniques cost-competitive and/or quality-competitive with rival processes (Ref. 3).

The forging business, although a fairly basic industry, is spread across the entire spectrum of manufacturing: the automotive industry, the mining and energy industries, and the construction and aerospace industries. The development of forging in all of these industries will have a common thread, resulting in near-net shapes, closer manufacturing tolerances, and lower costs. Each industry will address this development in different ways, and, of course, there will be an overlap in the benefits from one industry to another.

Productivity will also be addressed through automation, including implementation of robots and computer-controlled

equipment. The computer will be the biggest tool in forging development, allowing the "art" to become a "science."

Producing parts close to finished configuration is the basic concept of near-net shape forming, as illustrated in Figure 1-24. Comparison of conventional forging with hot die forging shows the savings achieved in metal and machining. Isothermal forging could produce even closer dimensions.

Forging Products

Component analysis is an important factor in every industry. In industries where expensive raw materials are used, the goal has been to produce shapes as close as to the finished shape as possible. The automotive industry is producing smaller vehicles. In order to protect themselves against component failure, automotive companies are increasing the specification of parts to include smaller machining allowances, dimensional tolerances, and defect-free surface finishes. This puts a greater burden on the forger to improve his forging techniques, equipment, and, in particular, his quality-control system. Behind this trend are advances in both equipment and forging technology.

In the aluminum forging industry, the current practical limit for net aluminum precision forgings is 250 in.^{2} PVA (plan view area), with some parts, under very special conditions, reaching 300 to 350 in.^{2} Some recent novel research into aluminum alloy forging technology has provided the process capability that would extend the size of net aluminum forgings to 400 to 600 in.^{2} PVA. Furthermore, the technology will be scaled up to produce very large (up to 1,500 to 2,000 in.^{2} PVA) near-net shapes with a fraction of the machine stock typical of current conventional closed die aluminum forgings. The technology called superplastic aluminum forging will be particularly important to successful, cost-effective forging part fabrication in more expensive advanced aluminum alloys such as advanced P/M alloys, aluminum-lithium alloys, and discontinuous metal matrix composites.

In titanium technology (and superalloy technology as well, but to a lesser extent), advanced alloys with enhanced forgeability, advanced hot die forging process technology, and larger, specially equipped forging presses are contributing to both the enhancement of shape sophistication and the size of net and near-net shapes producible in these difficult-to-forge, expensive metals. Net shapes in titanium and superalloys are now practically limited to 150

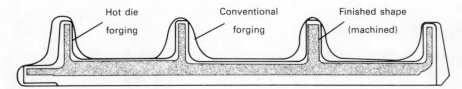

FIGURE 1-24. Capability of near-net shape forging concept. Courtesy of Wyman-Gordon Co.

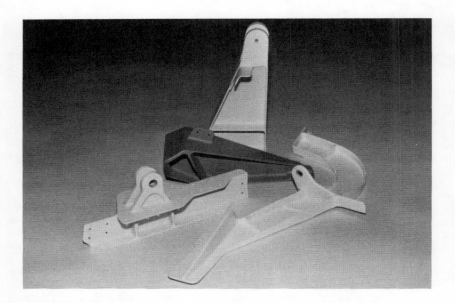

FIGURE 1-25. Net-shape aluminum aircraft forgings. Courtesy of Aluminum Precision Products.

(a)

(b)

(c)

FIGURE 1-26. Comparison of Ti-6Al-4V (a) conventional forging, (b) near-net-shape forging, and (c) machined part. Courtesy of Ladish Co.

to 200 in.² PVA. With advanced forging technology and equipment, this limit is being extended to 250 to 350 in.² PVA. In the future, titanium alloys are expected to be producible in near-net parts with PVAs two to three times this current limit.

Net and near-net forging technology is being combined with advanced materials to achieve shape sophistication at minimum cost. At the same time, the technology is being used with specially designed and controlled thermomechanical processing (TMP) technology to produce alloys with the best possible mechanical properties and optimum forging microstructural uniformity. Thus, advanced forging technology in aluminum-, titanium-, and nickel-base alloys will be employed not only to extend the size of net and near-net shapes, but also to achieve a degree of microstructural control and alloy optimization not achievable with conventional forging procedures.

The aluminum and titanium forgings shown in Figures 1-25 and 1-26 are examples of net shapes now available in difficult-to-forge materials.

The drive to net or near-net shapes in conventional ferrous forging is also prevalent. Another major net and near-net shape program in ferrous forgings includes bevel gears up to 20 in. in diameter for heavy-duty truck transmissions.

The limit to the near-net shape trend in the hot ferrous industry is the production of scale at the normal hot forging temperature. Therefore, the development of warm forming technology has filled the gap between the closer tolerance, but sometimes expensive, cold forging technology and the less accurate hot forging technology. Although still in its infancy, warm forging, which forms at temperatures between 1300 and 1800 °F (700 and 1000 °C), allows close-tolerance forging of steel alloys, which was previously economically unfeasible or impossible by the cold forging method. Figures 1-27 and 1-28 show a selection of warm forged parts.

The warm forming technique allows the production of parts that have been traditionally forged hot with large machining allowances to be forged very close to their final shape. Shafts, gears, and front-wheel-drive tulips are typical parts; as the technology develops, other components that are currently hot or cold forged will be warm formed. The process will also allow some parts that were previously forged by other processes because of lower costs or better quality to be recaptured by the forging industry.

FIGURE 1-27. Range of warm forged parts. Courtesy of E.P.A.G. (Walsall, England).

FIGURE 1-28. Warm forged sliding clutch gear, 7.25 kg. Courtesy of E.P.A.G. (Walsall, England).

Materials

Materials development for forgings will advance widely in virtually every base material—aluminum, titanium, alloy steel, superalloys, and even composites—with enhanced properties, more cost-effective parts, and superior performance. This is a double-edged sword to the forging industry. The development of new alloys allows traditionally forged parts to be manufactured by other processes, because acceptable mechanical properties are produced without the deformation caused by the forging process.

Advanced materials evaluated in terms of forgings are reviewed next. In many instances, these advanced materials are very difficult to successfully fabricate and present a challenge to the forger.

Aluminum Alloys. The development will progress along two major routes: ingot melted (I/M) and powder metallurgy (P/M) or rapid solidification, with both technologies producing advanced alloys to meet tomorrow's needs.

Aluminum-Lithium Alloys. Reduced density with excellent properties is the goal of I/M research and development. In forgings, 2xxx series aluminum alloys containing 1 to 3% lithium that possess tensile properties similar to 2024 or 7075, yet are 5 to 8% lighter, are now entering initial forging production and evaluation stages.

High-Strength Aluminum P/M Alloys. Alloys 7090 and 7091, the first generation of these types of materials, are now commercially available in forgings and offer 10 to 15% higher strengths than their I/M 7xxx alloy counterparts, with equivalent toughness and stress corrosion resistance. Second- and third-generation advanced 2xxx and 7xxx P/M alloys are now under development. These alloys will offer even better strength and toughness properties, with forgings as one of the primary products.

Elevated-Temperature Aluminum P/M Alloys. Emerging from recent research is an Al-Fe-Ce aluminum alloy (now called CU78). This alloy will demonstrate excellent elevated-temperature properties for service up to 550 to 650 °F (290 to 340 °C). CU78 and other elevated-temperature alloys based upon Al-Fe-Mo and Al-Fe-Ni are also being studied. However, they are dispersion-hardened and very difficult to fabricate where forgings have been found to optimize both part cost and alloy characteristics. These alloys may be competitive with some titanium alloys in gas turbines and other elevated-temperature applications.

Titanium Alloys. Beta and near-beta titanium alloys will be receiving increased research and development and production efforts. With their excellent combination of mechanical properties and superior amenity to hot die forging, they can be formed into cost-effective net and near-net shapes. Alloys such as Ti-10V-2Fe-3Al, Ti-17, and Transage 175 will be the subject of major production and research and development efforts. Ti-10-2-3 is one of the leading near-beta alloys and demonstrates particularly attractive properties—ultimate strength of 140 to 190 ksi and fracture toughness of 40 ksi $\sqrt{\text{in}}$. It has been successfully forged into several conventional and net forging shapes for commercial aircraft applications.

Other major alloy development efforts in titanium will be directed toward new alpha or near-alpha alloys (Ti-100, Corona 5, IMI 829) and titanium aluminides (TiAl, $TiAl_3$) for superior elevated-temperature or corrosion-resistance characteristics. Inherently, such alloys are extremely difficult to fabricate, require unique TMP schemes, and will be particularly demanding to produce in forged shapes.

Alloy Steels. Advanced melting technology has made both existing and new alloy steels with superior properties feasible for critical forging applications, particularly aerospace. Two examples of such grades which will find applications in forged components include HP310, a 310-ksi minimum ultimate strength alloy steel, and AF1410, a 240-ksi minimum ultimate strength specialty steel with fracture toughness of 120 to 150 ksi$\sqrt{\text{in}}$. These are but two alloys that are the product of alloy development that exploits advanced melting techniques, such as argon-oxygen decarburization, ladle metallurgy using rare earths and calcium treatments, vacuum induction melting (VIM), and vacuum arc remelting (VAR), to achieve superior alloys for severe service capabilities.

Superalloys. Advanced "super" superalloys, such as René 95, IN-100, and MERL76, are products of rapid solidification technology and unique consolidation and forging processes into net or near-net shapes. These materials, if made by conventional ingot techniques, are not forgeable; however, when processed through P/M routes, they can be made superplastic and forged under isothermal conditions in a vacuum to net shapes. Hot die and isothermal TMP forging processes will be used to expand the use of other superalloys, including nickel alloys 901 and 718, Waspaloy, Astroloy, and René 41, both to refine shape sophistication and to optimize properties via TMP.

Composites. Discontinuous metal matrix composites (DMMC), composed of silicon carbide or boron nitride whiskers, or a particulate in an aluminum alloy or titanium alloy matrix, are emerging from current research and development as potentially unique materials for a variety of applications requiring superior combinations of ultrahigh strength and high modulus of elasticity. Unlike most other composite types, DMMC is fabricable by all current hot working techniques, including forging. In fact, net and near-net DMMC is currently embryonic in its development, but will be a viable engineering material by the middle of this decade. Beyond DMMC are so-called metal glasses, solidified at such rapid cooling rates that normal crystal structures cannot form. The properties and fabrication of these unique "metal glasses" are not yet well understood.

Conclusion. With a few exceptions as noted, emerging advanced alloys and materials for forgings have much higher flow stresses but are much more difficult to successfully fabricate into forged shapes than are current alloys. This trend will continue, and it presents both a serious challenge and an opportunity for the forging producer. The opportunity lies in the application of new forging technologies to overcome this problem.

Difficult-to-fabricate alloys typically will demand slow or controlled strain-rate forging techniques and/or combinations of fabricating approaches, such as extrude plus forge, cast plus forge, hot isostatically press (HIP) plus forge, hot die, or isothermal. Most state-of-the-art press equipment being installed has such special capabilities that unique forging techniques, usually high temperatures and slow deformation rates, probably will emerge to meet the challenge of new alloys.

Productivity

The manufacturing industry in general is always seeking ways to improve productivity. Economic pressures dictate which segments of a manufacturing operation are highlighted for improvement or cost cutting—e.g., manpower, output per hour, uptime to downtime ratio, and number of operations. Because forging is a relatively

FIGURE 1-29. Automatic transfer system integrated into a forging press. Courtesy of National Machinery Co.

FIGURE 1-30. Robot-fed press setup. Courtesy of Letts Industries, Inc.

low-cost method of shaping metals and because there is an ever-increasing need to produce near-net shapes, the general trend toward more forging or forming operations is the basis for reducing the number of subsequent machining operations.

The pressure from alternative manufacturing processes on cost and quality will stimulate the need to reduce cost through improved productivity. The general emphasis in the Western industrialized areas will be to reduce manpower, raw material, and energy consumption while increasing output per hour, per machine, and per man. This will be achieved by the use of better design and technology and by mechanization. Design and technology will be formalized, with more emphasis being placed on the design, tooling, and forgings, which will in turn facilitate mechanized production.

The replacement of an operator, for cost reasons or because the task he performs is arduous or environmentally unacceptable,

is and will be the target for mechanization. Devices to automatically load furnaces, transfer forgings from the furnace to the first operating station, or even transfer forgings from all the forging stations are practical realities. Figure 1-29 shows an automated transfer system in a forging press. The combination of quick die-changing and setup techniques will allow the mechanized production of previously uneconomical small batch quantities (using programmable devices).

Industrial robot technology is no longer considered merely an interesting research project without any real applications. The technology has moved into the realm of actuality, with total acceptance in many production applications. Figure 1-30 shows a robot-fed press system.

The use of a robot or pick-and-place unit will eventually provide the mechanization the small batch forging producer requires. The key factor in this traditional area of the forging industry has been flexibility.

The development of reliable and simple handling devices will allow the producer of special, small-quantity forgings to improve productivity, quality, and hence cost effectiveness, while still remaining flexible. For forgings produced in large volumes, the trend toward high-speed and special-purpose equipment will continue to erode the traditional manual methods of production. Figure 1-31 shows highly automated high-production lines. Figure 1-32 shows a multiple punch and die system ready for replacement in a hot former.

The technology for the use of high-speed forging machines that can produce sometimes flashless, draftless, and close-tolerance forgings is continually developing, and methods are being devised to produce new parts. New equipment will be developed to produce small families of parts of similar size and shape. The forgings will be produced by utilizing several forming centers that are individually automated and synchronized by a central controller. Fig-

FIGURE 1-31. Hot former installation producing 100 wheel flanges per minute from hot rolled bar. Parts weigh up to 3.25 lb. Courtesy of Hatebur Metalforming Equipment, Ltd. (Basel, Switzerland) and Gerard Associates, Inc.

FIGURE 1-32. Four die and punch unit ready for hot former installation. Courtesy of Hatebur Metalforming Equipment, Ltd. (Basel, Switzerland) and Gerard Associates, Inc.

ure 1-33 shows an electrically controlled crankshaft and axle line.

Mechanized handling of large forgings will be seen for parts that have been traditionally handled by workers. Previously, forgings weighing up to 150 kg have been manipulated by hand with tongs and the aid of an overhead gantry. The unacceptable environment, fatigue, and difficulty in maneuvering heavy forgings will precipitate the use of programmable manipulating devices. The use of such devices will not only reduce the number of workers required to operate a forging line, but also allow an increased utilization of the equipment due to faster, more efficient movements. Figure 1-34 shows a mechanized line for bottle manufacture.

Computerization

The use of modern computer technology will provide the vehicle for forging technicians to scientifically solve forging problems. The previously identified trend toward net and near-net forging will in-

FIGURE 1-33. Fully automatic, electronically controlled forging plant for crankshafts (up to 250 kg) and truck front axles (up to 1,900 mm in length); output: 120 parts per hour. Courtesy of SMS Hasenclever (Dusseldorf, West Germany).

FIGURE 1-34. Two-press installation with automatic manipulators for upsetting, piercing, and ironing of closed-end cylinders. Courtesy of G. Siempelkamp GmbH & Co. (Krefeld, West Germany).

crease the demand on tool and die builders for more accurate tooling. The fast development of computer-aided design and manufacturing (CAD/CAM) systems will go a long way to fulfill this goal and, to a large degree, simplify the die-making process. In addition, the use of CAD/CAM systems will become almost mandatory in the forging industry as manufacturers design their products on a CAD system requiring them to be electronically transferred, as opposed to producing traditional hard copy drawings.

Future growth in more sophisticated systems is predicted. There is a growing belief in the future of advanced industrial robots and other computer-controlled devices throughout the world. This would include the development of plants on an unmanned concept. Unmanned does not necessarily mean no workers; however, it does mean that all productive functions will be carried out by automatic machines and supervised by a highly trained human work force.

Programmable controllers used for the equipment sequencing will be monitored by central processors and will, together with other monitoring devices, detect equipment failure and diagnose faults. Geometry verifiers will be used to automatically check the basic dimensions of forgings produced and automatically adjust die settings. Statistical analysis of quality-control data will then forecast die change times and facilitate automatic tool changing.

In conclusion, the forging industry must become much more cost effective, quality conscious, scientific, and innovative if it is to maintain its already eroded position as a major supplier of components to the manufacturing industry. It must rise to the challenge of the '80s and '90s with investment and development.

Purchasing of Forgings

By M. E. Rosberg, Administrator, Coordinated Purchases, Deere and Company, Moline, Illinois

As forging technology has grown, so too has the need for knowledge on the part of the forging buyer. Forgings, which are produced by a number of methods and from virtually all metals, are available in a wide range of sizes, shapes, and tolerances, and with a variety of physical and mechanical properties. It is therefore desirable that the purchaser of forgings have some understanding of the available forging methods. However, economically obtaining the component best suited for a particular ap-

plication also depends on a variety of additional considerations and requires an awareness of the manufacturing standards and practices offered by the forging industry.

This section provides general information that will be helpful in specifying and purchasing impression die forgings. With this knowledge, it is possible to take complete advantage of the opportunities presented by the forging process that result in the highest quality components, reduced total costs, greater reliability and strength, improved mechanical properties, uniformity from piece to piece, deliveries as scheduled, and effective service.

Types of Forge Plants

Over the years, the growth of forging technology and the increased complexity of forging processes have led the owners of forging plants to maintain maximum skill and efficiency in producing forgings with a degree of specialization. Today, there are several types of forge plants and a number of different ways in which they may be classified. One classification relates to the two basic forging methods: flat die forging and impression die forging.

Producers of flat die forgings (also referred to as open die or hand forgings) typically use heavy equipment to work hot stock into relatively large, symmetrical shapes between somewhat simple, smooth-surfaced dies. Flat die forging methods are often used to provide the advantages of the forging process where production quantities are extremely small or where the size or shape of the product to be forged makes other methods less practical. Flat die forging usually produces components that only generally approximate the final configuration desired and therefore often require extensive machining. Sometimes, flat die forging is used to preform stock in preparation for completion of the work in impression dies.

Producers of impression die forgings, on the other hand, use various types of equipment to produce quantities of uniform components in dies containing exact engineered impressions of the piece to be forged. Impression die forging is commonly but often imprecisely termed closed die forging and, by tradition, is loosely referred to as drop forging. Although the term "drop forging" correctly describes hammer forging operations, particularly where gravity hammers are involved, it does not properly apply to other types of forging methods.

In impression die forging, metal flow is rigidly controlled to ensure optimum mechanical properties in critical sections of the piece. Impression dies are employed to produce components that very often will require relatively little, if any, subsequent machining to achieve the desired shape, and are particularly suited for production of large quantities.

Forge plants may also be classified by the forging sizes they are capable of producing or by their typical quantities. For example, the term "production forge plant" is commonly used to describe facilities specializing in the production of large quantities of identical forgings, whereas the terms "short-run forge plant" and "job shop" are often used to designate companies whose primary capability is the efficient and economical production of a variety of small orders. "Commercial" or "custom" forge plants, of course, sell their products in the open market; "captive" forge plants produce forgings for the internal needs of their companies.

Forge plants are also identified with particular industries or by the types of forgings in which they specialize, such as aircraft or automotive forgings. Some forge plants may be identified by the special processes or equipment they use, such as warm forging, precision forging, cold forging, or extruding. A few forge plants are large, integrated organizations with broad forging capabilities that allow them to efficiently meet almost any forging requirement. Forge plants are most commonly classified by their manufacturing capabilities, such as hammers, presses, upsetters, ring rolling, cold forging, warm forging, precision forging, and open die forging.

Because production techniques and equipment vary considerably from plant to plant, no two forging producers are likely to approach the manufacture of a specific forging in exactly the same manner. For example, one forge plant might choose to preforge the material using equipment such as forging machines (upsetters) and forge rolls, while another might consider an entirely different series of preliminary forging operations, depending on shop practices and available equipment.

Forging costs vary accordingly. Therefore, to evaluate competitive quotations properly, the forging buyer must have at least a general knowledge of methods and equipment in each of the forge plants submitting a quotation.

A Service Industry

Regardless of the type of equipment and processes employed, a forge plant is essentially a service organization. Except for details of the manufacturing process, all considerations regarding a forging are finally established or specified by the purchaser.

Accordingly, the forging buyer, in conjunction with other company personnel, ordinarily participates in a number of important product decisions in addition to selecting a forging source. These decisions involve matters as basic as material selection, the degree of precision required, configuration of the component heat treat-

ing, quality certification and/or inspection procedures required, and other post-forging requirements.

The forging sales engineer can assist and advise his customers in these and other technical areas. Today, more than ever before, the technical services of the forging engineer are valuable to the manufacturer at all stages of production to ensure the integrity of the finished product. For example:

- *At the materials engineering stage,* where the merits of forgings are being compared with qualities of components produced by other methods, the forging engineer can help in determining what economic and technical advantages are obtained by using forgings for a specific application.

- *At the design stage,* the forging engineer may suggest modifications that will permit the designer to take full advantage of the benefits of the forging process to enhance the quality of the product and create maximum strength and fatigue resistance. Also, he may be able to suggest opportunities for significant production economics, such as zero or minimum draft, punched holes, or closer tolerances.

- *At the production stage,* where the design is fixed and the specifications are complete, the forging engineer is helpful in checking details of tooling, scheduling, and delivery commitments to ensure maximum economy and customer satisfaction.

- *At the heat treating stage,* he can be counted on for guidance in obtaining the optimum combination of properties required in the finished product.

- *At the machining and finishing stage,* his training is helpful in recognizing opportunities for savings in machining time, labor, and material.

The value of the forging engineer's services increases with his knowledge of the exact nature of the customer's requirements. Thus, a forging engineer should know about the application requirements of the product to be forged, its operating environment, the nature of mating parts, finishing and machining operations contemplated, anticipated quantities, and the degree of precision required. He will then be able to offer the best design and process suggestions, which ultimately will result in product improvement and manufacturing economies and the optimum finished part provided at the least total cost to the consumer.

Inquiries and Orders for Forgings

The industrial purchasing agent is keenly aware that the products of his company must be competitive, readily available, and of the highest quality consistent with service requirements and cost limits. In buying components for his company's products, he is intent upon obtaining confirmation that products comply with requirements and appropriate specifications and receiving maximum purchase values in terms of price, service, long-term materials availability, and just-in-time deliveries. By taking full advantage of the available forging services and by providing all pertinent information to the forgings supplier, the perceptive forging buyer can do a great deal to ensure excellent results.

Engineering Consultation

The responsibility for the design of a new component to be forged or the evaluation and redesign of an existing forging ordinarily falls on the customer's design and/or engineering department. It is the designer and the engineer who usually establish the function of the component, determine the service conditions to be imposed upon it, develop corresponding property requirements, make a tentative selection of material, and consider the general appearance of the component from the standpoint of utility and sales appeal. Because design is the most important single area where cost is influenced, it is at this point that the buyer wisely encourages the designer to augment his skill and knowledge with the specialized training of the forging engineer.

Often, slight changes in the shape will simplify forging requirements, reduce die costs, and speed production. Similarly, slight dimensional modifications can sometimes be made to permit greater precision and smaller draft angles and radii, resulting in reduced requirements for subsequent machining. In almost every case, the forging engineer, given basic design information, can work to good advantage with the designer in developing a design that meets functional requirements and maximizes production economies.

The forging engineer studies a new design from the standpoint of its tooling and processing requirements. He considers the complexity of the design and whether it can be forged in relatively simple dies or whether more elaborate (and more costly) tooling is required. He evaluates alternative positions of the parting line and whether it can be maintained in a single plane, or whether a flat top die can be used to provide economy. He must consider whether the distribution of metal will permit the development of the desired configuration and proper grain flow in a relatively simple sequence of operations, or whether special preforming operations, reheats, or additional dies and equipment are required.

The determination of required property levels and the selection of material to provide these properties economically are considerations that are closely interrelated with other aspects of the design problem. They are also important areas where close liaison between designer and forging engineer can reduce costs significantly. Once property requirements are established and agreed upon, selection of the most suitable material from the standpoint of economy and performance can be made.

Because there are often alternative grades and alloys of material that produce similar end results in mechanical properties, the forging engineer can be helpful in selecting, from among those that qualify, the material and grade that best combines satisfactory property levels with forgeability, heat treatability, machinability, and economy. Often, by using an alternate grade that is regularly forged, the buyer can benefit by permitting the forge plant to fully utilize its production experience and consolidation of common steel grades.

Requiring precision or property levels beyond the functional demands of the design increases forging costs. Best economies are achieved when mechanical property requirements (tensile, hardness, impact) are realistically based on the service requirements of the component being designed. This can often be accomplished by the selection of an appropriate ASTM specification covering the mechanical test requirement. Under these circumstances, the forging engineer can suggest the proper combination of material and heat treatment for the job. This often leads to the selection of less expensive materials that can be processed more economically.

Providing Complete Information

In addition to engineering consultation, it is extremely advantageous for the buyer to provide complete information on all requirements, including the application, for the component to be forged.

Specifications for forging can be grouped into two classifications: those affecting the material and those covering dimensional considerations. Material specifications deal with chemistry, strength, ductility, impact resistance, conductivity, soundness, and grain flow. These specifications can apply to forgings that have undergone no further processing, but they generally apply to conditions existing after heat treatment. There are many standard specifications that may be applied to the physical and mechanical qualities of a forging, such as those provided by the American Society for Testing and Materials, the Society of Automotive Engineers, the American Standards Association, and various offices of the U.S. government.

Dimensional specifications refer to the configuration of the forging and the degree of precision to be attained. Practical experience over a number of years has allowed the forging industry to prepare standards that set forth limits for such characteristics as size (length, width, and thickness), die match, and straightness—limits that have been found to be both practical and economical for most industrial applications. The standards are included in *Tolerances for Impression Die Forgings,* published by the Forging Industry Association, and are available to forgings buyers from company members of the Association. These tolerances are generally applied unless the purchaser, as indicated on the forging print, feels that more stringent dimensional controls are worth the additional cost, in which case special tolerances are agreed upon in advance of production.

Today, many components are being designed as forgings having surfaces that require little or no subsequent machining. Ordinarily, however, an "envelope" of material ≤0.32 in. thick is placed around the outline of the finish machined part, or on surfaces requiring machining, when a forging is designed. Envelope stock is added as required, and its location and thickness depend on factors such as the size, material, and complexity of the forging, as well as tolerances and heat treatment. At the same time, of course, many forgings are made to extremely close tolerances through coining operations or special metal-removal techniques such as electrochemical milling.

Often the customer writes his own specifications based on his particular needs. Even if the material and dimensional specifications to be used are standard or are developed by the customer, it is important that all aspects of the specification be clearly and thoroughly covered and all exceptions noted. Then, as the forging engineer reviews the applications—based on his experience or the experience of the supplier from whom he will purchase material—he can properly accept or request modification of the specification.

Physical property guarantees made by the owner of the forge plant are commonly based on experience with the particular alloy and the property level the alloy can be expected to meet when properly processed at the producing mill, skillfully forged, and adequately heat treated. The guaranteed mechanical properties are based on the alloy specified and are influenced to varying degrees by the size and shape of the forging and the particular forging location where tests are to be made. When no alloy is specified and only minimum physical properties are known, the forging engineer will ordinarily determine the proper material based on his forging experience. Mechanical properties can be certified by the forge plant through the destructive evaluation of a forging after heat treatment and prior to shipment. It is therefore important that provisions covering frequency and methods of testing be clearly spelled out so they may be agreed upon at the time the quotation is written.

Application and service requirements, which are provided by the forging buyer, should include comprehensive information on the application, function, operating environment, and mechanical properties for which the component is being designed. Advance information on how the component is to be processed after forging is also helpful. A thorough understanding of how and where the forging is to be stressed will often enable the forging engineer to make design suggestions, as well as recommendations on material, and will provide a frame of reference for decisions on processing. Information on the type of service (such as load-bearing, power-transmitting, impact, hydraulic pressure, service temperatures, the nature of mating parts, and corrosive atmosphere) will guide the forge plant staff in its approach to the complete processing cycle. Again, by supplying such information, the forging buyer increases the likelihood of receiving helpful suggestions that will increase quality and efficiency and reduce costs.

Delivery Requirements. An inquiry or order for forgings should indicate the delivery schedule required and anticipated future requirements for the forging on an annual basis. This information permits the forging vendor to anticipate immediate needs. At the same time it allows him to plan and design the most economical dies and tooling for the overall quantities required. The costs of forging and shipping a quantity of parts at one time are significantly different from the costs incurred in shipping to a monthly, weekly, or even daily quantity schedule.

Often overlooked in planning an incremental shipping schedule are elements of cost such as additional preparation for subsequent production runs and the carrying of completed forgings in inventory. When anticipated future requirements are known, these costs can be minimized and accurately reflected in the forge plant's quotation, and forgings can be adequately supplied at the increment level required by the forging consumer to meet manufacturing requirements and company ''just-in-time'' goals.

Also, by knowing the purchaser's future requirements, the forging company can often adjust its purchases of material to take advantage of lower cost large-quantity steel orders directly from the mill rather than ordering smaller quantities more frequently from warehouses or service centers at higher costs. Substantial reductions in material costs are often attainable through this type of advance planning.

Engineering drawings are key sources of information for the component to be forged. It is vitally important that drawings be legible and sufficiently detailed to communicate the customer's requirements clearly and accurately. Usually, several views of the component to be formed are desirable. Full-scale drawings are especially helpful (though not necessary) to the forging estimator in determining the net weight of the piece to be forged. The type of information that should be included on every drawing is discussed in the next section.

Checklist for Forging Inquiries and Orders

When necessary information is not available, the forger must make assumptions—which may increase the cost unnecessarily, cause production delays, or result in a forging that is less satisfactory. The following checklist may be used as a convenient guide in providing complete information on forging requirements. This will allow the forge plant owner to write an accurate quotation, and may be instrumental in highlighting the opportunities for cost reduction.

Identification

- Name of component
- Drawing number
- Part number
- Company name and address
- Name and title of person initiating the inquiry or order
- End use

Engineering drawing—forging print and machining print

- Name of component
- Drawing number
- Part number
- Position of locating points and/or chucking bosses for subsequent machining operations
- Surfaces to be machined and finish allowance desired
- Type of finishing operation to be used
- Location and nature of part numbers and trademarks (raised or indented numbers and letters)
- Identification of drawing as to issue status or number
- Test bar location, analysis, and specification number
- Heat treatment (if required)

Quantity

- Total quantity required (in pieces) for initial orders
- Number of pieces per release (if subject to release)
- Estimated annual quantity requirements
- Any limitations on application of FIA quantity tolerances. (Special quantity tolerances are usually quoted separately.)

Delivery schedule

- Initial delivery date and number of pieces
- Subsequent schedule (pieces required per delivery—monthly, daily, weekly, etc.)
- Date order is to be completed

Service data

- Maximum design stress (ksi)
- Description of stresses in service (impact, cyclic loading, or pressures)
- Nature of wear or abrasion to be encountered
- Operating environment (corrosive agents, maximum service temperatures)

Surface condition

- Surfaces to be machined (marked on drawing)
- Nature of finish (polish, plating, paint, other)
- Whether alternate quotation is desired, with machining and other operations included

Material

- Alloy or carbon grade by name, composition, and specifications
- Alternate materials permitted

Properties

- Standard specification which applies (additional requirements and/or exceptions)
- Minimum tensile strength (ksi)
- Hardness (maximum and minimum at specified locations)
- Other applicable properties

Heat treatment

- Nature of heat treatment
- Property levels required

Dimensional tolerances

- Standard tolerances (FIA)
- Critical dimensions where special tolerances apply

Special inspection requirements

- Inspection methods required (dye penetrant, magnetic particle, sonic)
- Customer's incoming inspection (complete, 100%; statistical; average quality level, AQL; or other)
- Government agency inspection
- First-piece inspection samples required

Shipping

- Special packaging specifications or crating requirements
- Type or name of carrier preferred

Government, customer, or technical society specifications

References

1. Kalischer, P.R., *Precision Metal*, Vol. 26 (No. 9), Sept. 1968, p. 55.

2. Scofield, C.G., "Influences Affecting the North American Forging Industry," 11th International Drop Forging Congress, Cologne, West Germany, June 1983.

3. Lester, R.D.N., The Future of the Drop Forging Industry, Part 1, *Metallurgia and Metal Forming*, Vol. 39 (No. 11), 1972, p. 385.

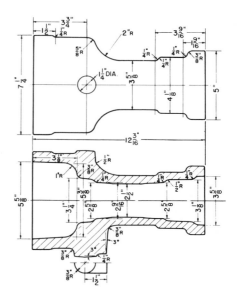

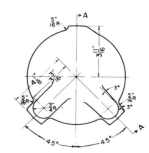

2 FORGING DESIGN

Designing a mechanical component is essentially the process of creating a useful shape that contains the needed properties and meets economic criteria. Occasionally alternative materials and processes can be used to meet these requirements when designing with metals. It is the designer's task to find that single combination of material and process that optimizes the factors of configuration, properties, and costs. Forging design ranks high on the growing list of design practices.

Forging design, similar to other metalworking processes, is influenced by the nature of the metal being processed and the capabilities and limitations of the available production equipment. Basic to the design function is a forging drawing of the desired finished part, such as those shown in Figures 2–1 and 2–2.

Types of Forgings

All forgings fall into two general classes: hand forgings and die forgings. Hand forgings are sometimes called open die forgings, because the metal is not confined laterally when being forged to the desired shape. The forger manipulates the stock between repeated squeezes of the hydraulic press or ring roller, or between blows of the hammer, and progressively shapes the forging to the desired form.

Die Forgings

Die forgings, the more common forging type (also called closed die forgings), receive their accurate and uniform shapes from hammering or pressing the forging stock in cavities or impressions cut into a set of dies. The primary equipment used to make die forgings includes hammers, mechanical presses, and hydraulic presses; each type possesses its own particular advantages. Forging design, forging tolerances, quantities required, and alloy selected must all be considered in determining the best and most economical equipment to use in making a specific part. Die forgings can be broadly classed as blocker type, finish only, conventional, or near-net.

A blocker-type forging is designed with large fillet and corner radii and with thick webs and ribs so that it can be produced in a set of finishing dies only. Producing such a forging may typically require a unit pressure of 20 to 30 ksi of projected plan area, depending on the alloy and the complexity of the design. This is less pressure than is necessary to make a more intricate forging. The projected plan area of the forging is used to arrive at the estimated total tonnage required. A blocker-type forging generally requires machining on all surfaces. Economics may dictate such a design if quantity requirements are limited or if the finished part tolerances necessitate complete machining. A blocker-type forging is an end product and should not be confused with a blocker forging, which is a preliminary shape requiring a subsequent finish forging operation to attain its final shape.

Finish-Only Forgings. As the name implies, forgings of this type require only one set of dies. The design is similar to a conventional forging, except that fillet ra-

FIGURE 2–1. Forging drawing of a hydraulic cylinder cap. Courtesy of the Drop Forging Association.

dii and the die closure tolerance may be increased. The part may require more than one forging operation, with processing in between forging operations. A typical unit

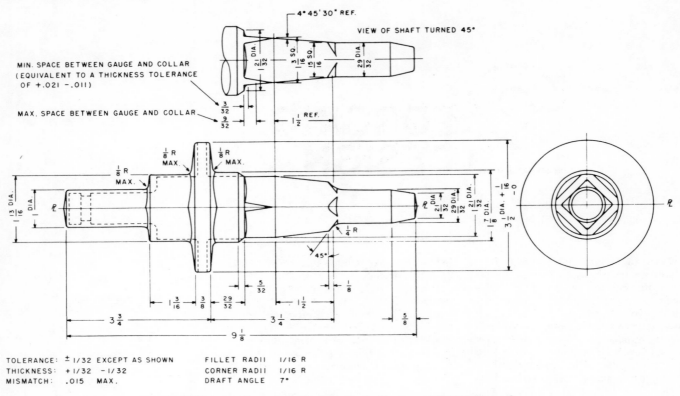

FIGURE 2–2. Forging drawing of a shaft. Courtesy of Cornell Forge Co.

pressure of 30 to 50 ksi is required for this operation.

A conventional forging, the most common of all die forging types, is more intricate in configuration than a blocker-type forging and has proportionately lighter sections, sharper details, and closer tolerances. Thus, it is more difficult to forge. A conventional forging normally requires only partial final machining. A typical unit pressure of 30 to 50 ksi of plan area is required, and usually a blocking operation is required prior to the finishing operation.

The designer usually evaluates the cost differences between blocker-type and conventional forgings. A blocker-type forging has a lower die cost but will be heavier, requiring more extensive machining; a conventional forging has a higher die cost but will be lighter, requiring much less machining. Only a cost comparison by the customer can determine which type of forging will give the lowest total cost.

Near-net forgings have lighter sections, sharper details, closer tolerances, and possibly more net surfaces than conventional forgings. A typical unit pressure of 40 to 70 ksi of plan area is required, and usually a blocking operation is required prior to the finishing operation.

Forging Design Principles and Practices for Aluminum

By Louis J. Ogorzaly, Design Supervisor, Alcoa Forging Division, Cleveland, Ohio

Parting Line

The planes of separation between the upper and lower parts of closed die sets are called parting lines. The parting line is usually, but not always, located through the maximum periphery of the forged part. It must be designated on all forging drawings. Its position can measurably affect the initial cost and ultimate wear of dies, the ease of forging, the grain flow, related mechanical properties, and the machining requirements for the finished part. Thus, within limits established by the geometry of the final part, the location of the parting line is determined after economic and technical considerations have been weighed. The parting line may be straight or irregular, depending on requirements; however, as a general rule, it is preferably positioned in one plane, wherever possible. Various methods of positioning the parting line are shown in Figure 2–3.

Maximum Periphery. It is preferable to place the parting line around the largest periphery of the forging. It is easier to force metal laterally in a spreading action than it is to fill deep, narrow die impressions.

Flat-sided forgings present an opportunity to reduce die costs, because the only machining is in the lower block, the upper being a completely flat surface. This simplifies production by eliminating the possibility of mismatch between the upper and lower impressions. Although a top die is always essential, in flat-sided designs the forge shop can use a stock or "standard" flat-top die to mate with the impression die. Obviously, there can be no integrally forged characters appearing on the flat surface when a standard flat-top die is used.

Inclination of the Parting Line. Forgings in which the parting line is inclined

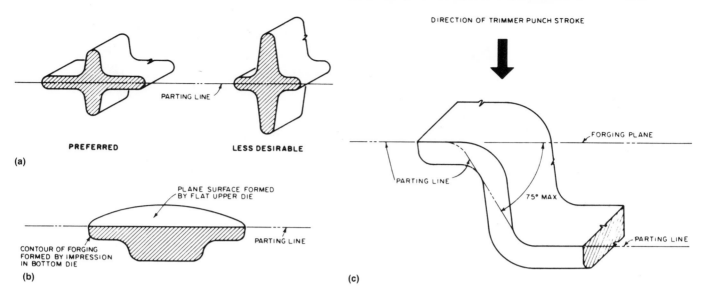

FIGURE 2-3. Parting line location. (a) Place the parting line through the largest cross section of the forging to avoid narrow, deep die impressions. (b) A flat surface at the parting line reduces die costs, simplifies the trimming operation, and eliminates any die mismatching. (c) A ragged trimmed edge on the forging may result if the flash on the parting line is inclined more than 75°. Source: Ref. 4.

to the forging plane may present difficulties in trimming if the inclination is too great. It is generally good practice to limit the inclination to 75°, thereby avoiding ragged trimmed edges.

Parting Line Effect on Grain Flow. The location of the parting line has a critical bearing on grain flow and the directional properties of the forged piece (Figure 2–4). During the forging process, excess metal flows out of the cavity into the gutter as the dies are forced together. In this flow toward the parting line, objectionable flow patterns may be created if the flow path is not smooth. A parting line on the outer side of a rib should be placed either adjacent to the web section or at the end of the rib opposite the adjoining web section. The placement of the line at an intermediate step can result in irregularities.

In forgings with a web of varying planes enclosed by ribs, the parting line should follow the center lines of the enclosed webs (Figure 2–5). The same forging could also have a satisfactory parting line at either the top or bottom surface. Since both are plane surfaces, a flat parting line would result. A die plug protruding above the parting line to form the forging recess is required, but would not be objectionable.

Putting the large flat surface on the parting line, with all of the impression in one die, is an economical way to produce a part. An undesirable approach is to exert unbalanced lateral forces on the die. A counterlock would have to be machined in

the dies to resist the side thrust, and this would mean more expensive dies. A good technique is to forge on an incline to minimize the side thrust. The part, relatively light in weight, will have good grain flow and provide a natural draft that will permit forging of the end flange at a true right angle. A method that is highly desirable from the standpoint of a unit cost per forging is to produce the part as a double forging that can be sawed apart into two units. Considerable economy is achieved, particularly if the end surfaces will have to be machined anyway. Examples of how parting line choices can affect forging designs are shown in Figure 2–6.

Die Side Thrust

While working clearance at the press or hammer guides is necessary for proper functioning, it can be a cause of match errors, especially when die side thrust is induced by an irregular parting line. Side thrust increases as the parting line inclines from parallel to the forging plane. One solution to the side thrust problem is to sink counterlocks into the dies, but this is expensive and potentially troublesome. The preferred solution is to place the part at an incline with respect to the forging plane, thus balancing the side thrust and eliminating the need for counterlocks. When the parting line is inclined with respect to the forging plane, or the principal dimensions of the piece are laid out in another set of axes, the draft angle is usually in the same

direction as the stroke of the hammer or the press (Figure 2–7).

Intersection of Parting Line and Forging Plane

Draft angles cannot be laid out until the forging plane is first established. A simple method of positioning the forging plane with respect to an irregular parting line is to select points approximately 0.50 in. or less beyond the ends of the forging (Figure 2–8). The intersection of the parting line and forging plane can be located easily with definitely established even dimensions outside of the forging proper.

Draft Angles

Draft refers to the taper given to internal and external sides of a closed die forging to facilitate its removal from the die cavity. Draft is normally expressed as an angle from the directions of the ram travel. Draft not only ensures good forgings with a minimum of production difficulty, but also reduces die sinking expense by adding rigidity to the die cutter. Normally, draft angle callouts are specified under the general notes, but sometimes they are omitted from the drawing if they are uniform throughout the part.

Standard draft angles refer to those most commonly used. A fully equipped die shop must have die cutters not only in all standard draft angles, but in a variety of diameters as well. The standard draft angles

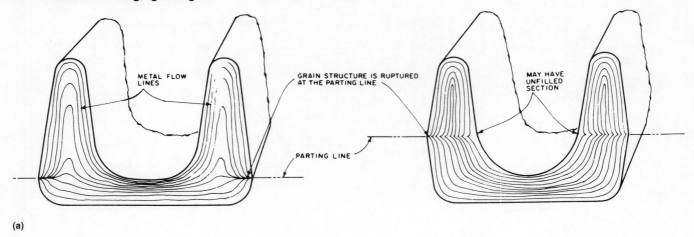

(a)

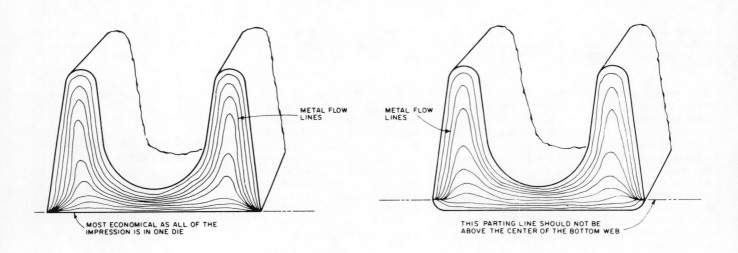

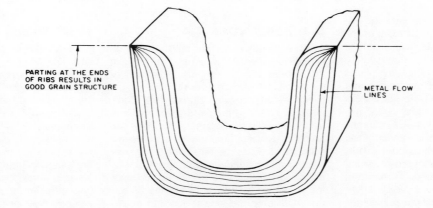

(b)

FIGURE 2-4. Various parting line locations on a channel section have differing effects. (a) Undesirable: These parting lines result in metal flow patterns that cause forging defects. (b) Recommended: The flow lines are smooth at stressed sections with these parting lines. Source: Ref. 4.

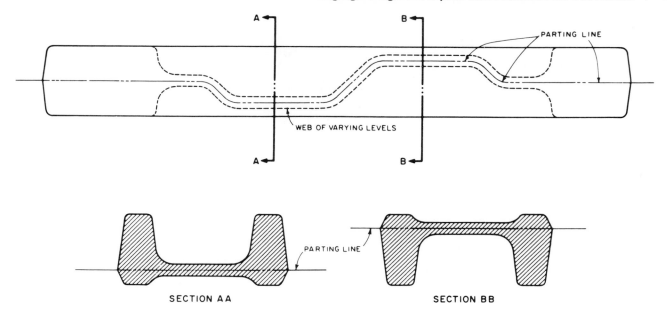

FIGURE 2–5. A parting line outside a rib or wall should follow the web inside the forging. Source: Ref. 4.

are 1°, 3°, 5°, 7°, and 10°. If other draft angles are specified, special die cutters have to be made. In most cases, draft angles less than 5° preclude the use of hammer equipment because additional mechanical aids are required to eject the forging from the die cavity. Table 2–1 shows standard draft angle dimension information useful for die-making and drawing purposes.

Cylindrical sections are assumed to have standard draft at the parting line. The customary die sinking practice is to leave the narrow tangential flats in these areas. It is not necessary to specify draft on drawings at such detail; only cylindrical contours need to be indicated.

Matching Parting Line Contours

Matching parting line contours at the parting line on the outer periphery is recommended where the top and bottom die contours do not coincide from normal die sinking. This can be achieved by the following three methods: When the matching portion is small, the method shown in Figure 2–9 is recommended. Weight increase is insignificant and die sinking is the simplest possible. Adding the 5° as shown in Figure 2–10 removes any doubt as to the intended construction. The second method is recommended when the matching portion is substantial and weight saving is important. Die sinking is routine in this case. The pad thickness may vary from 0.06 in. on small forgings to 0.50 in. on very large

forgings. Another method (Figure 2–11) adds less material than the first, but more than the second; it also adds to the die sinking. The original 5° draft cut leaves a jog or step at the parting line. An additional cut removes this step by a matching straight line tangent to the bottom radius.

Blending of draft angles can be troublesome and should be avoided where possible, but it can be done around the entire periphery of a boss (Figure 2–12) by applying either of the first two methods. However, if the lower portion had matching draft around the entire boss, there would still unavoidably remain a portion of the web that must be blended from the normal to the matching draft. The drawing should include a specific note defining the distance in which this transition is to be made.

Constant Draft Angles for Best Die Economy

The blending of varying draft angles results in an additional die sinking expense. There are important economies to be achieved by designing forgings with constant draft angles, because surfaces with draft are generally formed by conical cutters (Figure 2–13). A radius at the tip of a tool blends the drafted surface with the rest of the forging or forms a full radius at the edge of a rib. The size of the tool body determines the minimum fillet radius in a die between adjoining surfaces that have the same draft. Ribs, sidewalls, and bosses, with constant draft but changes in

depth, will vary in width at the base as a result of the draft.

If the die sinking cutter is traversed in a straight line, the top of the rib on a forging will be straight (in plan view) and have a constant width. However, the base of the rib will vary in width and will be curved or irregular (Figure 2–14). Making a rib vary in height with top and base widths parallel to each other requires blending of draft angles; die costs will also increase, so this should be done only in cases where it is imperative, not as a random choice. Figure 2–15 shows various choices and their defects.

Undercuts and Back Drafts

Undercuts on forgings are possible, but require careful consideration in their applications. Undercuts may increase cost slightly, because forging removal from the dies may be retarded. A forging with undercuts on both sides of the parting line (Figure 2–16) cannot be made. Because the die halves must separate in order to remove the forging, the die separation would mutilate the forging in overcoming the back draft resistance on both sides.

Undercuts in the bottom die can only be forged within limits. The forging must be removed from the bottom die at an angle because of the back draft. The draft angle opposite the back draft, as shown in Figure 2–17, must be larger by a recommended 6° to allow the forging to be removed. This applies to all similar surfaces.

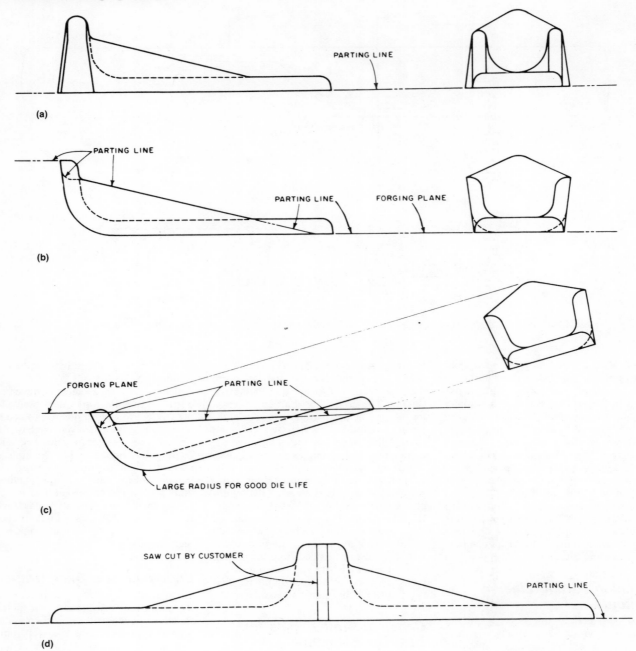

FIGURE 2–6. Illustrations of how parting line choices can modify a forging design. (a) Economical: Forging impression is only in one die, but forging sections are slightly heavier. (b) Undesirable: Design requires expensive counterlock to take side thrust forces. (c) Good: Inclining the forging in the die permits square end surfaces with natural draft. (d) Preferred: A double forging offers the most advantages in production. Source: Ref. 4.

An undercut can also be produced by inclining the forging plane to eliminate the back draft (Figure 2–18). Some types of forgings can be fabricated with back draft. However, these parts require a secondary operation (Figure 2–19) with the use of additional tooling. The extent of the back draft of these parts should, at the time of original design, be thoroughly discussed with a competent forging vendor. The increased cost of secondary tooling and increased piece price may outweigh the costs of conventional forging and machining costs. It should be noted that the parts in Figure 2–19 have greater strength due to an uninterrupted grain flow.

Finish Allowances for Machining

A designer must be familiar with both the magnitude and the application of the tolerances required on a forging to determine the necessary amount of finish allowance on the machined surfaces. Sur-

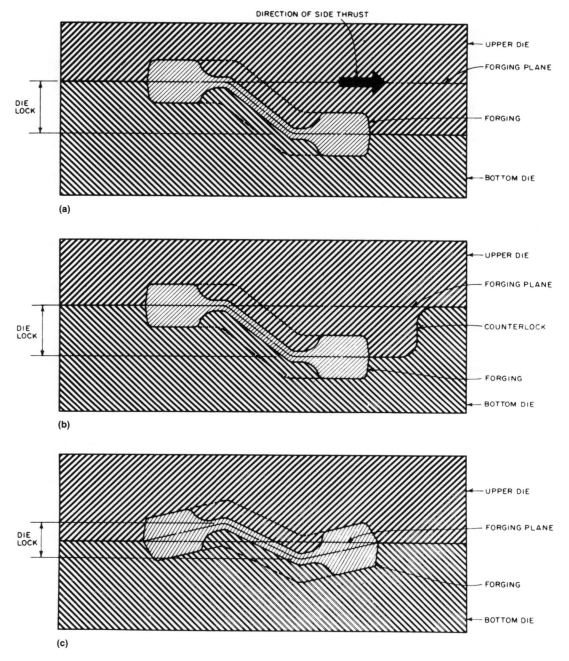

FIGURE 2–7. Forging plane positions to eliminate side thrust when locked dies are required. (a) Impractical: Side thrust makes it difficult to hold the dies in match accurately. (b) Not recommended: Dies with counterlocks are expensive to build and troublesome to maintain. (c) Preferred: The best method is to incline the forging with respect to the forging plane. Source: Ref. 4.

faces parallel to the parting line (Figure 2–20) are affected by the die closure and straightness tolerances; surfaces perpendicular to the parting line are affected by dimensional, straightness, and match tolerances. Only the adverse effect of these tolerances need be considered, and this amount should be added to a minimum clean-up allowance on each surface to ensure satisfactory machining. The minimum machining allowance is arbitrary and varies from 0.02 in. on very small forgings to as much as 0.25 in. on large forgings.

To lessen any possible adverse effect of forging tolerances, the designer must know how the forging will be positioned in the initial machining setup. For example, a forging fixtured on one side of the parting line is not affected by either the match tolerance or the minus die closure tolerance on that side, but these same tolerances do affect the opposite side of the parting line and must be considered. Not knowing how

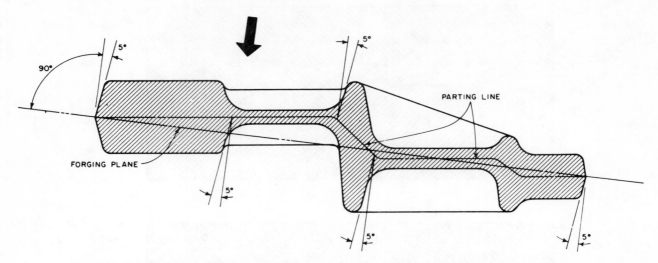

FIGURE 2–8. Minimum draft measured from a perpendicular to the forging plane. Source: Ref. 4.

TABLE 2–1. Standard Draft Dimensions, in.

Draft depth, in.	Draft angles				
	1°	3°	5°	7°	10°
1/32	0.0005	0.0016	0.0027	0.0038	0.0055
1/16	0.0011	0.0033	0.0055	0.0077	0.011
3/32	0.0016	0.0049	0.008	0.0115	0.0165
1/8	0.0022	0.0066	0.0109	0.015	0.022
3/16	0.0033	0.0098	0.016	0.023	0.033
1/4	0.0044	0.013	0.022	0.031	0.044
5/16	0.0055	0.016	0.027	0.038	0.055
3/8	0.0065	0.020	0.033	0.046	0.066
7/16	0.0076	0.023	0.038	0.054	0.077
1/2	0.0087	0.026	0.044	0.061	0.088
5/8	0.011	0.033	0.055	0.077	0.110
3/4	0.013	0.039	0.066	0.092	0.132
7/8	0.015	0.046	0.077	0.107	0.154
1	0.017	0.052	0.087	0.123	0.176

Source: Ref. 4.

a forging is to be fixtured makes it necessary to allow the full machining allowance for the adverse forging tolerances on both sides of the parting line. It is further recommended that tooling points be shown on drawings or that other means of indicating the machining fixturing points be used.

On small forgings, to make it possible to machine a forging to sharp corners, sufficient material must be provided; normally these corners are rounded on the forging. In many cases, by adding a ma-

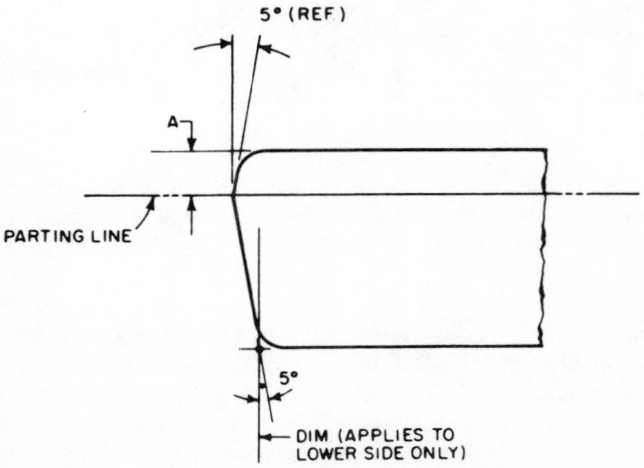

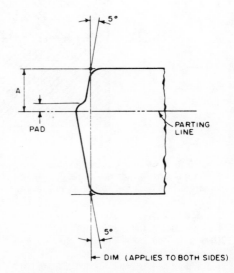

FIGURE 2–10. Matching parting line contours: recommended method when A is substantial and weight saving is important. Source: Ref. 4.

FIGURE 2–9. Matching parting line contours: recommended method when A is small. Source: Ref. 4.

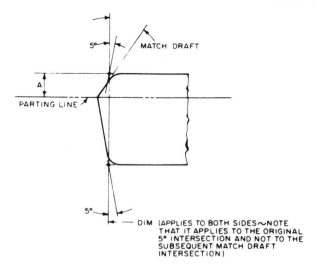

FIGURE 2–11. Matching parting line contours: recommended method when A is substantial. This method adds less material than in Figure 2–9, but more than in Figure 2–10. Source: Ref. 4.

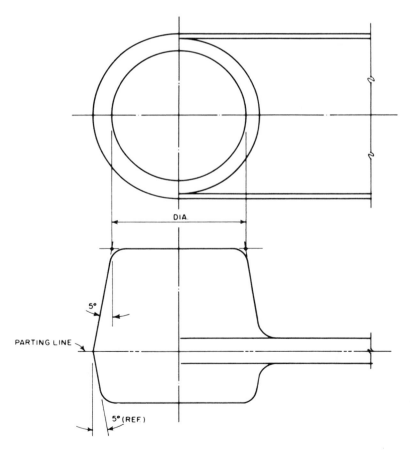

FIGURE 2–12. Illustration of how to avoid blending of draft. Source: Ref. 4.

chining allowance equal to the required corner radius, adequate material is provided to compensate for forging tolerances as well as to machine sharp corners. This practice will usually work out well on small forgings (maximum length of 8 in.). However, to be certain, the necessary finish allowance should be calculated from the applicable forging tolerances.

On large forgings, it is imperative that the finish allowance on all surfaces be calculated on an individual basis to ensure completely satisfactory machining. The finish allowance per surface varies considerably on the same forging because of the relation of different surfaces to the parting line and to the datum planes.

Corner Radii

A corner radius is formed by the intersection of two surfaces with an included angle (within the forging) of less than 180° or excluded angle (outside the forging) of greater than 180°. When minimum dimensions for corner radii are established, two factors are considered: the radius as a stress concentrator in the die and the pressure necessary to fill the die cavity. A sharp edge or corner on a forged piece requires a corresponding sharp recess in the die, because the forging is a positive shape and the die cavity is a direct opposite or negative shape. Thus, a forging's corner radius is formed by a corresponding die fillet radius.

When die fillet radii are excessively sharp, the force of metal under pressure and the stress-raising corner effect can cause checking or the development of small cracks in the die after only a few pieces have been forged. This adverse effect on die life is compounded, because a greater forging pressure is required to fill sharp die fillets versus large die fillets. Die life becomes short, and hence uneconomical.

Minimum corner radii limit the use of very high, thin ribs. With a full radius at its edge and standard draft on the sides, a rib should not be any thinner than twice R_R. A thicker rib or flange may have a flat edge with two corner radii, each equal to R_R. However, it is more desirable to use a full radius equal to one half the rib thickness. (See Figure 2–21 for the recommended corner radii sizes.) The height from the parting line usually determines the size of the corner radius, as shown in Figure 2–22.

The recommended corner radii for three classes of die forgings are: the smallest radii for precision-type forgings, the larger for conventional-type forgings, and the largest for blocker-type forgings. When

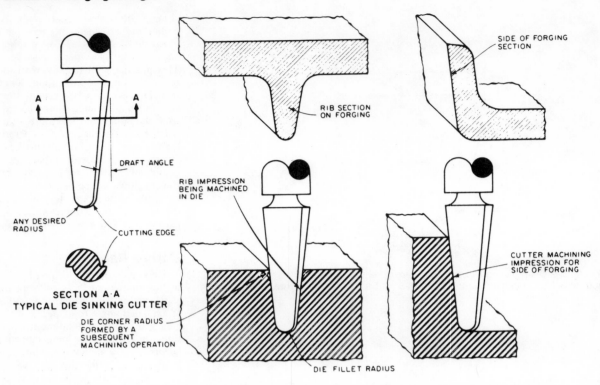

FIGURE 2–13. Die surfaces with draft machined with conical die sinking cutters. Source: Ref. 4.

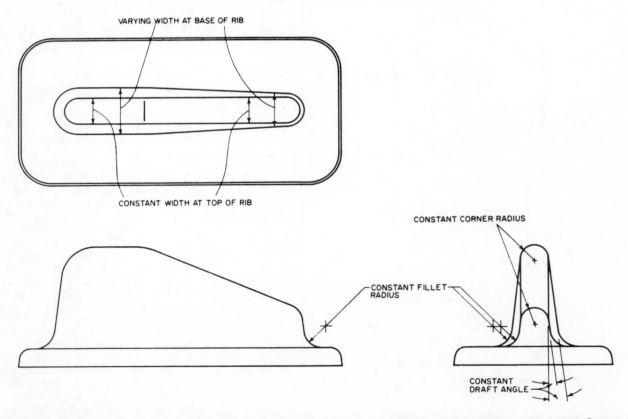

FIGURE 2–14. Draft angle, fillet radius, corner radius, and top width all constant along a rib with tapered or varying height. Source: Ref. 4.

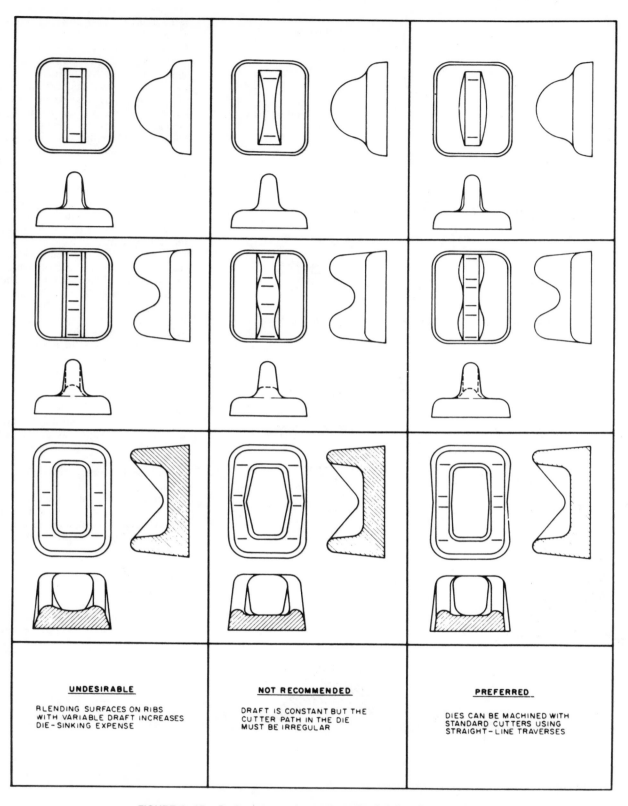

FIGURE 2–15. Draft contours shown at varying heights. Source: Ref. 4.

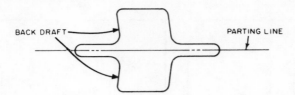

FIGURE 2–16. Illustration of back draft in both top and bottom dies, which does not permit the dies to be separated. Source: Ref. 4.

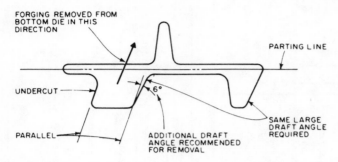

FIGURE 2–17. Illustration of undercut and opposite surfaces with larger draft angle. Source: Ref. 4.

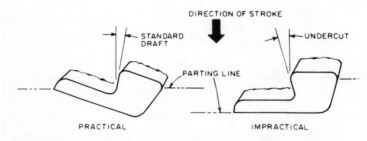

FIGURE 2–18. Illustration of rotating parting lines. Source: Ref. 4.

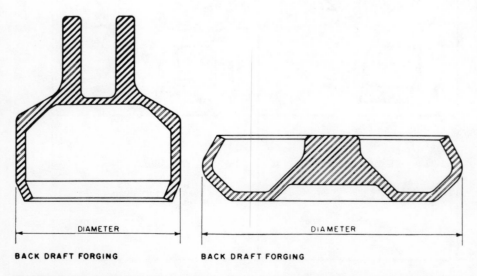

FIGURE 2–19. Back draft forgings showing lip sections preformed by a secondary operation. Source: Ref. 4.

designing any of the three types, every effort should be made to specify the maximum radii permissible, in order to keep forging costs down to a minimum.

The blending of corner radii on forgings should be carefully avoided to an even greater degree than the blending of fillets (discussed in the next section). Forging corner radii, as expressed physically in the die cavity, constitute fillets sunk by milling cutters of similar sized radii and, accordingly, are much more difficult to blend. Constant corner radii, as shown in Figure 2–23, will reduce die costs.

At rib or flange ends, a larger radius than the corner radius is recommended to permit forging fill without excessive difficulty. The die cavity is sunk by moving the die milling cutter in the same arc as the large radius at such an end. (This is not the same as the blending of radii, which requires difficult die sinking.) The suggested R_X radius (Figure 2–24) may be (1) equal to edge distance plus hole radius, (2) three times R_R, or (3) as large as possible. While it is recommended that R_X be as large as possible, the size of the radius in the plan view is also important. A full radius, as shown in Figure 2–25, is recommended. Such a die cavity is sunk by moving a full width diameter cutter in the same arc as R_X, a simple die sinking operation. If the arc is not a full radius, additional die sinking cuts are required to blend these radii throughout the distance of R_X.

A rib or flange edge that is rounded with a full radius will reduce die problems. The top or edge of a rib should be rounded (Figure 2–26) with a minimum radius of R_R, the value depending on the height from the parting line. Die life is reduced when

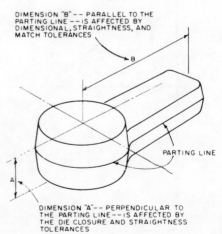

FIGURE 2–20. The relationship of a dimension to the parting line. Source: Ref. 4.

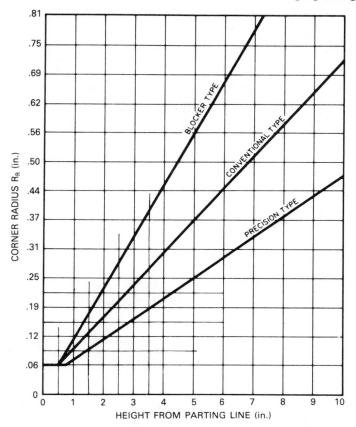

FIGURE 2–21. Recommended corner radius sizes. Source: Ref. 4.

deep, thin ribs are forged. The force required to fill a deep rib cavity, particularly when it is adjacent to a very thin web, causes severe die stresses to develop. Therefore, ribs that are next to very thin webs should be made as low as the design will permit.

Fillet Radii

A fillet radius is at the intersection of two surfaces with an included angle (within the forging) of greater than 180° or an excluded angle (outside the forging) of less than 180°. One of the most important contributions a designer can make to ensure the most economical quality forging of a blocker, conventional, or precision type is to allow generous radii for fillets and corners. Liberal fillets on forgings permit the forging stock to follow the die contours more easily during the forging process. Small fillets may cause momentary voids beyond the bend in the flow path of the metal. These voids are subsequently filled, but the interrupted flow results in a flaw in the forging (Figure 2–27).

A fillet radius equal to R_C (minimum) should be used when the flow of metal into the die gutter at the parting line is confined (Figure 2–28). If the forging stock must fill ribs on the outside periphery, web fillets at least as large as R_C should be used; when the web is very broad, larger fillets

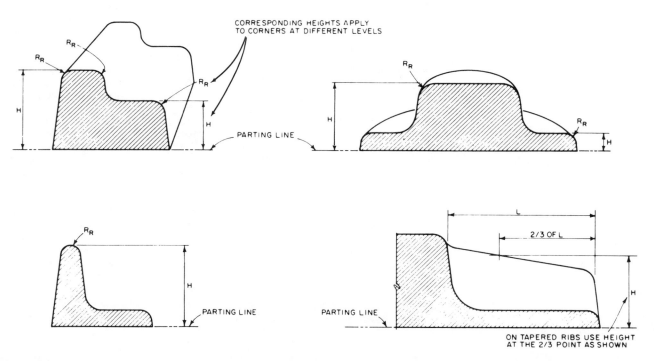

FIGURE 2–22. Illustration of how the corner radius is determined by the height from the parting line. Note: See Figure 2–29 for value of R_R. Source: Ref. 4.

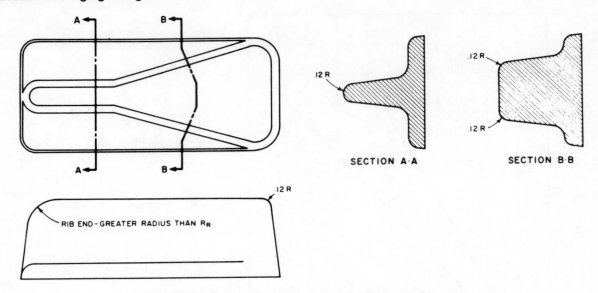

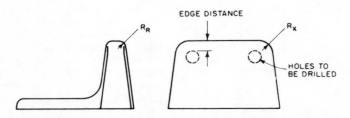

FIGURE 2–23. Constant corner radius. Source: Ref. 4.

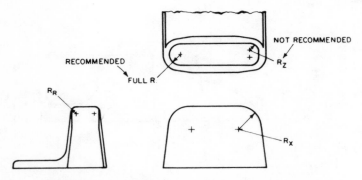

FIGURE 2–24. The R_x radius. Source: Ref. 4.

FIGURE 2–25. A full radius at rib ends. Source: Ref. 4.

may be advisable. When opposing ribs do not confine the metal flow to the gutter, values of R_U fillets should be used.

An area may be considered nonconfined if its outside periphery is confined less than 50% of the total periphery. The sizes of fillet radii required are governed by the step height from the surface where the fillet occurs to the adjoining surface level. There are recommended fillet radii for the three different classes of die forgings—precision, conventional, and blocker. Adherence to the standards is generally essential in good forging design. Significant decreases from these standards could result in an increase in forging costs. Figure 2-29 gives recommended fillet radii.

Extra die sinking costs result from designs requiring the blending of different fillet radii. However, when it is necessary to blend different radii, the change should be made at the plan view intersections, as shown in Figure 2–30.

At intersections of less than 90° and with rib heights of 1 in. or more, larger fillet radii are suggested (Figure 2–31) in order to avoid forging flaws. These larger radii are taken normal to the angle bisector.

Larger than normal fillets may be necessary to prevent "flow-throughs" when the web is enclosed by ribs and either web or ribs are excessively thin, and also when the width between ribs is more than ten times the rib height. An inadequate fillet at the base of the rib causes the metal to flow through into the die gutter rather than into the rib cavity. The result is a shear of the grain structure.

An inside radius in the plan view at the base of the web equal to or larger than the fillet radius is recommended from the forging standpoint, as it results in a uniform condition at this base. A smaller inside radius results in a nonuniform condition—an actual valley or trough that may be detrimental in forging. It also requires a blending of radii. The fillet radius continues into the die cavity from both directions, where these radii intersect at a sharp die edge. The inside radius where the sharp edge begins is used to break the edge constantly along the intersection into the web (Figure 2–28).

Grain Flow

The forge shop will usually produce the forging using the most economical forging technology. However, if a given forging

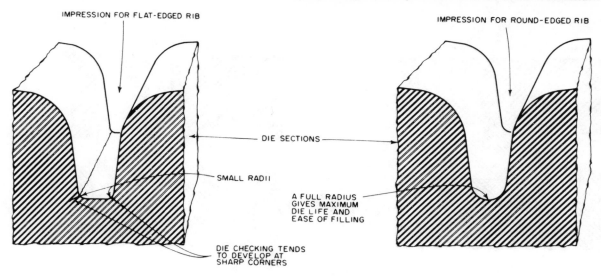

IMPRESSION FOR FLAT-EDGED RIB

IMPRESSION FOR ROUND-EDGED RIB

DIE SECTIONS

SMALL RADII

A FULL RADIUS
GIVES MAXIMUM
DIE LIFE AND
EASE OF FILLING

DIE CHECKING TENDS
TO DEVELOP AT
SHARP CORNERS

FIGURE 2–26. A rib or flange edge rounded with a full radius. Source: Ref. 4.

can be produced with several alternative grain flow patterns (fiber structures) and the grain flow direction is important because of the nature of loading, the preferred grain flow arrangement should be specified.

Grain flow is characterized by aligning the crystal structure of the base metal or the flow pattern distribution of secondary phases in the direction of working. Metals in the solid state are a crystalline aggregate and are therefore anisotropic to a certain degree regarding certain properties.

The mechanical properties may vary according to the grain flow direction and depending on the alloy and fiber direction. Transverse properties (perpendicular to the fibers) are generally lower; tensile strength and yield strength decrease slightly, if at all. Elongation, contraction, impact strength, and fatigue strength are generally lower. The differences increase with increasing amount of work in the material. In forging design, this anisotropy or directionality can be used to advantage by orienting the metal in the direction requiring maximum strength and toughness.

Flash

Conventional forging dies are designed with a cavity for flash, as shown in Figure 2–32. Important dimensions in the flash region of a die are the land width and flash thickness. For flash thickness, a reasonable guide to follow is to use about 3% of the maximum forging thickness. Increasing constraint is placed on the metal being forged by increasing the land width and/or by reducing the flash thickness. Re-

stricting formation of flash by such changes in the die design, however, results in higher forging load requirements. Tapered saddle designs place considerably more constraint on metal flow than do parallel saddle designs and thus reduce flash metal losses. The greater forging pressures required, however, impose greater stresses on forging dies. A tapered saddle design is generally used when the material savings justify the use of larger forging equipment. To a degree, therefore, the choice of flash design becomes a matter of economics.

Webs

The minimum dimensions of webs depend for the most part on the size of the forging (expressed as area at the parting line) and on the average width, as shown in Figure 2–33. Punchout holes are not included in the plan area when determining the minimum web thickness. However, these punchout holes should be of sufficient size to facilitate the forging operation. Recommended web thickness values applicable to small conventional-type forgings are listed in Table 2–2. Figure 2–34 is a nomograph that can be used in determining web thickness values for large conventional and high-definition forgings. On blocker-type forgings, the web thicknesses are usually considerably greater than the recommended values due to the natural summation of a finish allowance per surface—an amount to take care of the adverse effect of the forging tolerances—and the base web thickness. In any event, the minimum web thickness on a blocker-type

forging should not be less than one and a half times the recommended value.

It is difficult to maintain the dimensional tolerance on a thin web because, being thin, it cools and shrinks faster than other parts of the forging. Abnormal dimensional relationships may develop in such cases. Because of the tendency of thin sections to warp in heat treatment, additional straightening operations may be necessary and costs will increase. Thin webs and ribs tend to cause die checking, unfilled sections, and flow-through. Forging thin webs may also cause a temporary deflection in the dies, resulting in a web thicker in the middle than at the outer periphery. When forging excessively thin webs, this die deflection can become permanent.

The difficulty in forging thin webs may be lessened by web tapers or punchout holes, which are especially recommended for wide confined webs enclosed by high ribs, as shown in Figure 2–35. Punchout holes are the more desirable alternative, because they reduce the total projected plan area and the pressure required to make the forging. Even greater benefit is derived if the punchouts are located in central areas where metal flow resistance is at its greatest. Incorporating a flash and gutter cavity in the dies at these interior points provides a very useful means of escape for surplus metal.

A few large punchouts are more economical and helpful in forging than many smaller holes with the same total area; round punchout holes are preferred for ease of trimming. Small holes may even add to the cost without improving the forgeabil-

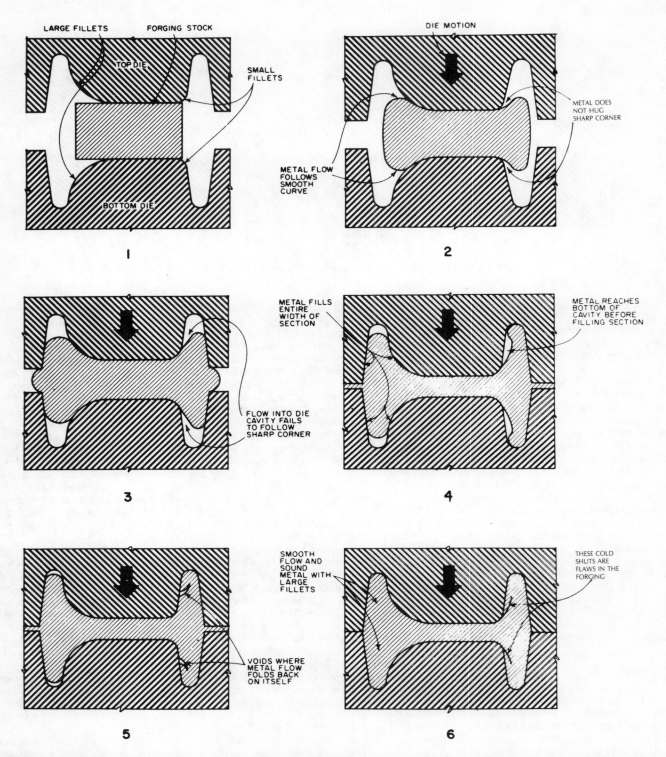

FIGURE 2–27. Progressive forging steps to demonstrate how small fillets may cause unfilled sections, laps, and cold shuts. Source: Ref. 4.

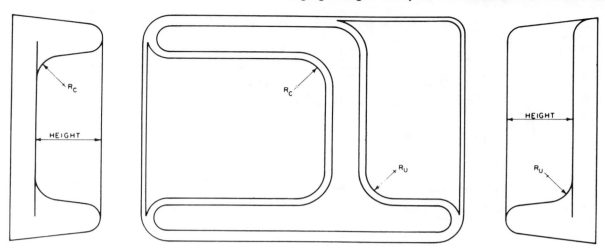

FIGURE 2–28. A fillet radius. Source: Ref. 4.

ity. If the web must be designed without punchouts, the best alternative is to incorporate web tapers by increasing the web thickness at the outer periphery. This design procedure also reduces the forging pressure required, although less than with the use of punchouts. Tapers improve the metal flow and permit easier filling of the rib cavities.

Die Deflection

Die deflection occurs on all forgings, regardless of the equipment used, but on most forgings it is so minute that it is not apparent. However, it is noticeable on large-area forgings that are exceptionally long and narrow and have very thin webs. The result of this die deflection is a thickening at the center of the forging, as indicated in Figure 2–36.

Die deflection can be anticipated, but cannot be predicted precisely. In sinking the die cavity, it is possible to compensate for the estimated deflection, but the special sinking involves more cost and time. To avoid compensation by die sinking, larger die closure tolerance is recommended. It may also be advisable to increase the straightness tolerance. Incorporating large punchouts in the central areas of these forgings also helps reduce or avoid die deflection.

Pockets and Recesses

Recesses are utilized in order to obtain desirable grain direction and, more frequently, to obtain better mechanical properties by reducing section thickness. Recesses in forgings are formed by corresponding raised sections or "plugs"

in the dies. It is good practice, in the interest of die life and forging reliability, to hold the depth of pockets and recesses to a minimum (Figure 2–37). Simple contours and generous fillets are recommended (Figure 2–38) wherever possible,

because they permit the use of smooth and durable die shapes that induce better metal flow.

The depth of a recess is normally restricted, because of die punch limits, to the size of its diameter or, for irregular or

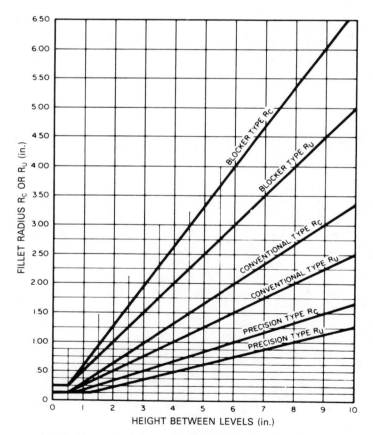

FIGURE 2–29. Recommended fillet radius sizes. Source: Ref. 4.

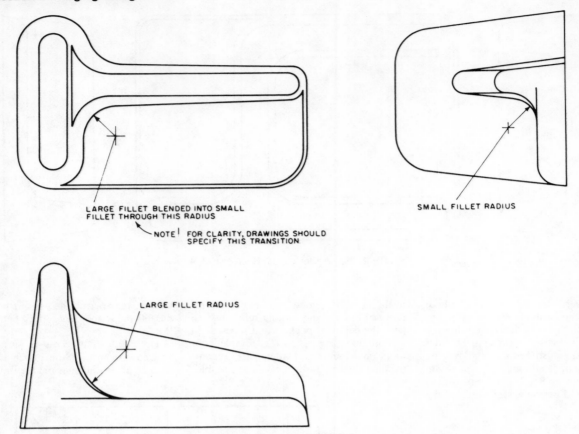

LARGE FILLET BLENDED INTO SMALL
FILLET THROUGH THIS RADIUS

NOTE! FOR CLARITY, DRAWINGS SHOULD
SPECIFY THIS TRANSITION.

SMALL FILLET RADIUS

LARGE FILLET RADIUS

FIGURE 2–30. Differing fillet radii blended through the plan view intersection radius. Source: Ref. 4.

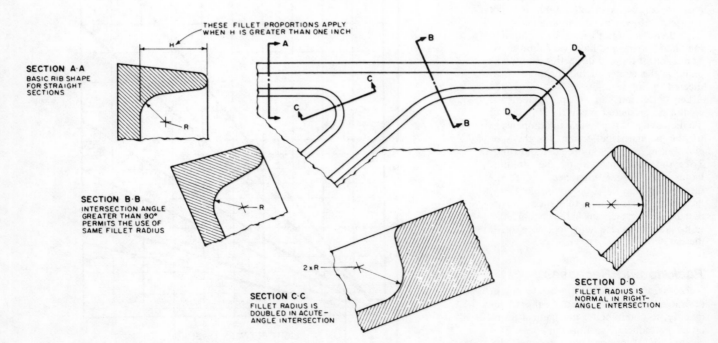

THESE FILLET PROPORTIONS APPLY
WHEN H IS GREATER THAN ONE INCH

SECTION A·A
BASIC RIB SHAPE
FOR STRAIGHT
SECTIONS

SECTION B·B
INTERSECTION ANGLE
GREATER THAN 90°
PERMITS THE USE OF
SAME FILLET RADIUS

SECTION C·C
FILLET RADIUS IS
DOUBLED IN ACUTE-
ANGLE INTERSECTION

SECTION D·D
FILLET RADIUS IS
NORMAL IN RIGHT-
ANGLE INTERSECTION

FIGURE 2–31. The fillet radius location on various sections of a forging. Source: Ref. 4.

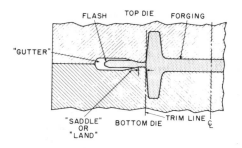

FIGURE 2–32. Example of conventional flash gutter design. Source: Ref. 5.

rectangular recesses, to the minimum transverse dimension (Figure 2–39a). The maximum depth may be one and a half times the diameter when a full radius (hemispherical shape) is allowed at the bottom, as shown in Figure 2-39(b). Holes of the finished component are normally preforged from both sides, and, as a result, a confined web is formed in the parting plane (Figures 2–39c and d), which is later removed in a separate punching operation. The thickness of this web is determined for deep holes by the permissible

depth of the recesses, as previously discussed. For shallow recesses, the thickness of the web (t_{min}) should never be less than the minimum specified for the given geometrical configuration.

"Bathtub" Sections

Dished or "bathtub"-type forgings become impractical when the wall thickness exceeds the web thickness. The walls should be no thicker than the web, as shown in Figure 2–40. If the final part re-

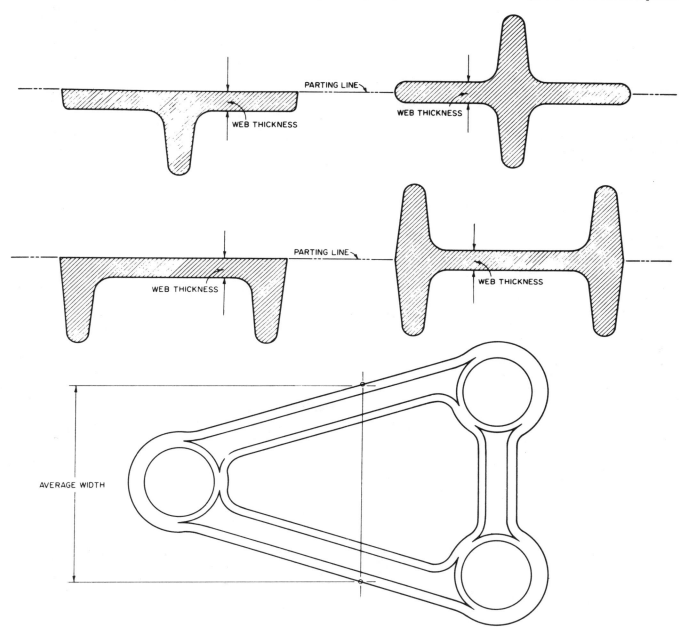

FIGURE 2–33. Minimum web thicknesses. Source: Ref. 4.

TABLE 2–2. Recommended Minimum Web Thickness Values for Small Precision or Conventional-Type Forgings

Within average width, in.	Within total area, in.^{2}	Recommended web thickness, in.
3	10	0.09
4	30	0.12
6	60	0.16
8	100	0.19
11	200	0.25
14	350	0.31
18	550	0.37
22	850	0.44
26	1200	0.50
34	2000	0.62
41	3000	0.75
47	4000	1.25

Note: Use the larger web thickness when width and area are not on same line. Source: Ref. 4.

quires a thinner web, it can be forged heavy and subsequently machined to size.

Punchouts

Besides aiding in the production of forgings with thin webs, as previously stated, punchout holes are used for functional reasons, such as lightening or providing clearance, or minimizing subsequent machining. As in web centers, punchouts may be beneficial in bosses or hubs, with the greater advantages resulting from large hole areas. The proportions of forged recesses govern the depth of the forged pockets left for piercing.

Punchout holes are incorporated by various means, such as die punching, machining, routing, or sawing. Where high strength is required, punchout holes may be reinforced with a bead, as indicated in Figure 2–41. A reinforced bead is not beneficial to the forging operation and should be generously blended into the web with a very large fillet or suitable taper. This is necessary to prevent flow-through, because the metal tends to flow more

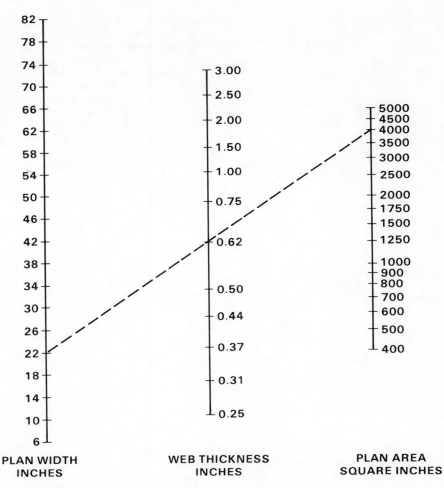

PLAN WIDTH INCHES **WEB THICKNESS INCHES** **PLAN AREA SQUARE INCHES**

FIGURE 2–34. Nomograph of recommended minimum web thicknesses for conventional and high definition forgings. Source: Ref. 4.

readily into the gutter basin than into the bead cavity.

Pads Added at Parting Line

When the top and bottom die impressions do not match at the parting line periphery, pads are recommended to match the protruding die impression. This permits the flash to be split equally about the parting line and simplifies die sinking. If both dies need matching, jogging the die flash would require more work. If only one die requires matching, incorporating all of the flash in the other die would make the forging appear to be out of match at the parting line. Adding small pads easily solves the problem. Such pads vary in thickness from 0.06 in. on small forgings to 0.50 in. on very large forgings.

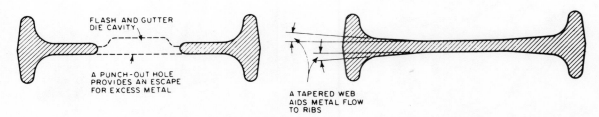

FLASH AND GUTTER DIE CAVITY

A PUNCH-OUT HOLE PROVIDES AN ESCAPE FOR EXCESS METAL

A TAPERED WEB AIDS METAL FLOW TO RIBS

FIGURE 2–35. Punchouts and web tapers. Source: Ref. 4.

FIGURE 2–36. Thickening at the center of thin and wide webs. Source: Ref. 4.

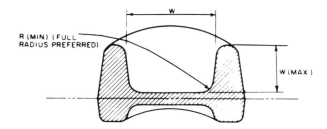

FIGURE 2–37. Shallow and rounded recesses. Source: Ref. 4.

Match Marks and Vent (Knockout) Marks

Match marks are useful to the inspector to quickly determine whether the forging is being made within the match tolerance. If not, production is stopped and the dies are adjusted to meet the tolerance. Match marks are very slight ridges placed in a line across the parting line, generally on the drafted surfaces; they are approximately 0.015 in. high by 0.375 in. long on each side of the parting line. The forging vendor determines the match marks, which need not be shown on the customer drawings.

The vendor also decides whether a specific forging requires that vents or knockouts (ejectors) be incorporated in the dies.

The customer's drawings need not indicate these marks.

Drafting Conventions for Forgings

A forging drawing should be prepared so that the designed forging can be produced economically and machined satisfactorily; it should illustrate the forging explicitly and completely to avoid misinterpretation. A few special conventions generally used in the forging industry are discussed next.

A full-scale drawing minimizes possible confusion and errors, because it depicts the part in true proportions. Drawings that are not made to full scale frequently cause mistakes and should be avoided. A small-scale drawing may be handy to use, but this convenience is costly if errors result. Enlarged views may be necessary when details are so minute that even full-scale views would be difficult to read.

Simplified drawings should be used, showing neither more nor less than is essential. They should have complete, clear, detailed information. General notes, preferably near the title blocks, should be used to avoid cluttering the drawing. The following note can be used to avoid duplicating dimensions, especially if these apply repeatedly throughout the drawing:

Note: Unless otherwise indicated:
    Draft angles   _____
    Corner radii   _____
    Fillet radii   _____

A similar note will eliminate the need to show tolerances on every dimension:

Note: Forging tolerances unless otherwise indicated:
    Die closure   _____
    Dimensional   _____
    Match   _____
    Straight within   _____
    Flash extension   _____

Perhaps the principal difference from the conventional drafting technique is the delineation of lines in the plan view. The method of showing all plan view lines (Figure 2–42), as if the corner and fillet radii were all theoretically sharp corners, is common on forging drawings. Thus, these radii are ignored, and the lines shown result from continuing the draft angles and terminating them at the top and bottom surface planes. This method depicts the

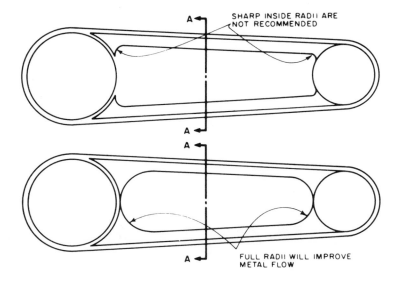

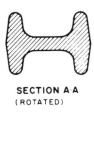

SECTION A-A
(ROTATED)

FIGURE 2–38. Illustration of full radii versus sharp inside radii. Source: Ref. 4.

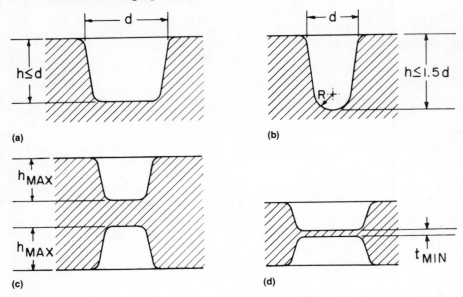

(a) **(b)** **(c)** **(d)**

FIGURE 2–39. Schematic examples of dimensional proportions for recesses and preforged holes. See text for discussion of (a) through (d).

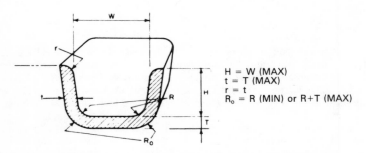

$H = W$ (MAX)
$t = T$ (MAX)
$r = t$
$R_o = R$ (MIN) or $R+T$ (MAX)

FIGURE 2–40. "Bathtub"-type forgings. Source: Ref. 4.

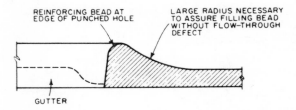

REINFORCING BEAD AT EDGE OF PUNCHED HOLE

LARGE RADIUS NECESSARY TO ASSURE FILLING BEAD WITHOUT FLOW–THROUGH DEFECT

GUTTER

FIGURE 2–41. Reinforced bead at edge of punchout holes. Source: Ref. 4.

draft more vividly in the plan view and simplifies the work of the draftsman. All other intersections are depicted by lines indicating a change of direction.

Dimensions to Points of Intersection. The dimensions are likewise taken to the theoretical points of intersection or sharp corners. In most cases, they are taken to the tops of the forging, with the draft material added on. A note and sketch similar to Figure 2–43 will also ensure that the point is understood.

Most drawings are now specified in decimal dimensions, and the trend is increasing. It is most important to use a proper method of dimensioning. Dimensions should be related to the subsequent machining operations and should even follow the sequence of these operations, if possible. In making his own drawing, the

forging vendor must follow the customer's drawing exactly to avoid inspection controversies. The responsibility rests with the customer to dimension properly. If the vendor finds any dimensional irregularities, the customer is advised and must resolve them before the die impressions are begun.

Tooling Points Shown on Drawings. Every effort should be made to have the drawing dimensioned so explicitly that the customer and the vendor can agree precisely in their dimensional findings when making a layout inspection of a forging. To prevent such controversies, tooling points (actual points of fixture contact when machining or when inspecting layouts) should be indicated on the drawings.

These tooling points can also be used to establish datum planes from which all possible dimensions originate. By having a common starting setup when checking a forging dimensionally, it is possible to avoid inspection differences that sometimes occur. A tooling point or similar dimensioning system reduces layout time because it eliminates many preliminary setups. It also gives more positive verification that adequate finish allowance has been provided before machining begins. A tooling point system or a similar system of dimensioning is strongly recommended, and more companies are adopting such methods.

Avoid Chain Dimensioning. Chain dimensioning should be avoided to prevent an accumulation of tolerances (see Figure 2–44).

Reference dimensions (normally labeled "Ref.") are adjacent to the dimension and used for reference purposes only, not for manufacturing or inspection. They may result from other dimensions or be duplicated elsewhere to reference other drawings. Reference dimensions should be used discriminately and only if required for clarification.

Variables With Differing Dimensional Methods. The following illustrates two different methods of dimensioning and the variable effect on the resultant dimensions. In Figure 2–45(a), if the outside and inside diameters are at the extreme tolerances, the resultant wall thickness may vary ±0.12 in. In Figure 2–45(b), if the outside diameter and the wall thickness are at the extreme tolerances, the resultant inside diameter may vary ±0.36 in. In both (a) and (b), the referenced dimensions should not be verified for size tolerance, because these are only resultants from the given dimension and tolerances. This example is given only to show how the results may vary on referenced dimensions

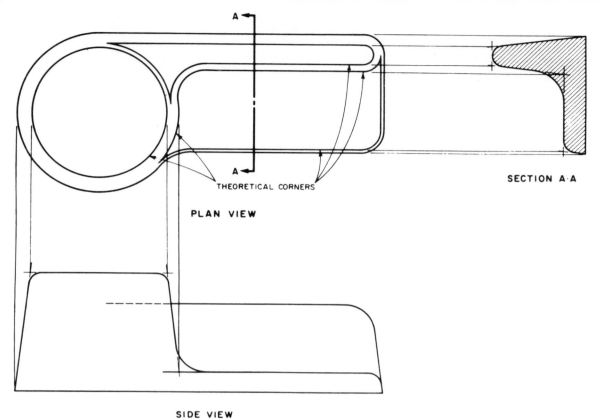

FIGURE 2–42. Plan view lines as the theoretical intersection points of the corner and fillet radii of the side views. Source: Ref. 4.

when different base dimensions are used. The customer should select the method of dimensioning that produces the best results.

Combine Tolerances on Side Walls. A side wall with normal draft angle formed by a bottom die cavity and a protruding top die is affected by both dimensional and match tolerances. Checking these tolerances separately is difficult and causes inspection controversies. To avoid this difficulty, both tolerances should be combined and their sum added to the side wall di-

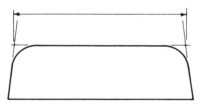

ALL DIMENSIONS TAKEN TO
POINT OF INTERSECTION

FIGURE 2–43. Example of a sketch with note that is conventional on forging drawings. Source: Ref. 4.

mension with an explanatory note, as illustrated in Figure 2–46.

Designated Forging Plane When Parting Line Is Irregular. The parting line should always be specified on the drawing, but the flash extension at the parting line periphery should not be pictured. When the parting line is irregular, the forging plane must also be designated (Figure 2–47) for die-making purposes and for drawing completeness, because draft angles are usually taken normal to the forging plane. There is no need to indicate the forging plane when it coincides with the parting line.

Uniform Draft Indicated by Concentric Radii. At plan view intersections of plane surfaces, at different levels but parallel to each other, concentric radii indicate uniform draft, as shown in Figure 2–48. These are incorporated easily and naturally in sinking the die cavity with conical cutters. When not shown as concentric radii, these areas require difficult and more costly hand-worked die blending.

Dimensioning Bosses Inclined to Forging Plane. Bosses inclined to the forging plane should be dimensioned at the points of tangency to simplify die sinking and to obtain round bosses, as shown in Figure 2–49. When so dimensioned, the die may be initially inclined to the required angle and the boss cavity sunk with a straight-sided cutter. After the die is returned to its normal position, a conical cutter cuts the normal draft tangent to the already incorporated bottom radius, but only around that part of the periphery where it is necessary. If this same boss cavity were sunk to the usual point of intersection dimensioning, the bottom radius would actually take on an oval contour instead of the intended round contour, and the forged boss would appear to be discrepant.

Dimensioning Acute Angle Intersections. By not specifying the radius when ribs intersect at an acute angle (Figure 2–50), the variations within the standard tolerances remain at a minimum.

Intersections of Webs and Tapered Ribs. The recommended delineation (Figure 2–51) to show intersections of webs and tapered ribs follows the concept of disregarding radii and showing the forging as though it has sharp corners.

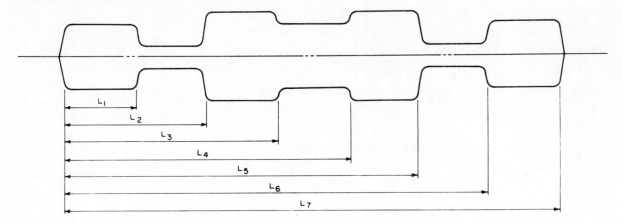

FIGURE 2–44. Illustration of a common reference used to avoid chain dimensions. Source: Ref. 4.

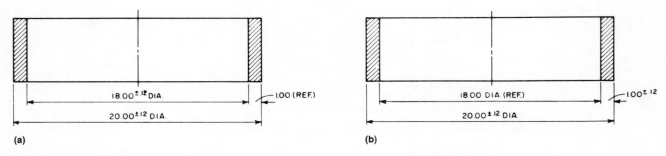

FIGURE 2–45. Two methods of dimensioning. See text for discussion of (a) and (b). Source: Ref. 4.

Dimensioning Plan View Radii at Inside Intersections. The radii in the plan view at inside intersections are not considered fillet radii and must be specified separately (Figure 2–52). These are incorporated in the die by moving the cutter in a radial path, which results in the previously described concentric radii, indicating uniform draft. It is preferred practice to dimension these radii at the bottom of the draft, making them at least as large as the general fillet radii, to ensure adequacy where fillets are most important.

Incorporating Fillet Radii at Rib Intersections. There are various ways of depicting blends of transitions of differing fillet radii at rib intersections, and the drawing should indicate exactly what is required. The four methods in Figure 2–53 are used commonly, and each produces a different physical effect.

Dimensioning an Offset in Plan View. A recommended method of dimensioning a plan view offset is to the tangency of the outside radius (Figure 2–54). It is also satisfactory to locate the center of the same radius. Dimensioning to the intersection of the outside and inside radii should be avoided, because this approach is ambig-

uous, especially if the two radii are unequal.

Dimensioning Shallow Steps Between Two Surfaces. The general point of intersection rule applies on a shallow step between two surfaces and locates the corner radius; this is constructed first, followed by the corner radius. Any exception to this must be specifically shown on the drawing—for example, when the fillet radius is to be positioned first. Figure 2–55 illustrates both a drawing and the step-by-step die instruction.

Deviation From Normal Dimensioning. There are cases where it is preferable to deviate from the normal practice of dimensioning to the point of intersection of the corner radius. Thus, where the corner radius is much larger than the thickness of web, as illustrated in Figure 2–56, the drawing should clearly indicate that the dimension is taken to the parting line. This is more measurable and helps to avoid layout controversies.

Draft Projections at Rib Ends With Large Radii. The draft contour is a func-

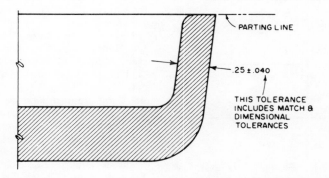

FIGURE 2–46. Tolerancing on side walls. Source: Ref. 4.

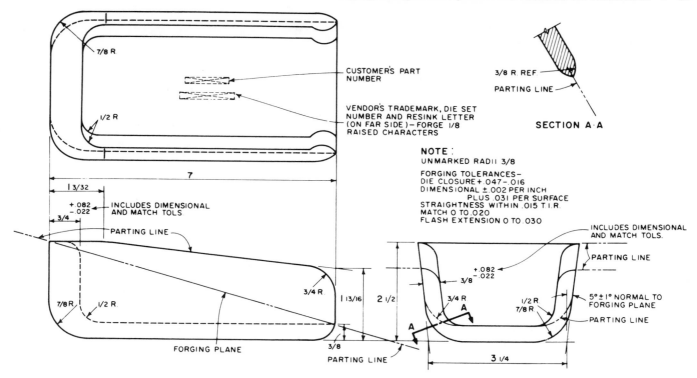

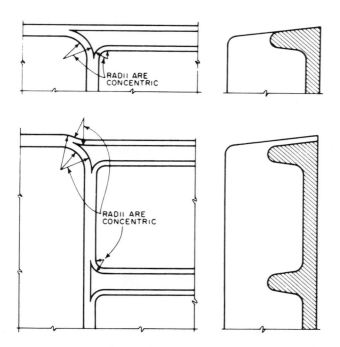

FIGURE 2–47. Designation of forging plane. Source: Ref. 4.

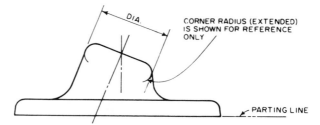

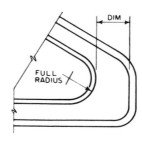

FIGURE 2–48. Concentric radii. Source: Ref. 4.

FIGURE 2–49. Round bosses inclined to the forging plane. Source: Ref. 4.

FIGURE 2–50. Specifying radius dimensions at acute-angle intersections. Source: Ref. 4.

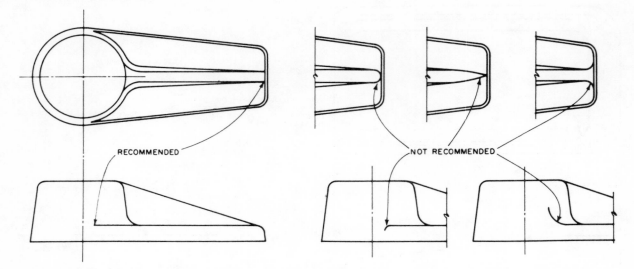

FIGURE 2–51. Proper and improper methods of showing intersections of webs and tapered ribs. Source: Ref. 4.

tion of the depth of the impression from the parting line. At a rib end with an end radius larger than the corner radius, the draft develops naturally when the ball die cutter follows the path of the large radius (Figure 2–57). Note that the top lines are projected from the theoretically sharp corners of the die cutters and the large end radius is taken into consideration. On cylindrical parts with the parting line through the diameter, the draft configuration is developed similarly.

Draft Projections on Tapered Webs. The draft in the plan view should indicate the change in web thickness, as shown in Figure 2–58. The projections reflect the web thickening with the taper or developed curve lines.

Identification characters on forgings may be either permanent or temporary. The permanent types include integral, impression-stamped, vibra-stamped, and electroetched characters; temporary characters are ink stamped.

Integral characters, either raised or depressed, are formed by opposite or negative characters in the forging die. The most common are raised letters from depressed lettering stamped into the die impression. These are also the most economical, because the die stamping is routine and performed only once. If distinctive and superior characters are preferred, these can be provided by a more expensive die engraving process and may be either raised or depressed.

Integral characters are forged and therefore must be located on a surface parallel or nearly parallel to the forging plane and in an area where the metal flow is relatively retarded, such as near the center. The area must also be flat and large enough for the necessary lettering. The standard sizes of lettering stamped in the dies are: 0.06, 0.09, 0.13, 0.16, 0.19, 0.25, 0.31, and 0.38 in. The lettering size should be proportional to the forging size.

The height of integral characters above the adjoining surface is from approximately 0.015 in. on small forgings to 0.031 in. on large forgings. Die stampings in areas exceeding these heights may invite difficulty in obtaining uniform lettering or may cause lettering washouts, especially in areas of excessive metal movement.

Impression-stamped, vibra-stamped, or electro-etched characters must be added on each idividual forging. This increases costs because it demands more handling and more careful attention. Of the three methods, impression stamping is strongly preferred. The desirable location for lettering is a flat, adequately large, accessible surface. The normal depth of these characters is a maximum of 0.015 in. If a specific depth is designated, special equipment becomes necessary to control the operation. Impression stamping is frequently located on raised pads deliberately used where the stress-raising effect of such

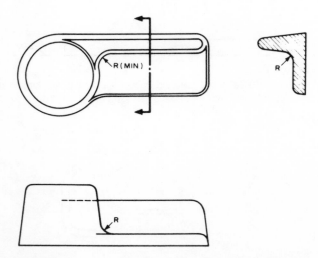

FIGURE 2–52. Individually dimensioned radii at inside intersections in plan views. Source: Ref. 4.

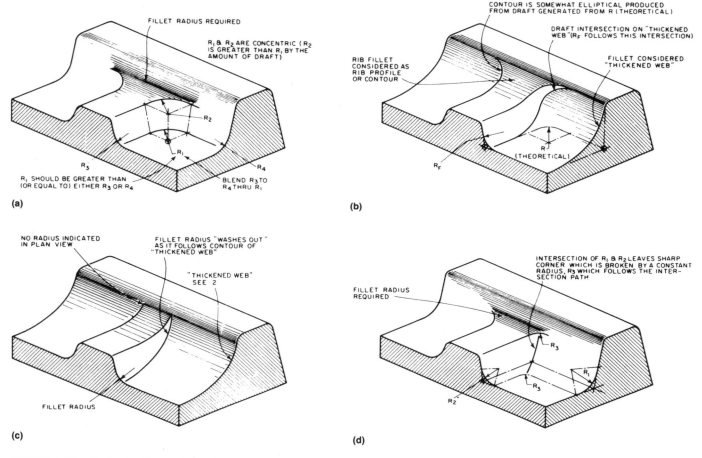

FIGURE 2–53. Methods of illustrating fillet radii at rib intersections. (a) Most common application. (b) When rib top fillet area is a constant profile. (c) When no plan view radius at the rib intersection is indicated. (d) Constant plan view radius also applies when differing draft angles must be blended at the intersections. Source: Ref. 1

stamping is feared. These pads should be large enough to accommodate all the stamping easily.

Ink-stamped characters are temporary and may be easily obliterated. Because of its impermanent nature, ink-stamping can be done only at the completion of all forging operations. This procedure is not routine and also demands special handling. If the forging configuration is such that it is not possible to obtain integral characters, or if integral characters are illegible, rubber ink-stamping may be used.

The essential information that must appear on a forging is determined by the customer. Typical drawing identification callouts are shown in Figure 2–59. Part of the identification data is also helpful to the vendor in processing forgings through his shop. If identification is omitted on the original customer drawing, the vendor may propose identification, but it must be approved before being applied. The identi-

fication may include the customer's part number, the vendor's trademark and the die set number, the alloy designation, the lot number, an inspection symbol, and a serial number. The part number, vendor's identification, and alloy designation should preferably be in integral characters, because they do not change. The lot number (sometimes called melt, heat, or heat-treat

number), the inspection symbol, and the serial number change must be applied individually on each forging.

Contours Defined by Ordinates. The layout of a contour defined by ordinate dimensions is done with a flexible spline or numerical control to obtain the truest possible curvature, unless otherwise specified. If the contour takes a decided change,

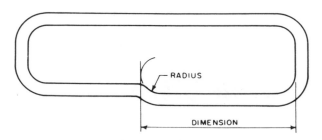

FIGURE 2–54. Dimensioning offsets to the tangent of the radius. Source: Ref. 4.

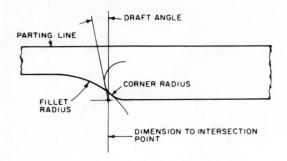

DRAWING LAYOUT SEQUENCE

1. INTERSECTION POINT
2. DRAFT ANGLE
3. CORNER RADIUS
4. FILLET RADIUS

DIE CONSTRUCTION SEQUENCE FOLLOWS

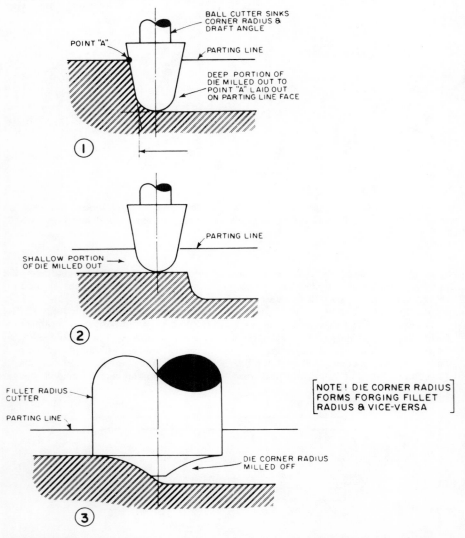

NOTE! DIE CORNER RADIUS FORMS FORGING FILLET RADIUS & VICE-VERSA

FIGURE 2–55. Step-by-step drawing and die sinking sequence in incorporating a shallow step. Source: Ref. 4.

this portion should have sufficient ordinate points to define the desired curve adequately.

Outside Corner Radii in Plan View. When the general corner radius to an outside corner radius is in the plan view, it is drawn smaller because of the draft angle and in keeping with the practice of showing plan view lines as intersections of the draft and top surfaces (Figure 2–60).

Dimensioning in Crowded Areas. To avoid crossing the solid drawing lines with extension and dimension lines (this makes a drawing difficult to read, especially when details are compacted), a flag system of dimensioning is suggested, as shown in Figure 2–61. The dimension is placed inside the flag, which has one side coinciding with the extension line and the opposite point directed toward the datum line from which the dimension is taken. The extension line is kept as short as possible. A general drawing note should be added to explain this method of dimensioning. The drawing will be relatively clear and easy to interpret. Another suggestion would be to follow the same suggestions, but omit the flag; however, such a drawing would not give assurance of proper interpretation.

Drawing Callouts. A drawing callout such as "break all sharp edges" is normally interpreted to be applicable to all surfaces except at the flash trim. Thus, all marks caused by die checking, die venting, die knockouts, or handling are to be removed.

A similar drawing callout that specifies the amount of radius or chamfer required (either maximum or minimum) can increase costs unnecessarily, especially if the tolerances are restrictive. Unrestrictive tolerances such as 12 in. maximum or 0.005 in. minimum would be fairly easy to control. Actually, radius or chamfer requirements would apply only to knockout marks, as other marks would normally be removed flush with the forging contour. Removing the sharp edges or the knockout marks, which would require very careful grinding, machining, or controlled tumbling to comply with the dimensional requirement, would be difficult. If the callouts, "break all sharp edges," are intended to apply also to the flash trim, the drawing should specify that removing the sharp edges from the flash extension presents a greater problem because of the much larger periphery involved.

Engineering Terms

A basic dimension defines the theoretically perfect size or location of a feature

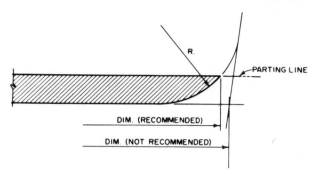

FIGURE 2–56. Illustration of where the usual point of intersection dimension rule should be avoided. Source: Ref. 4.

and is without tolerance. It establishes a base from which allowable deviations are to be measured.

A datum establishes the exact feature from which other features are located or measured. It may be a centerline, plane, or diameter from which dimensions originate.

True position (T.P.) signifies the theoretically perfect position of a feature.

MMC stands for "maximum material condition" of a feature and would define a maximum shaft diameter and a minimum hole diameter. This expression is used with form tolerance, which is minimum when the feature is at maximum size, but increases the form tolerance when the feature is smaller by the difference from MMC.

LMC stands for "least material condition" of a feature and would define a maximum hole diameter and a minimum shaft diameter. This expression is used with form tolerance, and when the feature contains the least amount of material, the form tolerance increases accordingly.

RFS stands for "regardless of feature size." It is used with form tolerances and with MMC callouts. It signifies that the form tolerance must not be exceeded, regardless of the actual size of the feature.

T.I.R. stands for "total indicator reading" and signifies the maximum tolerance. It may represent a diametral tolerance zone or a tolerance zone between two parallel planes. The callout "F.I.R." means the same and stands for "full indicator reading." The callout "F.I.M." means "full indicator movement" and when appropriately applied to a surface measures its variations.

Dimension "Max." defines the high limit of the dimension, with zero being the low limit.

Dimension "Min." defines the low limit of the dimension, with infinity (or whatever is practical) as the high limit.

Dimension "Typ." implies that the dimension is applicable to all similarly sized features.

Typical forging drawings are illustrations of forgings in true proportions.

Tolerances for Die Forgings

Tolerances, on the whole, represent a compromise between the desired accuracy and that obtainable within economical reason. Therefore, tolerances cannot be re-

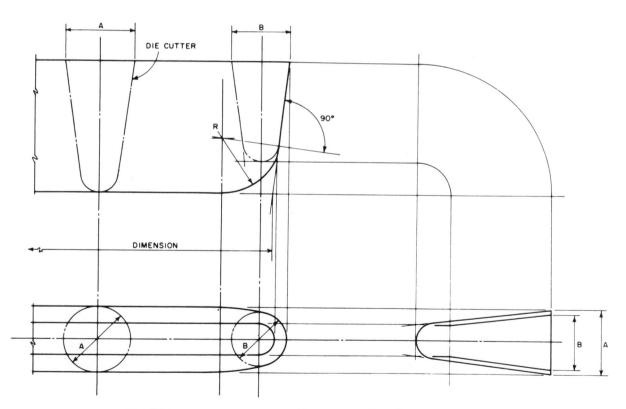

FIGURE 2–57. Draft projection rib end with a large radius. Source: Ref. 4.

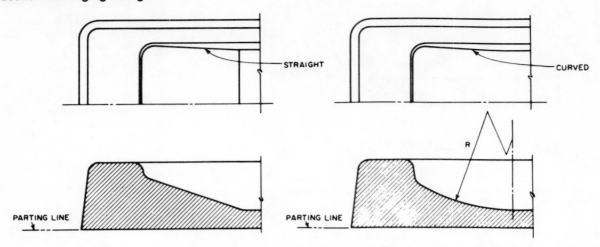

FIGURE 2–58. Draft line projections of reflecting web thickening patterns. Source: Ref. 4.

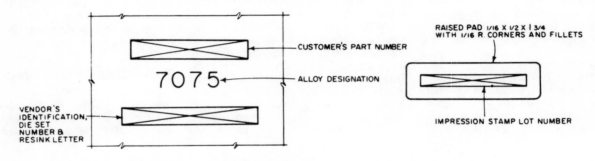

FIGURE 2–59. Typical drawing identification callouts. Source: Ref. 4.

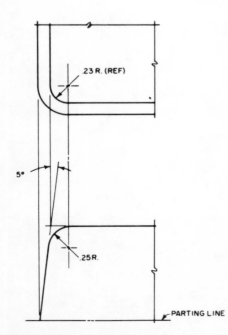

FIGURE 2–60. The referenced plan view radius and the corner radius. Source: Ref. 4.

garded as rigid standards. Their value changes with the development of technology and design concepts. Because the number of factors influencing dimensional accuracy is so great, one cannot expect tolerance systems to be applicable to all situations. At best, they represent the reasonable average performance that can be expected from a forge plant in a given industrial climate. Tolerances closer than this obviously require greater care, skill, more expensive equipment, or a more expensive technology and therefore carry a premium. When tolerances are excessively tight, cost of production may become disproportionately high. An experienced and capable producer can offer valuable assistance in handling dimensional control problems.

Dimensional Accuracy Influenced by Design. A forging's proportions have a great deal to do with its dimensional accuracy. Shrinkage, warpage, and other deviations usually can be anticipated and their effects minimized by appropriate design precautions. Shrinkage is the dimensional contraction that occurs when a forg-

ing cools to room temperature after it is removed from the die. The extent of shrinkage depends on the temperature of the part as it comes from the die and on the forging alloy's coefficient of thermal expansion. By providing a simple, well-proportioned shape and ample fillets, the designer can be assured that the average forged part will be within its straightness tolerance. When it is necessary to employ odd shapes, radical section changes, or extremely long, thin members, the likelihood of warpage is increased. Figure 2–62 is an example of a potentially troublesome part.

When producing die forgings that have to meet tolerance limits, forging plant operators will combine reasonable tooling and part costs with adequate dimensional accuracy. In individual cases, tolerances smaller than standard values may be justified. Certain dimensions often require close limits to suit fixtures used in subsequent processing, to ensure clearance for mating parts, or to satisfy other conditions stemming from a forging's application. The vendor's engineering departments should

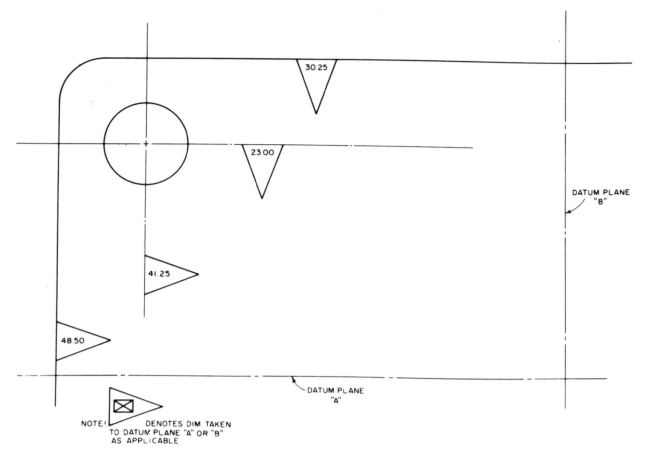

FIGURE 2–61. A flag dimensioning system. Source: Ref. 4.

be consulted when changes are desired in tolerance values. Their estimate of the cost increase will help decide whether very close tolerance limits, or other measures such as additional machining, will result in the most economical forging.

In most cases, tolerances vary for the different types of forgings (precision, conventional, and blocker). Tighter-than-necessary tolerances should be avoided as these invariably result in higher forging costs. A properly dimensioned drawing will result in the minimum possible tolerances, whereas poor dimensioning can result in a buildup of tolerances. Unless a drawing explicitly specifies otherwise, each class of tolerance is separate and independent of every other class of tolerance. All applicable tolerances should be specified on the drawing. Some suggested die forging tolerances follow.

Die Closure Tolerances

Die closure tolerance pertains to thicknesses across and perpendicular to the fundamental parting line and is affected by the closing of the dies. This tolerance normally includes the initial die sinking limits, the subsequent die polishing necessary to maintain smooth die cavity surfaces in production (correcting for "die wear"), and an allowance for die deflec-

tion (which creates a thickening at the middle, especially with thin webs). In addition to thicknesses, there are dimensions perpendicular to the fundamental parting line that are covered by varying tolerances. They fall into categories with tolerances applicable as follows:

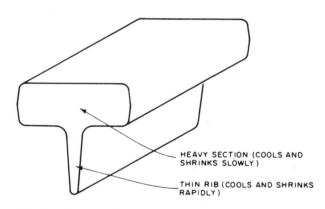

FIGURE 2–62. Forged part with heavy and light sections. Source: Ref. 4.

TABLE 2–3. Recommended Die Closure Tolerances (in.)

Weight, lb (within)	Plan area, in.² (within)	Forging Precision type		Forging Conventional type		Forging Blocker type	
0–½ 10		+0.020	−0.010	+0.020	−0.010	+0.031	−0.016
½–1 30		+0.020	−0.010	+0.031	−0.016	+0.047	−0.031
1–5 100		+0.031	−0.010	+0.047	−0.016	+0.062	−0.031
5–20 400		+0.047	−0.016	+0.062	−0.031	+0.093	−0.062
20–50 750		+0.062	−0.016	+0.093	−0.031	+0.125	−0.062
50–100 1000		+0.093	−0.016	+0.125	−0.031	+0.187	−0.062
100–200 2000		+0.125	−0.016	+0.187	−0.031	+0.250	−0.062
200–500 3500		...		+0.250	−0.031	+0.375	−0.062
500 up 5000		...		...		+0.500	−0.062

Note: Use the larger tolerance when weight and plan area are not on same line. Source: Ref. 4.

Full-die closure tolerance

- All thicknesses, whether the forging impression is all in one die half or in two die halves
- Center dimensions across parting line or from center to parting line when the impression is all in one die

One-half die closure tolerance plus a symmetry tolerance of ±0.015

- Dimensions from surface to parting line when impression is in both dies

- All center dimensions to parting line when impression is in both dies

All other dimensions covered by dimensional tolerances

Recommended die closure tolerances are listed in Table 2–3. Note, however, that the applicable tolerance is determined by the weight or the plan area of the forging—whichever is greater. The larger tolerance should be used when weight and plan area are not on same line.

Coining Tolerances

Coining tolerance is ±0.005 in. minimum. Similar to die closure tolerance, it is applied to the surface across and parallel to the parting line. It is limited to small areas only and usually to portions of forgings, such as boss faces. The coining operation is performed on cold forgings after heat treating and prior to aging.

When two or more sets of bosses are coined simultaneously (Figure 2–63), each set of bosses will have the faces parallel to one another and within coining tolerance. However, the coining operation may not straighten the forging, and it may spring back to its original distorted shape if any warpage existed before the coining.

Dimensional Tolerances

Dimensional tolerances (sometimes called length and width tolerances) usually apply to dimensions essentially parallel to the fundamental parting line. They also actually apply to all dimensions not otherwise covered by the die closure tolerances. Dimensional tolerances consist of two factors, a shrinkage tolerance and an

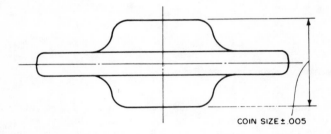

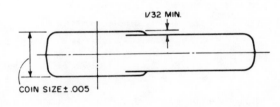

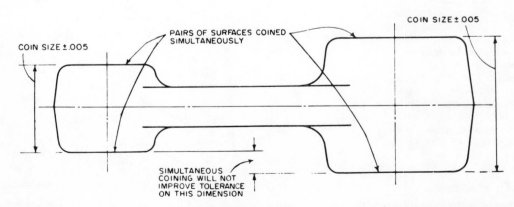

FIGURE 2–63. Example of coin sizing to refine dimensional relationships between opposing flat surfaces. Source: Ref. 4.

TABLE 2–4. Suggested Dimensional Tolerances

Forging type	A Shrinkage variation per in.		B and C Per each surface on forgings with maximum dimensions to 60 in.		Per each surface on forgings with maximum dimensions over 60 in.	
Blocker	+0.002	−0.002	+0.047	−0.015	+0.078	−0.015
Conventional	+0.002	−0.002	+0.020	−0.010	+0.047	−0.015
Precision	+0.0015	−0.0015	+0.010	−0.010	+0.020	−0.010

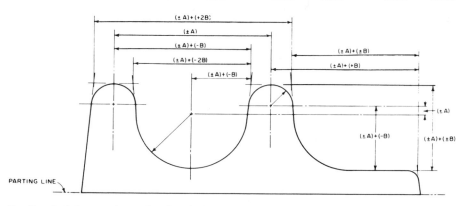

Note: Dimensional tolerances apply to any dimensions other than those covered by die closure tolerances, even to dimensions perpendicular to the parting line. Dimensions are (1) between opposite outside surfaces, (2) between opposite inside surfaces, (3) between centerlines, (4) between center and outside surface, (5) between center and inside surface, and (6) between steps. Figure shows only how dimensional tolerances apply with various types of dimensions, but the method of dimensioning is unfavorable and not recommended. Source: Ref. 4.

TABLE 2–5. Suggested Match Tolerances

Weight, lb (within)	Overall length, in. (within)	Forging Precision type	Conventional and blocker types
0–1	10	0.010	0.015
1–5	17	0.015	0.020
5–20	25	0.020	0.030
20–50	50	0.030	0.045
50–100	75	0.045	0.060
100–200	100	0.060	0.080
200–500	150	· · ·	0.100
500 up	250	· · ·	0.120

Note: Use the larger tolerance when weight and overall length are not on same line. Source: Ref. 4.

allowance per surface. The shrinkage tolerance is solely for the oversize or undersize variations that may occur because of shrinkage differences. The per surface allowance is a "plus material" amount on the forging, frequently called "die wear allowance." This allowance includes the die sinking limits, die wear, and die dressouts; it results in larger outside dimensions and smaller inside dimensions.

Suggested dimensional tolerances that depend on both the size and type of dimensions are listed in Table 2-4. The individual dimension may be affected by shrinkage only or by both shrinkage and die wear; center dimensions are affected only by shrinkage. Dimensions from a center line to a surface are affected by both shrinkage and die wear allowance. Dimensions from a center line to a surface are affected by both shrinkage and die wear allowance. Dimensions to two surfaces are affected by shrinkage and two die wear allowances. Die wear is principally the result of abrasion by the forging or die cavity polishing that is used whenever it is necessary to maintain the smoothness of die surfaces required for proper metal flow. On cold-worked die forgings, greater-than-normal dimensional tolerances may be necessary and are determined on an individual basis.

Match Tolerances

Match tolerance is the maximum shift or misalignment variation allowed between the two die halves at and parallel to the parting line, as shown in Figure 2-64. The features on one side of the forging are slightly out of line with those on the other side because of this shift. Mismatching is caused when the forging forces are exerted parallel to the forging plane. To counteract these forces and maintain match on the forging, guide holders, guide pins, and counterlocks are employed. Match tolerances are applied separately from, and independently of, all other tolerances. They depend on the weight or overall length of the forging, whichever is greater, as indicated in Table 2-5.

Straightness Tolerances

Straightness tolerance is a deviation applicable generally to flat surfaces and has a total indicator reading (T.I.R.) limit. On a continuous flat surface, it is the total maximum deviation from a plane surface. On noncontinuous surfaces, the deviation is a total flatness relationship of all parallel surfaces; however, it does not include the step tolerance that may exist between any two surfaces, nor is it applicable to any surfaces that are inclined to the major surfaces being measured. Contoured or tapered surfaces are not covered by the specified straightness tolerance, but must be within the specified dimensional tolerances, which should be large enough to allow for any warpage existing at these areas. The contour envelope tolerance

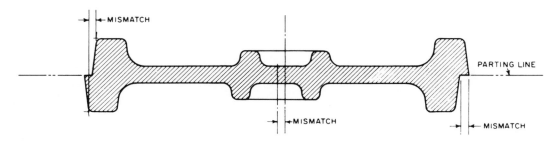

FIGURE 2–64. A misalignment between the dies. Source: Ref. 4.

TABLE 2–6. Suggested Straightness Tolerances

Overall length, in. (within)	Tolerance (T.I.R.), in.
10	0.015
20	0.030
30	0.045
40	0.060
50	0.075
60	0.090
90	0.120
150	0.180
Over 150	0.250

Source: Ref. 4.

method is sometimes used. On cylindrical forgings, the straightness tolerance is applicable to the axis of the part.

Straightness tolerances are also measured separately from, and independently of, all other tolerances. Warpage in a forging, caused by differential cooling of varying sections, occurs both after the hot forging operation and (especially) during quenching after solution heat treating. Straightening is done when the forgings are cold, as soon as possible after solution heat treating, but prior to aging treatment. Sometimes an additional set of dies is used for straightening, but the total production requirements must be high enough to warrant the cost. In most cases, however, forgings are straightened in hydraulic arbor presses, utilizing V-blocks or other simple supports. This process is not as positive as die straightening and must sometimes be repeated to bring the part within allowable straightness tolerance. However, it may be the most economical method.

The use of stress relief dies on cold forgings is effective, not only in straightening the forging, but also in substantially reducing residual stresses. The overall length of the forging determines the amount of straightness tolerance (Table 2-6).

Although the straightness tolerance is independent of, and in addition to, all other tolerances, it may be difficult or even impossible to measure it separately from other tolerance deviations because of part configuration. This applies particularly to ring, tubular, and contoured shape forgings.

However, the straightness tolerance must be considered, and an allowance must be made for it. On such shapes, a 0.0015-in. total tolerance per inch of dimension should be added to the dimension tolerance as an allowance for the straightness deviation. On ring-type and tubular-type forgings, this larger tolerance would be applicable only to the diameters, with the additional straightness allowance allowed for a possible ovality condition. The general straightness tolerance would otherwise apply to the flat surfaces.

An ovality is permissible even if a drawing omits any mention of ovality, but it must not exceed the diametral tolerance limits. If the ovality is specified and is restricted to only a portion of the total diametral tolerance, both tolerances can easily be checked by conventional methods. On extremely thin sectioned ring forgings, which may present severe ovality problems, it may be necessary to specify an ovality tolerance even greater than the diametral tolerance. This would require that the diameters be measured with "PI" tapes to determine if they are within the dia-

metral tolerances. On a contoured shape, there is no practical method of checking the straightness deviation separately; it must therefore be included in the dimensional check.

Methods and Procedures for Checking Straightness. The method of measuring straightness should verify that the surface or surfaces do not deviate by more than the tolerance from the required plane surface. On cylindrical parts, the method must ensure that the actual axis does not exceed the straightness limit with respect to the true axis. If a forging is flat on one side, or nearly so, a straightedge can be used to measure the straightness. However, experience and careful inspection are required to detect twist conditions that may exist, or to make certain that a whole area is within the T.I.R. reading, since only a line contact can be checked in sliding the straightedge over the entire surface.

Another method of verifying straightness is to place the forging, if one side is entirely flat, against a surface plate and use a feeler gage to check the straightness deviation (Figure 2-65). Extremely narrow parts can also be checked in this manner. The middle portion of a forging of any appreciable width cannot be checked, because feeler gages can only verify that the outside periphery is within tolerance limits.

The recommended method is to position the forging on a surface plate on three supports, at least two of which are adjustable, and adjust them so that the top surfaces of the forging are parallel with the surface plate. Checking the top rather than the bottom surfaces is suggested because

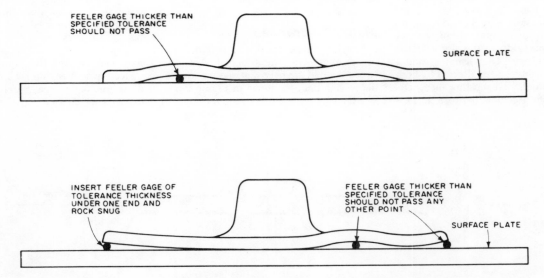

FIGURE 2–65. Methods of checking straightness on very narrow parts. Source: Ref. 4.

these are more accessible. Since the straightness tolerance is separate from, and independent of, all other tolerances, and the forging is supported on the bottom side while the top side is checked, all other forging tolerances must be excluded from the straightness check. In this case, for example, the forging may have a die closure variation, even for an allowable taper. Such variations are avoided by using the adjustable supports and working to a plane established by the top surfaces.

Getting the best straightness reading on the top surfaces may require a few adjustments to the supports before a satisfactory checking plane, from which all the top surfaces can be checked, is established. This plane may vary from forging to forging of the same lot. Most forgings do not have large, continuous areas to establish the checking plane.

Any offsets, from one level to another, needed to determine the plane must be set up to the actual forging step measurements rather than to the drawing dimensions. The dimensional variations in such steps are covered by dimensional tolerances and should not be absorbed into the straightness allowance.

Once the horizontal plane has been established, the dial indicator used to check straightness should be run over the top surface area. Whenever the dial indicator is at the end of one level, it should be started at the same dial reading on the adjacent level in order to continue measuring the straightness relationship. Narrow or small areas such as ribs or pockets can be disregarded.

The total dial reading variation should not exceed the straightness tolerance. In all methods described, the straightness check is typical for the common drawing callout stating that the forging must be straight within a specified amount. If the drawing calls out a plus and minus straightness tolerance, the drawing must establish the plane from which the tolerance can be measured by specifying the location dimensions that define the plane. If such dimensions are omitted, the applicable straightness tolerance is considered to be the total of the plus and minus amounts.

When a drawing callout specifies that the straightness (or flatness) must be within x amount in y in., an entirely different check is required. This callout defines the rate of change and requires that as a straightedge of y length is slid over the flat areas, the flatness must not exceed x in. This type of checking is relatively simple in comparison to that required by the more common straightness callout.

Contour Checking. Contoured shapes may vary from the theoretically perfect contours in part because of distortion, and also because the accuracy of contours is affected by die sinking limits, die wear, and shrinkage variations. The method of checking contoured shapes depends on the tolerances specified. A contour with a constant and uniform tolerance can be checked with a template and feeler gage; a contour with a variable tolerance may require a special fixture with appropriate features incorporated to check out the tolerance requirements.

Flash Extension Tolerances

Flash extension tolerances govern the accuracy with which flash must be trimmed from the forging. The length of projection (measured from the body of the forging) left after trimming may vary from zero (theoretical intersection of the draft angle at the parting line) to the dimension specified. Standard flash extension tolerances (Table 2-7) are based on the net weight or overall length of the forging, whichever is greater.

Negative Trim. Although most forgings are supplied with a slight extension remaining, there are some requirements for flash trimming that prohibit any flash extrusions. In such cases, a negative trim tolerance is essential; this allows the trimming to cut into the body and remove part of the draft material. The tolerances for this negative trim should be specified on the drawing and should be at least one half of the tolerances for a protruding trim. Because flash trimming adds to costs, this requirement is usually applied only to specified portions, not to the entire periphery of the parting line.

Unless the requirement is specified otherwise, the trimmed surfaces will vary depending on the method of trimming, which may be either shearing, sawing, machining, or grinding. The flash extension or

TABLE 2-7. Suggested Flash Extension Tolerances

Weight, lb (within)	Overall length, in. (within)	Forging Conventional and precision types	Blocker types
$^1/_2$	5	0.015(a)	...
5	15	0.030(a)	0.120
25	30	0.060(a)	0.180
50	60	0.120	0.250
100	120	0.120	0.500
Over 100	Over 120	0.250	0.500

Note: Use the larger tolerance when weight and overall length are not on same line. Source: Ref. 4. (a) These tolerances apply for die trimming and should be doubled for saw trimming. All other tolerances shown apply for saw trimming.

TABLE 2-8. Standard Quantity Tolerances

No. of pieces	Plus, pieces	Minus, pieces
1–2	1	0
3–5	2	1
6–19	3	1
20–29	4	2
30–39	5	2
40–49	6	3
50–59	7	3
60–69	8	4
70–79	9	4
80–99	10	5

No. of pieces	Plus, %	Minus, %
100–199	10	5.0
200–299	9	4.5
300–599	8	4.0
600–1249	7	3.5
1250–2999	6	3.0
3000–9999	5	2.5
10,000–39,999	4	2.0
40,000–299,999	3	1.5
300,000 up	2	1.0

Source: Ref. 4.

negative trim tolerance should be taken at the parting line from the portion that protrudes because of the mismatch condition. The flash extension or negative trim tolerance applies in all cases, even if the drawing dimensions are specified as the parting line.

Draft Angle, Angular Radii, and Quantity Tolerances

The following draft angle tolerances are the permissible variations from the specified draft angles:

- Draft angles 3° or less, $\pm^1/_2$° tolerances
- Draft angles 3° and above ± 1° tolerances

Standard angular tolerances apply to angle dimensions other than draft angles. This tolerance is $\pm^1/_2$° for general applications and is also applicable to unspecified but implied 90° angles. Standard tolerances for both corner and fillet radii are ± 0.03 in. on dimensions up to 0.30 in. and $\pm 10\%$ on dimensions exceeding 0.30 in. Quantity tolerances are the shipping limits within which an order is considered to be complete. Table 2-8 lists standard quantity tolerances.

Knockout and Vent Mark Tolerances

Knockout pins or ejectors are necessary to extract forgings of small draft angles (3° or less) from the die cavity (Figure 2-66). Sometimes knockouts are used to facilitate removal of heavy forgings or simply to speed up production. The area enclosed

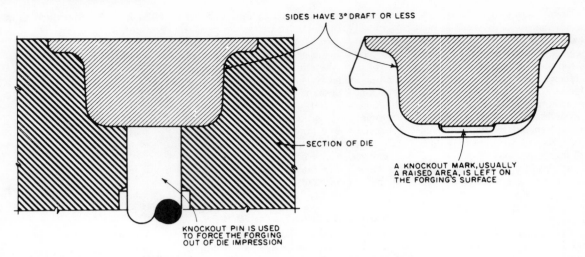

FIGURE 2–66. Die knockout pins or ejectors. Source: Ref. 4.

by the knockout is likely to be raised on the forging, but it may occasionally be depressed. The height or depth of such knockout marks is normally held to within 0.03 in. maximum. Metal may flow between the die and the knockout pin, creating a thin circumferential fin. This is also normally trimmed to within 0.06 in. maximum.

Vent holes may be incorporated into the die cavity (Figure 2-67) to facilitate and ensure forging fill in troublesome pockets. The vent holes allow trapped gas to escape and allow metal to fill the cavity. Protrusions formed in die vents are normally removed to within 0.06 in. maximum height. Tightening the normal maximum for knockout marks, knockout fins, and vent marks will increase costs. Specifying no restrictions regarding these marks may present advantages if the part is to be subsequently machined in these areas, and if

no handling problems arise. The location and size of knockouts and the number and location of vent holes are determined by the forging vendor at the time of quotation.

Additional Tolerances

The following tolerances often are specified:

The out-of-round tolerance is the allowable roundness or ovality deviation from a perfect circle. Generally, it is expressed as T.I.R. tolerance, which requires the actual diameter to fall within two concentric circles with a total difference equal to the tolerance. When specified as a radial tolerance, it defines the allowable space radially between two noncentric circles. In the absence of any drawing comment pertaining to out-of-roundness, it is assumed

that any such deviations must fall within the diametral tolerances.

The concentricity tolerance is the maximum eccentricity from the true axis between two or more diameters or features. It expresses the allowable diametral zone about the true axis or the T.I.R. limit. The drawing should specify the diameter or feature to which the concentricity tolerance applies and the diameter or feature establishing the datum axis. If the drawing specifies no concentricity tolerance, it is assumed that any such deviation must fall within the dimensional tolerances for the diameters or features.

The squareness tolerance, also called perpendicularity tolerance, is the allowable deviation from true squareness and represents the zone between two parallel planes. When specified, a drawing note should indicate the feature or features to

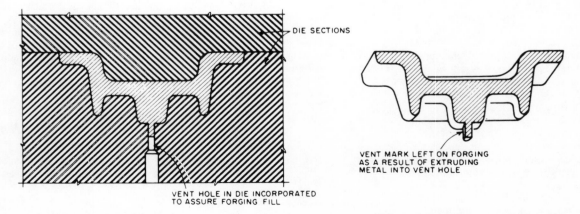

FIGURE 2–67. Die vent holes to allow trapped air or lubricant to escape. Source: Ref. 4.

which the tolerance applies and the feature establishing the datum plane.

The flatness tolerance applies to plane surfaces and has been covered under straightness tolerances.

Designs for Shapes Produced on Horizontal Forging Machines (Upsetters)

Generally, considerations for designing forgings to be produced on horizontal forging machines (upsetters) are basically similar to those for press forgings. The basic difference is that in the forging machine there are two mutually perpendicular forging planes: one is perpendicular to the ram (heading tool) motion, and the other lies in the parting plane of the gripping dies. For this reason, many components can be produced as closed die forgings, and a parting plane—in the sense employed in impression die forging—does not always exist. Other designs may require flash formation, in which case parting lines are chosen according to considerations previously discussed, with suitable modifications for the effect of split gripping dies.

Draft angles are required on all surfaces that are withdrawn from a confined cavity and on all surfaces from which the heading tool is removed. Actual values of draft vary considerably according to the design and the special capabilities of individual forging producers.

Corner radii contribute relatively little to the ease of metal flow and are correspondingly chosen to be as small as convenient. Often they are equal to the value of the machining allowance. Larger fillet radii do facilitate metal flow and can be taken to one fifth of the relevant height or depth.

Similarly, radii promote the development of sound metal flow in operations where recesses are formed by the heading tool. Radii should be chosen as large as possible, consistent with the need for keeping excess materials within economical limits.

The considerations described yield different radii for various sections of the forging. Die making will be more economical if the nearest round number is used and if radii are unified for sections made with the same milling tool.

Conventional Forging Design: Other Alloy Systems

By Dr. S.L. Semiatin, Principal Research Scientist, Battelle's Columbus Laboratories, Columbus, Ohio

The design of dies for the manufacture of forgings of aluminum alloys has been discussed at length in the previous subsection. Similar considerations apply to other materials. However, the exact design parameters are very dependent on specific material, determined largely by its deformation resistance and forgeability. Therefore, in this section, design guidelines for various other materials are summarized briefly. Additional information on these aspects is contained in Section 3 of this Handbook (Material Characteristics).

As for aluminum forgings, basic design considerations include specification of the following:

- Parting line
- Allowance for machining (forging envelope)
- Draft, corner, and fillet radii, and minimum section thicknesses and maximum rib heights

- Die closure/thickness tolerances
- Length and width/die wear tolerances
- Match/mismatch tolerances

Parting Line

The selection of the parting line is one of the most important decisions in die design for impression die forging. Obviously, the parting line should be selected to allow ease of part removal following forming. However, the parting line and accompanying flash design have a large impact on the overall metal flow during forging and thus the ability to forge parts successfully. Other considerations when choosing a parting line include (1) possible die breakage due to bending stresses when deep impressions are forged, (2) possible die breakage due to side thrusts when the parting line selection gives rise to an asymmetry in the forging, and (3) control of grain flow (discussed in Section 3). In addition, the choice of parting line may have important economic impacts insofar as it affects the ease of machining and die material losses (Figures 2-68 and 2-69).

Finish Allowance

Finish allowance is the amount of excess metal surrounding the intended final part shape that is included in the forging design for purposes of cleanup and machining. Also known as the forging envelope, finish allowance depends largely on the oxidation behavior of the metal, its forgeability, and the severity of the application of the final part. Aluminum and magnesium alloys that do not oxidize readily are often forged with minimal forging envelopes (Figure 2-70). Other metals, such as steels that are prone to substantial scale and decarburization losses at hot forging temperatures, require substantial finish allowances. Still other materials, such as titanium and nickel-base alloys, require large forging envelopes so that surface defects and other metal flow

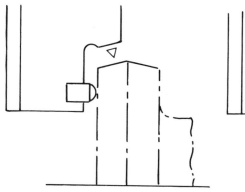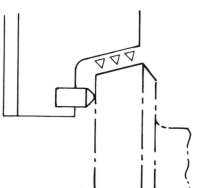

FIGURE 2–68. Choosing the parting line for ease of machining. Source: Ref. 6.

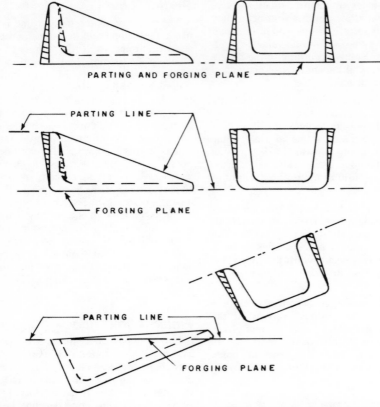

FIGURE 2–69. Parting line and material losses. The shaded areas should be removed by machining. Source: Ref. 6.

irregularities can be removed in order to produce high-integrity structural parts for jet engines and structural applications, for example. Finish allowances also generally increase with increasing forging size because of longer heating times, added operations, and a greater chance for the introduction of defects during handling.

Draft

Forging projections are typically tapered to allow easy part removal from the die cavity. The most common draft angles are between 5 and 7° for conventional steel forgings (Figure 2-71, Table 2-9). Smaller draft angles can be used, but will usually result in greater production difficulty. For steel, outside draft angles are usually smaller than inside ones (Table 2-9). This is done because outside surfaces shrink away from the die during cooling and permit removal of the part.

The primary factor in determination of draft angle is the depth of the die cavity (Table 2-9). This is by no means the sole criterion on which draft design can be based. At times, the height-to-width ratio of the impression may give a more reasonable means of estimating drafts (Figure 2-72). At other times, draft angles may have to be increased if the material is difficult to form or if the rate of deformation is slow, leading to greater cooling during

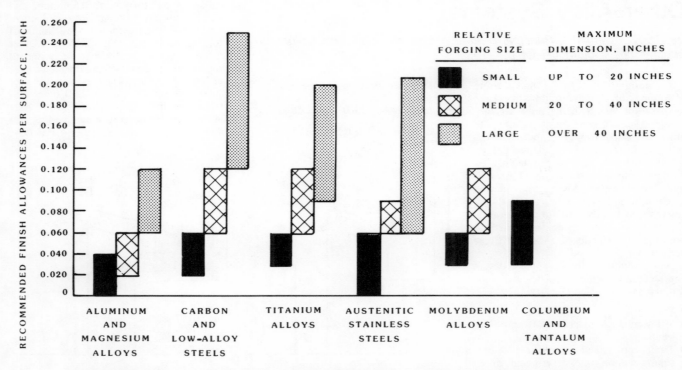

FIGURE 2–70. Recommended finish allowances for die forgings of various alloy systems. Source: Ref. 7.

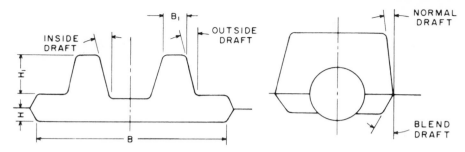

FIGURE 2–71. Draft angles. Source: Ref. 6.

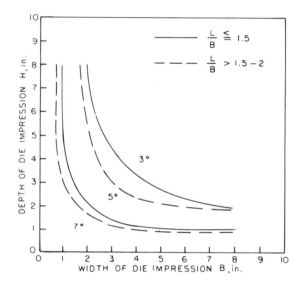

FIGURE 2–72. Draft angles chosen on the basis of die dimensions. Source: Ref. 6.

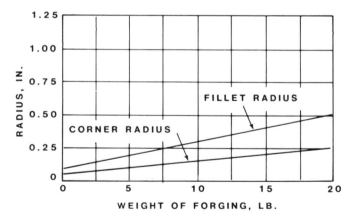

FIGURE 2–73. Minimum fillet and corner radii for steel forgings without deep ribs and bosses. Source: Ref. 6.

TABLE 2–9. Draft and Draft Tolerances for Steel Forgings

Height or depth of draft, in.	Commercial standard Draft, degrees	Tolerance(a) plus, degrees	Special standard Draft, degrees	Tolerance plus, degrees
Outside draft				
$1/4$–$1/2$	...	...	3	2
$3/4$–1 5		3	...	...
Over $1/2$, up to 1 ..	...	...	5	2
Over 1, up to 3 7		3	5	3
Over 3 7		4	7	3
Inside draft				
$1/4$–1 7		3	5	3
Over 1 10		3	10	3

(a) The minus tolerance is zero. Source: Ref. 6.

metal flow problems such as laps, cold shuts, and flow-through defects in structural rib-web forgings and other parts with deep cavities. Moreover, sufficient radii are necessary to avoid die breakage due to stress concentrations. In contrast, fillet radii that are too large are undesirable if a much smaller radius is subsequently machined. This is because the forged fiber is cut through and the part weakened. Forging to a smaller radius, even at the expense of additional forming stages, may be preferable.

As with draft angles, exact values of corner and fillet radii cannot be quoted, as they depend on part design and material. For instance, suggested radii for ferrous and nonferrous forgings are given in Figures 2-73 and 2-74, where it is observed that fillet radii are generally about twice as large as corner radii. In addition, radii for difficult-to-forge materials (e.g., high-temperature materials) are often substantially larger than those for easily forged materials such as aluminum alloys. Figure 2-74 shows that recommended corner and fillet radii also increase with the height of ribs or bosses in a forging geometry. As noted above, such a design trend is necessary to enhance metal flow and to avoid defects.

It is important to maintain consistent corner radii for any given shape. This practice minimizes the need for numerous tooling changes during die machining and reduces forging die costs.

Minimum Section Size/ Maximum Rib Height

One of the prime means of minimizing the forging weight and hence material costs is to forge as close as possible to final cross-

forming and more subsequent shrinkage. Titanium and nickel-base alloy forgings generally require 7° or greater drafts. On the other hand, ejectors often permit the use of lower draft angles (2 to 4° for steels).

Corner and Fillet Radii

Another important die design feature is that of corner and fillet radii. Proper selection of such radii is critical in avoiding

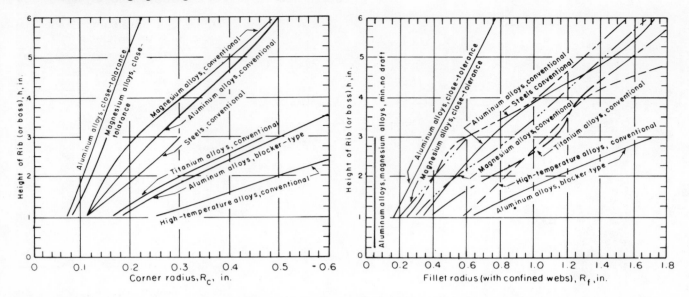

FIGURE 2–74. Corner and fillet radii for various materials recommended by forging users and producers. Source: Ref. 8.

section size and thickness. Small reductions in the forging envelope around tall ribs or on large webs can account for sizable savings. However, there are practical limits on minimum section thickness, because many alloys have narrow forging temperature ranges or require large forging pressures. Webs are difficult to forge because of their low mass and their tendency to cool rapidly upon contact with the dies during conventional hot forging. Similarly, rib formation requires high forming loads. Rib-web forging may also

be difficult when several ribs need to be formed adjacent to each other.

Minimum web and rib dimensions that can be achieved for steels by conventional hot forging are shown in Figures 2-75 and 2-76. In Figure 2-75, it is seen that the minimum achievable web thickness decreases as its minimum plan dimension decreases. A similar dependence of minimum thickness is found in steel forgings with unconfined webs, or webs in which material is free to flow in at least one lateral direction (Figure 2-76a). As ex-

pected, the maximum achievable rib height also decreases as the rib thickness decreases (Figure 2-76b). Metal flow is more difficult in forgings with confined webs. Thus, their web thicknesses must be greater and rib heights decreased.

Analogous trends in minimum section size apply for materials other than steels. Typical guidelines are given in Figures 2-77 and 2-78, and in Tables 2-10 and 2-11. From these data, it is obvious that thinner webs are attainable in aluminum alloys, but not in the less forgeable titanium and nickel-base materials. The difficulty of forging the high-temperature alloys is graphically illustrated in Figure 2-79, which depicts the relative forging design limits for comparable aluminum and nickel-base alloy rib-web parts.

Die Closure/Thickness Tolerances; Length and Width Tolerances

Once the forging design has been established, thickness (die closure), length and width, and match/mismatch tolerances must be specified. These specifications determine when forgings must be rejected or when dies must be removed from service because of excessive wear. Figure 2-80 shows that the thickness tolerance usually increases with part weight. The thickness tolerance may also be expressed as a function of plan area (Table 2-12) and typically varies with workpiece material. More forgeable materials not surprisingly have tighter die closure tolerances. Tolerances on extremities of forgings extending perpendicularly more than 6 in. from the

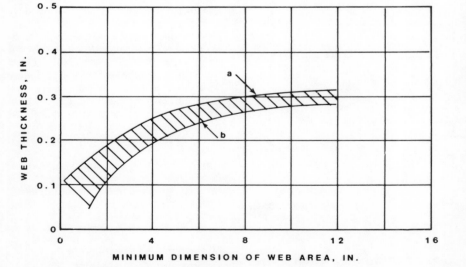

FIGURE 2–75. Recommended minimum web thickness for steels of good forgeability in relation to web dimensions. (a) Minimum for rapidly completed forgings. (b) Attainable usually at extra cost. Source: Ref. 6.

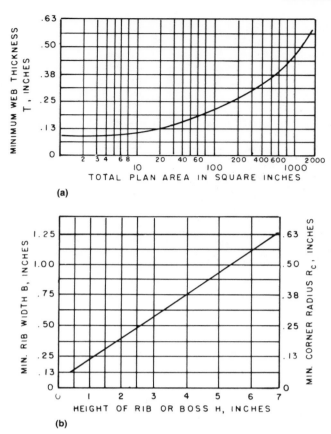

(a)

(b)

FIGURE 2-76. Recommended dimensions for web and rib thickness and corner radii of forgings with unconfined webs for steels with good forgeability. See text for discussion of (a) and (b). Source: Ref. 6.

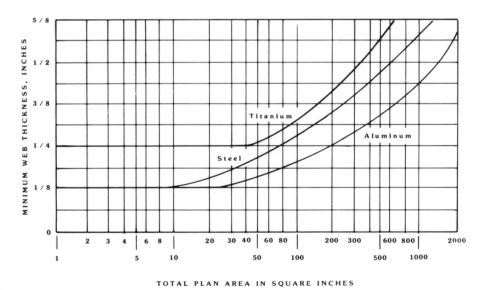

FIGURE 2-77. Relation of minimum web thickness to total plan area for three materials, showing how the materials affect web thickness and total plan area. Source: Ref. 8.

parting line typically include the die closure tolerance and an additional tolerance of 0.003 in./in. This tolerance is added to those in Figure 2-80 and Table 2-12, but only apply to such extremities.

As for thickness tolerances, there are standard length and width tolerances for conventional forgings (Figure 2-81, Table 2-13). These tolerances account for dimensional inaccuracies due to such causes as die wear and shrinkage from the hot-forging temperatures as well as variations due to die sinking and polishing. Both die wear and shrinkage tolerances are a function of part size, workpiece material, and forging shape. Because of this, die wear tolerances for various materials are often applied in addition to the length and width tolerances on dimensions pertaining to forged surfaces only. Die wear tolerances do not apply to center-to-center dimensions.

Experience with conventional closed die forging in presses and hammers has established that the tolerances quoted above provide adequate dimensional accuracy for most industrial applications. Furthermore, the tolerances for forgings made to conventional design can be met by all producers using equipment normally available. Narrower tolerances than those discussed can often be established by mutual agreement between producer and customer.

Match/Mismatch Tolerances

Match/mismatch tolerances (also known as "die shift" tolerances) refer to the axial alignment of two opposing impression dies. These tolerances are a measure of the lateral displacement of a point in one die from a corresponding point in the opposite die in any direction parallel to the forging plane. Because the various types of forging equipment are not perfectly rigid, there is never perfect alignment between the opposing dies. The amount of mismatch is measured as shown in Figure 2-82. Typical tolerances for this parameter are given in Table 2-14 for various materials. As for other forging tolerances, the allowable mismatch increases with increasing weight and decreasing material forgeability. The mismatch tolerances are applied independent of other tolerances, however. Where possible, the measurements are made at locations unaffected by die wear.

To decrease mismatch, integral matching "locks" or guide pins are machined into the opposing dies. They provide lead-ins that align the dies during forging. The use of such devices is particularly important when forging asymmetric parts for which side thrusts are developed during forming.

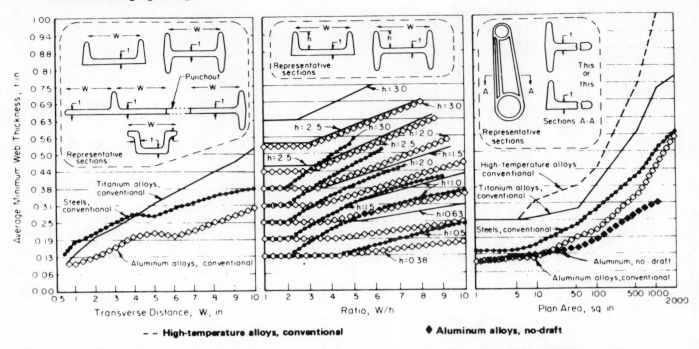

FIGURE 2–78. Average minimum producible web thickness for various materials and forging processes, as estimated by users and producers. Source: Ref. 8.

TABLE 2–10. Minimum Web Thickness

Total plan area(a), in.^{2}	Inches (min.)					
	Conventional forgings (7°, 5°, 3°)			Blocker forgings		
	Aluminum, magnesium	Carbon and alloy steels	Titanium, Inconel(b)	Aluminum, magnesium	Carbon and alloy steels	Titanium, Inconel(b)
To 10	0.09	0.13	0.20	. . .	. . .	. . .
30	0.12	0.16	0.25	0.19	0.25	0.38
60	0.16	0.20	0.32	0.22	0.28	0.43
100	0.19	0.25	0.40	0.25	0.31	0.47
200	0.25	0.32	0.48	0.31	0.38	0.56
300	0.31	0.37	0.58	0.38	0.45	0.64
500	0.37	0.44	0.70	0.43	0.50	0.75
800	0.44	0.50	0.80	0.50	0.56	0.85
1200	0.50	0.56	0.90	0.56	0.62	1.00
1600	0.56	0.62	1.00	0.62	0.69	1.13
2000	0.62	0.70	1.13	0.69	0.75	1.25
2500	0.69	0.80	1.25	0.75	0.81	1.38
3000	0.75	0.88	. . .	0.81	0.88	1.50
3500	1.00	1.13	. . .	1.13	1.19	. . .
4000	1.25	1.38	. . .	1.31	1.38	. . .
5000	2.00	2.25	. . .	2.13	2.50	. . .

(a) When required plan area falls between those listed, determine web thickness by interpolating in increments of 0.01 in. Specify thickness to two decimal places. (b) Also nickel- and cobalt-base superalloys. Courtesy of McDonnell-Douglas Aircraft Co.

TABLE 2–11. Design Guidelines for Superalloy Forgings(a)

Alloy	Type of forging	Minimum web thickness, in.	Minimum rib width, in.	Thickness tolerance, in.	Minimum corner radii, in.	Minimum fillet radii, in.
A-286, Incoloy 901, Hastelloy X, Waspaloy, Udimet 630, TD-Nickel(b)	Blocker	0.75–1.25	0.75–1.00	0.18–0.25	0.62	0.75–1.25
	Finish	0.50–1.00	0.62–0.78	0.12–0.18	0.50	0.62–1.00
Inconel Alloy 718, René 41, X-1900(b)	Blocker	1.00–1.50	1.00–1.25	0.20–0.25	0.75	1.00–2.00
	Finish	0.75–1.25	0.78–1.00	0.15–0.20	0.62	0.75–1.50
Astroloy, B-1900(b)	Blocker	1.50–2.50	1.25–1.50	0.25–0.30	1.00	1.25–2.50
	Finish	1.00–1.50	1.00–1.25	0.18–0.25	0.75	1.00–2.00

(a) For forgings over 400 in.^{2} in plan area and forgings of 100 to 400 in.^{2} in plan area, design allowables can be reduced 25%. For forgings under 100 in.^{2} in plan area, design allowables can be reduced 50%. Recommended draft angles 5 to 7°. Machining allowance for finish is 0.15 to 0.25 in. (b) Based on limited data. Source: Ref. 8.

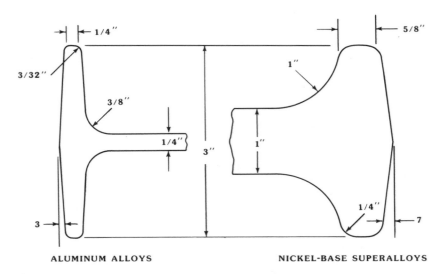

FIGURE 2–79. Comparison of typical design limits for type 4 rib-web structural forgings of aluminum alloys and nickel-base superalloys. Source: Ref. 8.

TABLE 2–12. Thickness (Die-Closure) Tolerances for Various Materials

Tabulated values are plus values, only, expressed in inches.

Materials	Area at the trim line, flash not included, in.^{2}						
	10 and under	Over 10 to 30 incl.	Over 30 to 50 incl.	Over 50 to 100 incl.	Over 100 to 500 incl.	Over 500 to 1000 incl.	Over 1000
Carbon and low-alloy steels ..	$1/32$	$1/16$	$3/32$	$1/8$	$3/32$	$3/16$	$1/4$
400 series stainless	$1/32$	$1/16$	$3/32$	$1/8$	$3/16$	$1/4$	$5/16$
300 series stainless	$1/16$	$3/32$	$1/8$	$5/32$	$3/16$	$1/4$	$5/16$
Superalloys, titanium	$1/16$	$3/32$	$1/8$	$3/16$	$1/4$	$5/16$	$3/8$
Aluminum, magnesium	$1/32$	$1/32$	$1/16$	$3/32$	$1/8$	$3/16$	$1/4$
Refractory alloys	$3/32$	$1/8$	$5/32$	$3/16$	$1/4$	$5/16$	$3/8$

Source: Ref. 8.

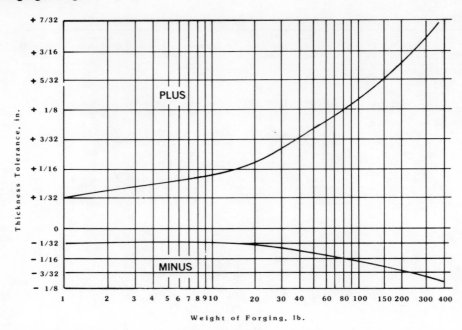

FIGURE 2–80. Thickness tolerance for conventional forgings as a function of the forging weight. Source: Ref. 8.

TABLE 2–13. Length and Width Die Wear Tolerances

Length or width(a), in.	Conventional forgings (7°, 5°, 3°)			Blocker forgings		
	Aluminum, magnesium, carbon steel	Alloy steel	Titanium(b)	Aluminum, magnesium, carbon steel	Alloy steel	Titanium(b)
0–5 . . . . 0.01		0.02	0.02	0.03	0.04	0.05
5–10 . . . . 0.01		0.02	0.03	0.03	0.04	0.05
10–15 . . . . 0.02		0.03	0.04	0.04	0.05	0.06
15–20 . . . . 0.02		0.03	0.05	0.04	0.05	0.06
20–25 . . . . 0.03		0.04	0.06	0.05	0.06	0.07
25–30 . . . . 0.03		0.05	0.07	0.05	0.06	0.08
30–35 . . . . 0.04		0.05	0.08	0.06	0.07	0.09
35–40 . . . . 0.04		0.06	0.09	0.06	0.08	0.10
40–45 . . . . 0.05		0.07	0.10	0.06	0.08	0.12
45–50 . . . . 0.05		0.08	0.12	0.07	0.09	0.13
50–55 . . . . 0.06		0.08	0.13	0.07	0.10	0.14
55–60 . . . . 0.06		0.09	0.14	0.08	0.11	0.15
60–65 . . . . 0.07		0.10	0.15	0.08	0.11	0.16
65–70 . . . . 0.07		0.11	0.16	0.09	0.12	0.17
70–75 . . . . 0.08		0.11	0.17	0.10	0.14	0.18
75–80 . . . . 0.08		0.12	0.18	0.11	0.15	0.19
80–85 . . . . 0.09		0.13	0.19	0.12	0.16	0.20
85–90 . . . . 0.09		0.13	0.20	0.12	0.17	0.21
90–95 . . . . 0.10		0.14	0.22	0.13	0.18	0.23
95 and up 0.10		0.15	0.23	0.13	0.19	0.25

(a) Use largest dimension of forging (length or width) to determine the applicable amount of die wear per side. The allowance thus found is expressed as a portion of the tolerance applied to all forging dimensions that are affected by die wear. (b) Also nickel- and cobalt-base superalloys. Courtesy of McDonnell-Douglas Aircraft Co.

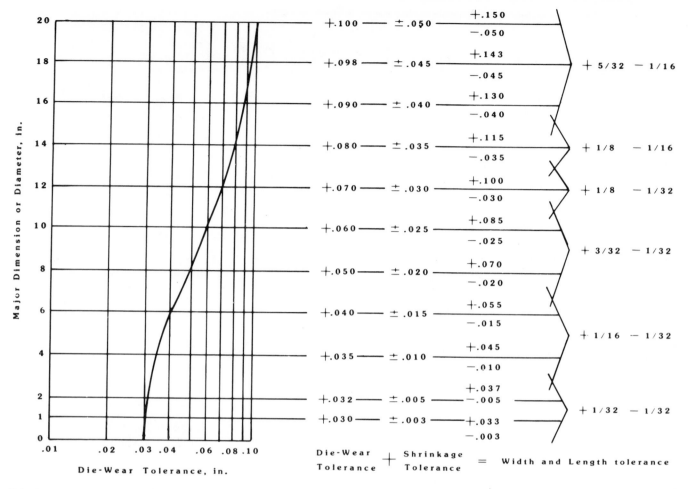

FIGURE 2-81. Length and width tolerances for conventional steel forgings as a function of the forging size. The width and length tolerance is the sum of the die wear and shrinkage tolerances, as individually calculated. Bethlehem commercial-close die wear tolerances are two thirds the commercial-standard values shown. Source: Ref. 8.

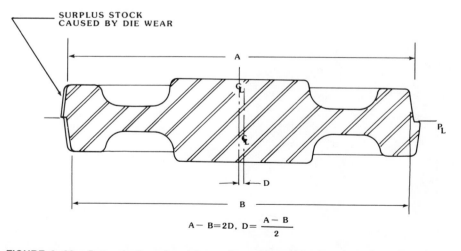

$$A - B = 2D, \quad D = \frac{A - B}{2}$$

FIGURE 2-82. Determination of match or mismatch on a forging. A, projected maximum overall dimensions measured parallel to the main parting line of the dies; B, projected minimum overall dimensions measured parallel to the main parting line of the dies; D, displacement. Source: Ref. 8.

TABLE 2–14. Match Tolerances for Various Materials as a Function of Forging Weight

Tabulated figures are amounts of displacement, expressed in inches, of a point in one die-half from the corresponding point in the opposite die-half in any direction parallel to the parting line of the dies.

Materials		Less than 2	Over 2 to 5 incl.	Over 5 to 25 incl.	Over 25 to 50 incl.	Over 50 to 100 incl.	Over 100 to 200 incl.	Over 200 to 500 incl.	Over 500 to 1000 incl.	Over 1000
Carbon and low-alloy steels	(a)	$1/64$	$1/32$	$3/64$	$1/16$	$3/32$	$1/8$	$5/32$	$3/16$
Stainless steels	(a)	$1/32$	$3/64$	$1/16$	$3/32$	$1/8$	$5/32$	$3/16$	$1/4$
Superalloys, titanium	(a)	$1/32$	$3/64$	$1/16$	$3/32$	$1/8$	$5/32$	$3/16$	$1/4$
Aluminum, magnesium	(a)	$1/64$	$1/32$	$3/64$	$1/16$	$3/32$	$1/8$	$5/32$	$3/16$
Refractory alloys	(a)	$1/16$	$3/32$	$1/8$	$5/32$	$3/16$	$1/4$	$5/16$	$3/8$

(a) Customarily negotiated with purchaser. Source: Ref. 8.

Computer-Aided Design and Manufacturing (CAD/CAM) Applications in Forging

By Dr. Aly Badawy, Principal Research Scientist, Battelle's Columbus Laboratories, Columbus, Ohio; and Dr. Taylan Altan, Senior Research Leader, Battelle's Columbus Laboratories, Columbus, Ohio

The practical design of a forging process involves:

- Conversion of the available machine part geometry into a forging geometry by using guidelines associated with design of forgings and limitations of the forging process.
- Design of finisher dies, including determination of flash dimensions, forging stresses, and forging load. (In some cases it may even be appropriate to calculate die stresses and modify the die geometry in critically stressed areas of the die to reduce the probability of premature die failure.)
- Design of blocker or preblocker dies; this includes calculation of forging volume, including flash allowance and the estimation of blocker and preblocker die geometry (including web thicknesses, rib heights, and fillet and corner radii).
- Design of the preform and estimation of stock size; this includes prediction of desired metal distribution in the stock

(by preforming or busting operations) prior to forging in the blocker die.

Traditionally, the above process and die-design steps are carried out using empirical guidelines, experience, and intuition.

Applications of computer-aided design (or drafting), engineering, and manufacturing (CAD/CAE/CAM) are continuously increasing in the forging industry. A number of companies are already using stand-alone CAD/CAM systems such as UNIGRAPHICS, ANVIL 4000, COMPUTERVISION, CALMA, and others in designing, drafting, and numerical control (N/C) machining of forging dies.

In introducing CAD/CAM for forging applications, it is useful to consider the following factors:

- A CAD/CAM graphics system consists of both hardware and software (Figure 2-83). It is imperative that support and maintenance of both be secured for the present and the future, because most forging companies do not have extensive in-house computer expertise.
- For most applications, a CAD/CAM system must be capable of (1) describing full three-dimensional geometries with complex sculptured surfaces, as shown in Figure 2-84; (2) drafting with dimensions, tolerances, and text (Figure 2-85); and (3) N/C cutter path generation (Figures 2-86 and 2-87). The utilization of CAD/CAM systems in the design and manufacture of forging dies is illustrated schematically in Figure 2-88.
- The trend is toward the use of graphics software hosted on 32-bit processors (such as VAX, Prime, or Perkin Elmer) that use unaltered, stand-alone operat-

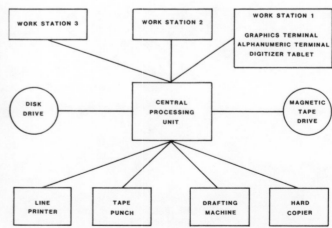

FIGURE 2–83. Typical CAD/CAM interactive graphics work station and system components.

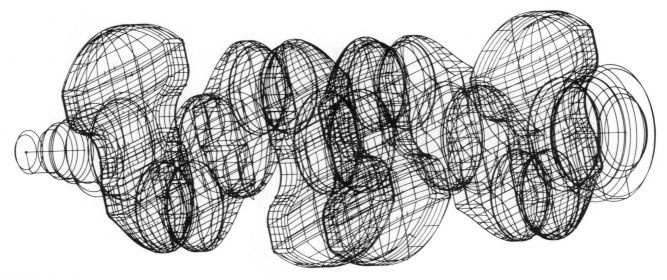

FIGURE 2-84. Three-dimensional surface representation of a complex crankshaft geometry. Generated on COMPUTERVISION's CADDS/4 stand-alone CAD/CAM system.

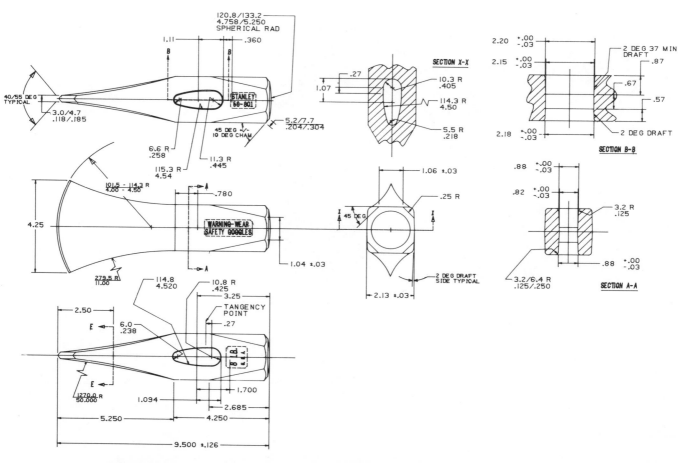

FIGURE 2-85. Engineering drawing of a maul drawn on the COMPUTERVISION stand-alone system.

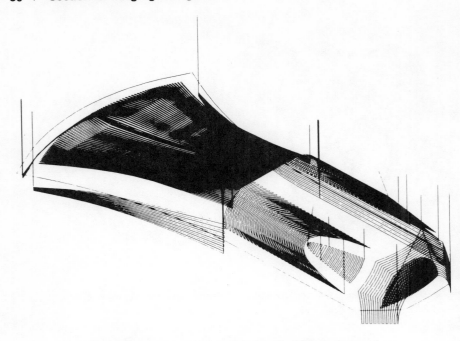

FIGURE 2–86. N/C tool paths for all surfaces of a maul.

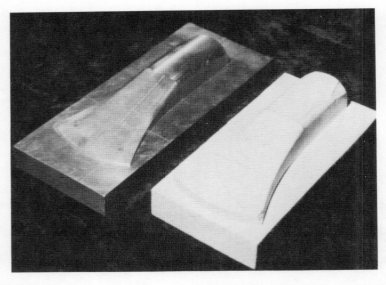

FIGURE 2–87. Machined models of the maul depicted in Figures 2–85 and 2–86 (left, aluminum; right, wood).

ing systems. These hardware/software configurations allow the computer hardware and operating system to be used for other engineering and analysis purposes. This feature can be quite advantageous under some conditions.

- For three-dimensional representation of solid objects—for example, forging dies or models—"solid modeling" software packages are available. Solid models can display complex parts on a graphics terminal by assembling solid building blocks. A graphics display by "solid modeling" looks like a photograph of the real object (with shades and depths). Such a display greatly enhances visualization of the object before it is manufactured. As a result, "solid modeling" is expected to be accepted also in the metalforming industry in the near future.

- Presently (early 1980s), the geometric capabilities of most solid modelers are not quite sufficient for design and manufacture of complex dies and molds. However, with the rapid improvements that are being made in graphics software, it can be expected that these limitations will be eliminated within the next few years.

Typical applications of CAD/CAM techniques in the forging industry are given in the following sections to illustrate the value of this new technology.

Close-Tolerance Forging of Spiral Bevel Gears

Traditionally, straight bevel and spiral bevel gears were manufactured by machining in special gear-cutting machines. Precision forging of bevel gears is now routinely used in production. This technology offers considerable advantages, such as reductions in machining (material and energy) losses and increases (up to 30%) in fatigue life.

Production methods for precision forging of truck differential-quality spiral bevel gears have been developed in the United States using CAD techniques. This method uses the following as input:

- The gear geometry, i.e., the coordinates that describe the tooth surface, which is obtained by computer simulation of the gear-cutting process (Figure 2-89)
- The overall dimensions of the die insert and of the shrink ring
- The process variables, such as forging temperature, die temperature, friction factor, material properties, and forging speed

With this information, local die corrections are calculated to account for thermal shrinkage and for elastic deflection due to forming stresses. Graphite electrodes, with calculated dimensional corrections, are then machined in a spiral bevel gear-cutting machine. The forging dies are manufactured by electro-discharge machining (EDM). The die assembly allows only internal flash and is contained in the die holders mounted on the top and bottom bolsters of a 2000-ton mechanical press. In the bottom die holder, the tooling consists of a die ring, a die insert, and an ejector mechanism.

The gears were forged to finish machining tolerances, thereby eliminating the need for rough machining. The results of this developmental work indicate that it may be quite feasible to produce some spi-

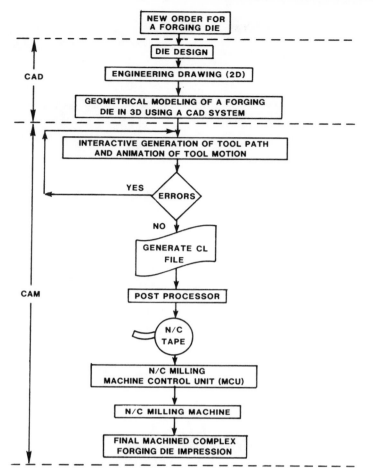

```
          NEW ORDER FOR
          A FORGING DIE
               │
          ┌────────────┐
          │ DIE DESIGN │
          └────────────┘
               │
    ENGINEERING DRAWING (2D)
               │
    GEOMETRICAL MODELING OF A FORGING
      DIE IN 3D USING A CAD SYSTEM
               │
    INTERACTIVE GENERATION OF TOOL PATH
       AND ANIMATION OF TOOL MOTION
               │
      YES    ◇ ERRORS ◇
               │ NO
          GENERATE CL
             FILE
               │
          POST PROCESSOR
               │
             N/C
             TAPE
               │
          N/C MILLING
    MACHINE CONTROL UNIT (MCU)
               │
       N/C MILLING MACHINE
               │
     FINAL MACHINED COMPLEX
      FORGING DIE IMPRESSION
```

CAD

CAM

FIGURE 2–88. Schematic of the use of CAD/CAM in the design and manufacturing of forging dies.

ral bevel gears with tolerances such that no machining of the tooth surface is necessary.

Design of Finisher Dies

In designing the finisher and blocker dies for hot forging, it is best to consider critical cross sections of a forging, where metal flow is plane strain or axisymmetric. Cross sectioning approximates, in two dimensions, the complex three-dimensional geometry of and metal flow in a practical forging; thus, stresses and pressures can be calculated.

The computerized "slab method" of analysis has been found to be the most practical method of analyzing stresses. The method can be used to estimate forging stresses for plane-strain as well as axisymmetric metal flow sections. The computerized "slab-method" for predicting stresses and loads was originally implemented in a large-frame CDC 6500 computer as well

as in a PDP 11/40 minicomputer. Recently, this software has also been developed for a COMPUTERVISION stand-alone CAD/CAM system.

By modifying the flash dimensions, the die and material temperatures, the press speed, and the friction factor, the die designer is able to evaluate the influences of these factors on forging stress and load. Thus, conditions that appear most favorable can be selected. In addition, the calculated forging stress distribution can be utilized for estimating the local die stresses in the dies by means of elastic finite element method (FEM) of analysis. After these forging stresses and loads are estimated for each selected section, the loads are added and the center of loading for the forging is determined. Figure 2-90 shows the stress distribution and the flow model (i.e., approximate shear surfaces) obtained with the new software developed for designing finisher dies.

Design of Blocker Dies

Designing the blocker and preform geometries is the most critical part of forging die design. Traditionally, blocker dies and preforms were designed by experienced die designers and were modified and refined by die tryouts. The initial blocker design was based on several empirical guidelines.

Presently, using computer-aided design techniques, blocker cross sections can be designed using interactive graphics. However, this method still employs the same traditional empirical guidelines, but stored in a quantitative manner in the computer memory. The main advantages of computer-aided design of blocker dies are:

- Cross-sectional areas and volumes can be calculated rapidly and accurately.
- The designer can easily modify geometric parameters such as fillet and corner radii, web thickness, and rib height and width and can immediately review the alternative design on the screen of the computer graphics terminal.
- The designer can examine any given portion of the forging on a graphics terminal and perform sectional area calculations at the workstation.
- If necessary, the designer can review the blocker positions in the finisher dies at various opening positions to study the initial die blocker contact point during finish forging.

These geometric manipulations and drawings are basically no different from what is being done manually today; however, with the aid of computers, they can be done much faster and much more accurately.

Computer-Aided Simulation of the Forging Process

The ultimate advantage of computer-aided design in forging is achieved when reasonably accurate and inexpensive computer software is available for simulating metal flow throughout a forging operation. Thus, forging "experiments" can be run on a computer by simulating the finish forging that would result from an "assumed" or "selected" blocker design. The results can be displayed on a graphics terminal. If the simulation indicates that the selected blocker design would not fill the finisher die or that too much material would be wasted, then another blocker design can be selected and the computer simulation (or the experiment) can be repeated. This computer-aided simulation will reduce the required number of expensive die tryouts.

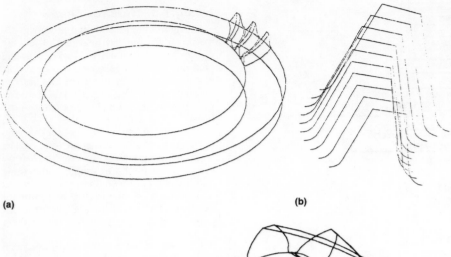

(a)

(b)

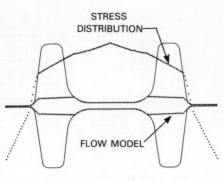

FIGURE 2–90. Stress distribution and flow model on one cross section of a connecting link. Source: Ref. 9.

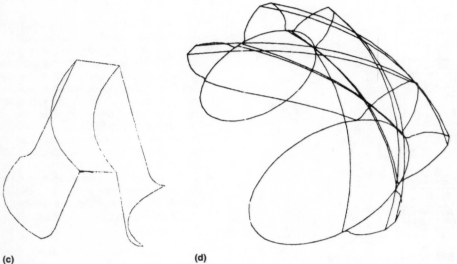

(c)

(d)

FIGURE 2–89. Gear geometry as displayed on CRT screen. (a) Gear. (b) Tooth sections. (c) Tooth form. (d) Pinion.

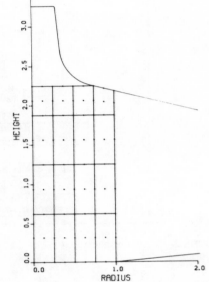

Recently, the finite element method (FEM) has been utilized to simulate the metal flow in various two-dimensional axisymmetric and plane strain hot and cold forging cross sections. Initial, yet highly promising, results were obtained by this advanced FEM-based program, known as ALPID (Analysis of Large Plastic Incremental Deformation), developed at Battelle's Columbus Laboratories under Air Force sponsorship. These results indicate that this software is able to simulate a large number of two-dimensional forging operations with reasonable accuracy and at acceptable cost. In simulating an axisymmetric forging operation (under assumed isothermal conditions, for titanium alloy Ti-6242 forged at 1740 °F, or 950 °C), the ALPID program was capable of predicting the strains, strain rates, and stresses at specific grid points in the deformed material (Figure 2-91). In fact, by using spe-

cial plotted software, the results could be displayed on a graphics terminal and plotted on paper in the form of contour plots.

Future Outlook

CAD/CAM is being applied in forging technology at an increasing rate. Using the three-dimensional description of a machined part, which may have been computer designed, it is possible to generate the geometry of the associated forging. For this purpose, it is best to use a stand-alone CAD/CAM system with software for geometry handling, drafting, dimensioning, and N/C machining. Thus, the forging sections can be obtained from a common data base. Using well-proven analyses based on the slab-method or FEM techniques, forging loads and stresses can be obtained and flash dimensions can be selected for each section where metal flow

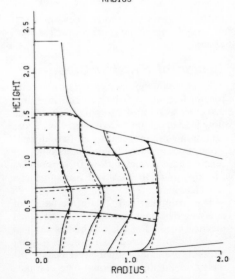

FIGURE 2–91. Simulation of axisymmetric spike forging. Source: Ref. 10, 11.

is approximated as two-dimensional (plane strain or axisymmetric). In some relatively simple section geometries, a computer simulation can be conducted to evaluate initial guesses on blocker or preform sections. Once the blocker and finisher sections are obtained to the designer's satisfaction, this geometric data base can be utilized to write N/C part programs to obtain N/C tapes or disks.

This CAD/CAM procedure is still in the development stage. In the future, this technology can be expected to evolve in two main directions:

- Handling of the geometries of complex forgings—e.g., three-dimensional description, automatic drafting and sectioning, and N/C machining
- Utilization of design analysis—e.g., calculation of forging stresses and of stress concentrations in the dies, prediction of elastic deflections in the dies, metal flow analysis, and blocker/preform design

Research and development is in progress, and preliminary results are already being obtained on the design and simulation of some practical forging operations.

References and Bibliography

4. *Aluminum Forging Design Manual,* 1st ed., Forgings Division, Aluminum Association, New York, Nov. 1967.
5. *Open Die Forging Manual,* 3rd ed., Open Die Forging Institute, Forging Industry Association, Cleveland, 1982.
6. Schey, J.A., *Principles of Forging Design,* American Iron and Steel Institute, Washington, DC, 1965.
7. Sabroff, A.M., Boulger, F.W., Henning, H.J., and Spretnak, J.W., *Fundamentals of Forging Practice,* Supplement to Technical Documentary Report No. ML-TDR-64-95, Contract No. AF 33(600)-42963, Battelle Memorial Institute, Columbus, OH, March 1965.
8. Altan, T., Boulger, F.W., Becker, J.W., Akgerman, N., and Henning, H.J., *Forging Equipment, Materials, and Practices,* MCIC-HB-03, Metals and Ceramics Information Center, Battelle's Columbus Laboratories, Columbus, OH, 1973.
9. Badawy, A., Billhardt, C.F., and Altan, T., Implementation of Forging Load and Stress Analysis on a COMPUTERVISION CADDS-3 System, *Proceedings of the Third Annual COMPUTERVISION Users Conference,* Dallas, Sept. 1981.
10. Oh, S.I., Lahoti, G.D., and Altan, T., ALPID–A General Purpose FEM Program for Metal Forming, *Proceedings of NAMRC IX,* May 1981, State College, PA, p. 83.
11. Oh, S.I., Finite Element Analysis of Metal Forming Problems with Arbitrary Shaped Dies, *International Journal of Mechanical Sciences,* Vol. 24 (No. 8), 1982, p. 479.

- Altan, T., Raghupathi, P.S., Badawy, A., and Ostberg, D., CAD/CAM Techniques Makes it Practical: Spiral Bevel Gears Precision Forged, *U.S. Army ManTech-Journal,* Vol. 8 (No. 4), 1983, p. 3–8.
- Kelley, M. and Davis, L., Precision Forged Gears, *Metals Engineering Quarterly,* Vol. 14 (No. 4), Nov. 1974, p. 20.
- Mages, W., Advantageous Application of New Forming Processes in Gear Drive and Manufacture (in German), *VDI-BERICHTE,* No. 332, 1979, p. 97.
- Semiatin, S.L. and Lahoti, G.D., Forging of Metals, *Scientific American,* Vol. 245, Aug. 1981, p. 82.
- Akgerman, N. and Altan, T., Modular Analysis of Geometry and Stresses in Closed-Die Forging: Application to a Structural Part, *ASME Transactions, Journal of Engineering for Industry,* Vol. 94, Nov. 1972, p. 1025.
- Subramanian, T.L. and Altan, T., Application of Computer Aided Techniques to Precision Closed Die Forging, *Annals of CIRP,* Vol. 27 (No. 1), 1978, p. 123.

3 MATERIAL CHARACTERISTICS

By Dr. S.L. Semiatin, Principal Research Scientist, Battelle's Columbus Laboratories, Columbus, Ohio

Contributing Authors: James R. Becker, Cameron Iron Works, Houston, Texas; and Richard Wood, Principal Research Scientist, Battelle's Columbus Laboratories, Columbus, Ohio

There are a number of special attributes that make metals suitable for shaping operations such as forging. Of prime importance in this regard is the crystalline structure and metallic bond that allow large permanent or plastic deformations to be imposed. The amount of deformation prior to fracture, however, depends on the particular metal being formed as well as on alloying that has been added to it to enhance its service properties. Such alloying often leads to strengthening and increased difficulty in working relative to the pure metals, factors that must be understood to optimize working operations. In this section, a discussion of the properties of metals serves as an introduction to alloy forgeability, which is addressed in several subsections.

The Metallic Bond and Crystal Structure

Metals are sometimes defined as substances having attributes such as good thermal and electrical conductivity, opacity to light, or ductility. Nevertheless, the concept of the metallic character can best be understood by considering the nature of the metallic bond itself. There are three technologically important types of bonding in solids; metallic is only one. Figure 3-1(a) illustrates another type called the ionic bond, using sodium chloride as an example. The lattice (atomic structure) consists of positive and negative ions. Because like changes repel one another, it is not possible to shift a group of ions to a position directly above their like-charged counterparts by the "slip" process. As will be shown later, "slip" is typical of conventional plastic deformation in metals. Many ceramic materials possess ionic bonding.

Figure 3-1(b) depicts a third type, the covalent bond, using the structure of diamond as an example. The bond consists of a pair of electrons resonating between two ion cores, giving rise to a strong, but highly localized bond with limited possibility of flow. Many organic compounds are covalently bonded.

Figure 3-1(c) depicts the metallic bond, which is composed of a lattice of ion cores held together by a free electron gas. The bonding is decentralized and flexible, giving rise to the capacity for plastic flow by slip and the extremely useful property of plasticity. Essentially all of the characteristics of the metallic state can be explained in terms of the metallic bond.

Although the bonding of atoms is similar in all metals, the particular arrangement of atoms may vary from one metal to another. These arrangements are known as crystal structures, or regular three-dimensional arrays that are often close-packed because of the nature of the metallic bond. The three most common crystalline structures (Figure 3-2) are face-centered cubic (FCC), body-centered cubic (BCC), and hexagonal close-packed (HCP).

Metals with an FCC crystal structure include aluminum, copper, nickel, silver, and gold. Those with a BCC structure include chromium, vanadium, molybdenum, and tungsten. Metals such as zinc and magnesium have an HCP structure.

Furthermore, some metals have a crystal structure that changes with temperature. These changes are called allotropic transformations, and the two most important metals that exhibit such behavior are iron (BCC → FCC at 1670 °F, or 910 °C) and titanium (HCP → BCC at 1620 °F, or 885 °C). Alloys of these two elements may also exhibit such transformations, and control of the transformations by processing (forging, heat treatment) offers an important method of varying the basic structure and hence the properties of metal parts.

85

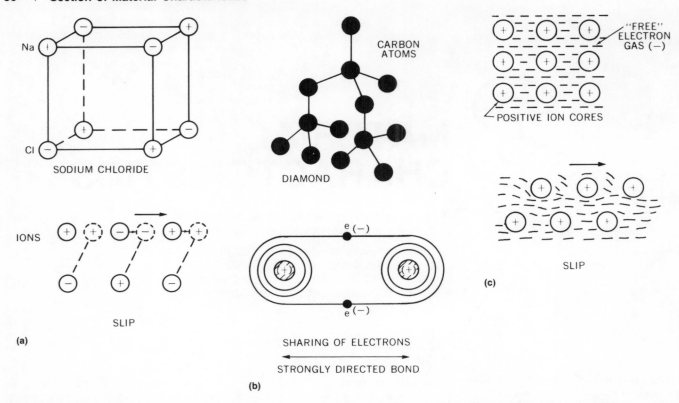

FIGURE 3-1. Three basic types of bonding in solids. (a) The ionic bond, involving Coulombic forces of attraction and repulsion. (b) The covalent bond, involving a pair of shared electrons resonating between two ion cores. (c) The metallic bond, involving a lattice of positive ion cores and "free" electron gas.

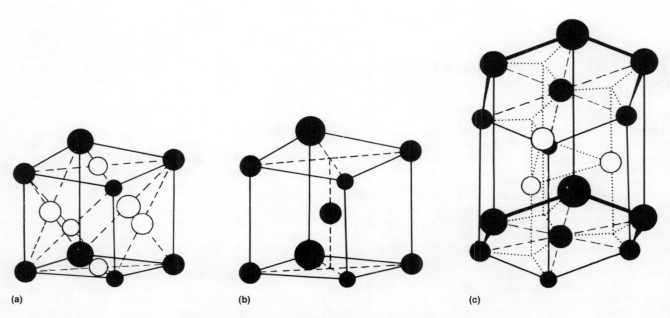

FIGURE 3-2. (a) FCC, (b) BCC, and (c) HCP crystal structures.

Plasticity in Metals

Aside from the factors of availability and economics, the two most important properties giving rise to the widespread use of metals and alloys are: (1) their ability to be formed into useful shapes by plastic deformation, and (2) their ability to develop a wide range of properties, particularly in terms of attractive combinations of strength, ductility, and toughness, after suitable thermal and mechanical treatments and alloying.

Of these two factors, the capacity for permanent changes in shape while maintaining coherency must be considered the most important. This is a particularly impressive property when one considers that metals are generally used as polycrystalline aggregates, or assemblages of individual crystals, characteristically of size 10 to 250 μm (10^{-3} to 2.5×10^{-2} cm). Each grain must adjust to the change in the shape of 6 to 13 neighboring grains during plastic deformation. Without this remarkable capacity of plasticity, the forming of metals would be restricted to casting, powder metallurgy techniques, metal removal processes such as machining and grinding, electrodeposition, and vapor deposition.

The basic science underlying the deformation of solids is called rheology, which is defined as the science of flow and deformation of matter. The classic body is considered to be composed of fundamental particles defined in an arbitrary manner. Flow and deformation are the results of the movement of particles relative to one another within the body. If there is no relative motion of the particles, the entire body will be displaced in a rigid manner under the action of a force.

The basic equations of rheology connect a kinematical quantity called "strain" or deformation with a dynamical quantity called "stress" by means of material constants or coefficients describing the rheological behavior of various materials. A material's behavior is dependent not only on stress and strain, but also on the rate of strain and the rate at which stresses are produced.

The mechanical behavior of real metals (Figure 3-3) is very complex, and no single model can adequately represent the phenomena of elasticity, elastic aftereffect, hysteresis, plastic flow, and creep. The Hookean elastic solid and the Newtonian viscous fluid are the best-known models, which are mathematical abstractions. The most important model for large plastic deformations is the perfectly plastic solid (called the Tresca solid, or the Mises solid) in which elastic response to loading occurs up to a critical stress (flow stress) at which flow continues at constant stress.

These models are useful in obtaining what is known as a phenomenological description of elastic and plastic deformation of metals. Underlying these models, however, is a large body of information on the microscopic mechanisms that characterize the plastic deformation of crystals. The specific mechanisms are reasonably well understood.

It is convenient to discuss deformation in terms of two temperature ranges, because ease of deformation is related to temperature: (a) temperatures below approximately one third the melting point (degrees Kelvin) of a material, and (b) temperatures above one third to one half the melting point (degrees Kelvin). Range (a) is usually referred to as the low-temperature range, and range (b) as the high-temperature range. This temperature division has a physical basis, since self-diffusion rates become appreciable at one half the melting point (Kelvin), which corresponds roughly with the temperature above which cold-worked metals are softened by recovery and recrystallization. Self-diffusion refers to the fact that atoms, even in pure metals, move about in the lattice by exchanging with vacancies, or empty sites.

Plastic deformation in the low-temperature range is referred to as cold working, whereas that in the high-temperature range is warm or hot working, depending on whether the temperature is relatively low or high. At cold-working temperatures, plastic deformation occurs by processes *within* the crystal grains and the stress-strain

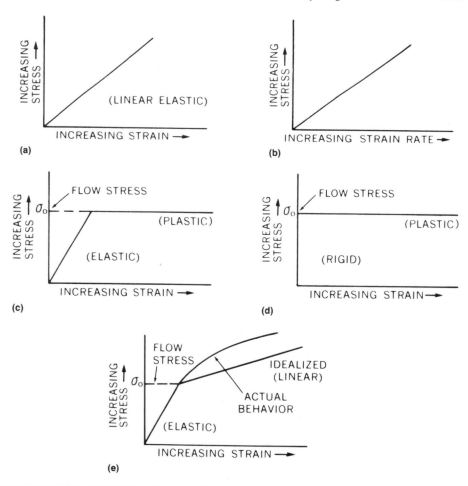

FIGURE 3-3. Mechanical behavior of classical bodies. (a) Hookean solid. (b) Newtonian fluid. (c) Perfectly plastic solid. (d) Rigid plastic solid. (e) Strain-hardening solid.

curves exhibit strain hardening (Figure 3-3e). In contrast, at warm- and hot-working temperatures, deformation processes within the grains as well as at the grain boundaries are occurring. At the higher hot-working temperatures in this regime, a large amount of softening accompanies deformation. This softening may balance the strain hardening, leading to a stress-strain response such as that depicted in Figure 3-3(c). The mechanisms controlling these behaviors are discussed next.

Deformation Mechanisms at Cold-Working Temperatures

The deformation mechanisms in the low-temperature range (below one third the melting point in Kelvin degrees) include slip, twinning, deformation banding, and kinking. Slip and twinning are the most important modes of deformation and account largely for the plastic strain that is generated before fracture.

Slip. Deformation by slip can be said to take place by the movement of lamellae (planes of atoms within the crystals) over each other, much like the shearing of a deck of playing cards. The elements of the slip process for a single crystal are illustrated in Figure 3-4. The slip translation takes place on a definite slip plane in a definite crystallographic direction. The combination of these two elements constitutes a slip system. In a given crystal structure, the slip direction remains the same despite the fact that new slip planes may become operative at elevated temperatures. Slip usually occurs on close-packed planes along close-packed directions in the crystal structure.

Slip is depicted in Figure 3-4 according to the classical model in which a portion of a crystal moves rigidly over another portion. The force required to do this is far in excess of that required to cause slip in real metals and alloys. It is well established that slip occurs by motion of dislocations, which are atomic disregistries in the crystal occurring in lines and loops. The basic notions of the "edge" and "screw" types of dislocations are illustrated in Figure 3-5. The difference in force requirements for the slip process by the classical model (rigid sliding of part of a crystal over the remaining part) and the dislocation model (motion of a dislocation) can be envisioned by the force required to move a large rug across the floor as a unit compared to that required to move it by passing a wrinkle along the rug.

Strain hardening as explained by dislocation concepts is due to the increased density of dislocations generated in plasic flow; softening requires the reduction of dislocation density. A limited amount of dislocation reduction occurs during deformation by a process of "dynamic recovery," which involves annihilation of individual dislocations. However, this is not enough to overcome the rate at which dislocations are multiplied by interactions with other dislocations or nondeformable barriers such as alloy precipitates. As the temperature is increased, nevertheless, the rate of dynamic recovery increases, leading to a reduction in the overall rate of strain hardening, as evidenced in stress-strain curves such as that shown in Figure 3-3(e). Several other phenomena such as twinning, the yield point, and strain aging are rationalized by dislocation concepts as well.

Twinning. The other important deformation mechanism at cold-working temperatures is twinning, which results in a limited amount of plastic deformation. Twinning is particularly important in hexagonal close-packed crystals, which normally slip only on the basal plane. It results when a lamellae within a crystal takes up a new orientation related to the rest of the crystal in a definite symmetrical manner. The lattice within the twin is a mirror image of the rest of the crystal (Figure 3-6).

There are two classifications of twins according to the mode of formation: deformation twins and annealing twins. Deformation twins are those that develop as a result of mechanical loading. They form very rapidly and are usually thin lamellae within a grain. Their formation is favored by increasing the strain rate and by lowering the temperature, both of which mitigate against the slip process.

Mechanical twinning occurs in body-centered cubic and close-packed hexagonal metals. Face-centered cubic metals normally do not form deformation twins. They do, however, form twins upon annealing after plastic straining, which accordingly are called annealing twins. Annealing twins are generally much thicker than deformation twins.

Deformation Mechanisms at Warm- and Hot-Working Temperatures

The high-temperature range was defined previously as the range above one third to one half the melting point in Kelvin degrees. The temperature at one half the melting point has been named the "equicohesive" temperature, below which the grain boundaries are more resistant to deformation by grain-boundary sliding than to slip inside the grain. The converse is true above this temperature.

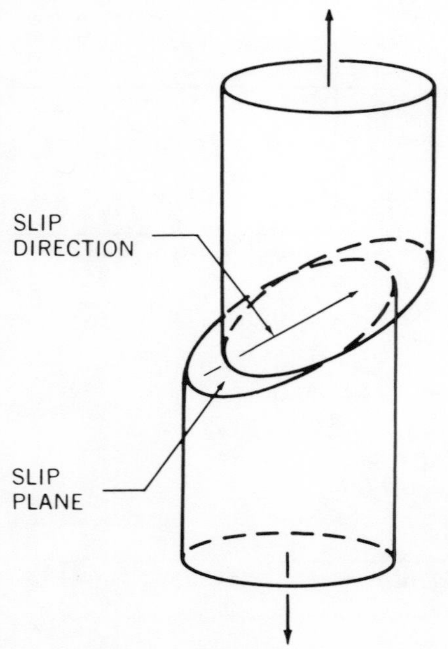

FIGURE 3-4. Slip sytem for a single crystal.

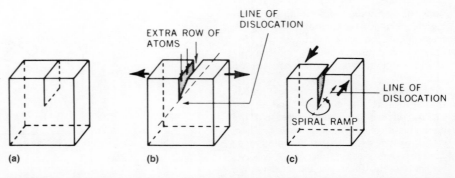

FIGURE 3-5. Creation of edge and screw dislocations. (a) Block with cut. (b) Edge dislocation. (c) Screw dislocation.

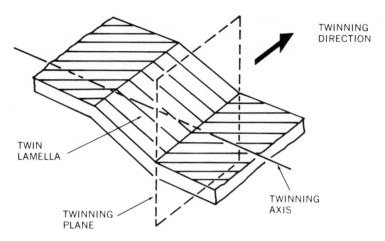

TWINNING DIRECTION

TWIN LAMELLA

TWINNING PLANE

TWINNING AXIS

FIGURE 3-6. Schematic of the elements of a twin lamella within a grain.

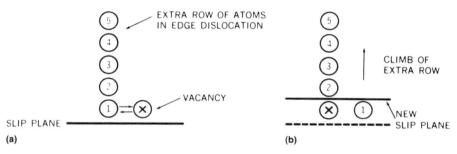

EXTRA ROW OF ATOMS IN EDGE DISLOCATION

VACANCY

SLIP PLANE

(a)

CLIMB OF EXTRA ROW

NEW SLIP PLANE

(b)

FIGURE 3-7. Concept of climb of a dislocation out of the slip plane. (a) Before and (b) after exchange with vacancy.

Perhaps the most significant factor at the equicohesive temperature and at higher temperatures is that the rate of self-diffusion in metals becomes significant. This atomic movement takes place by the exchange of atoms with vacancies. The exchange makes possible the phenomenon of "climb of dislocations" out of their slip planes, as illustrated in Figure 3-7. Each exchange process lifts the dislocation one interatomic distance out of the slip plane. This climb process allows dislocations to climb over obstacles and to recover from strain hardening.

The processes occurring in the high-temperature range can be summarized as follows:

- Breakdown of grains into subgrains through dislocation climb
- Grain-boundary sliding and grain-boundary migration
- Fine slip, difficult to resolve by ordinary optical-microscopy techniques

These processes are manifestations of the increased atomic mobility in the high temperature range. The characteristics of this range are:

- Recovery from strain hardening caused by plastic deformation
- Recrystallization, which is the formation of a new strain-free grain structure

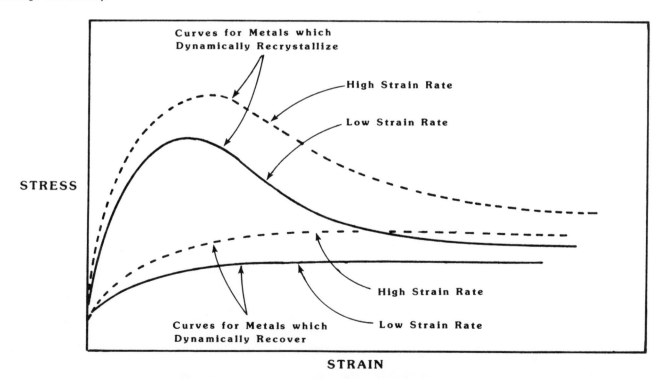

Curves for Metals which Dynamically Recrystallize

High Strain Rate

Low Strain Rate

STRESS

Curves for Metals which Dynamically Recover

High Strain Rate

Low Strain Rate

STRAIN

FIGURE 3-8. Flow curves showing strain hardening response.

that may then grow in size by migration of grain boundaries

● Viscous flow in which the rate sensitivity (the dependence of flow stress on rate of straining) becomes appreciable. The degree of viscous flow increases as the deformation temperature approaches the melting point.

When the above processes are operative, the flow curves show negligible, or even negative, strain-hardening response (Figure 3-8). If dynamic recovery is the principal softening mechanism, the metal strain hardens to a certain level, after which it exhibits a steady-state flow stress. In metals in which dynamic recrystallization occurs at hot-working temperatures, the flow curve passes through a maximum at which recrystallization is initiated, "flow softens," and then achieves a steady-state flow stress. Both dynamic recovery and dynamic recrystallization are affected by the rate of straining. With an increase in strain rate, the deformations at which dynamic recovery leads to a steady-state flow stress, or at which dynamic recrystallization is initiated, both increase (Figure 3-8).

Besides the change in the shape of the flow curve, an important effect of increasing the deformation temperature is the generally lower flow stresses compared to those at cold-working temperatures. The obvious advantage of increasing the deformation temperature is, namely, the reduction in required working loads. The dependence of flow stress on homologous temperature T_H (= test temperature T divided by the melting point T_M) is illustrated in Figure 3-9 for several common alloys. As mentioned, the flow stress becomes much more dependent on strain rate at hot-working temperatures. This is shown schematically in Figure 3-8, and the dependence of tensile strength on strain rate in high-purity aluminum is depicted for a variety of cold- and hot-working temperatures in Figure 3-10.

The effect of increasing strain rate on the increase in flow stress can be mitigated to a certain extent by the effects of deformation heating, a phenomenon that occurs at both hot- and cold-working temperatures. This is a result of the fact that more than 90% of the deformation work (area under the stress-strain curve) is converted into heat and only 10% is retained in the metal (in the form of dislocations or subgrains).

At high strain rates (>1 s^{-1}, typically), this heat cannot be dissipated into the tooling and thus raises the temperature of the metal workpiece. An estimate of the tem-

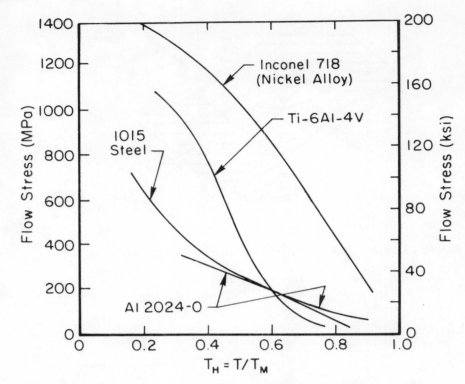

FIGURE 3-9. Flow stress versus $T_H = T/T_M$ for several common engineering alloys.

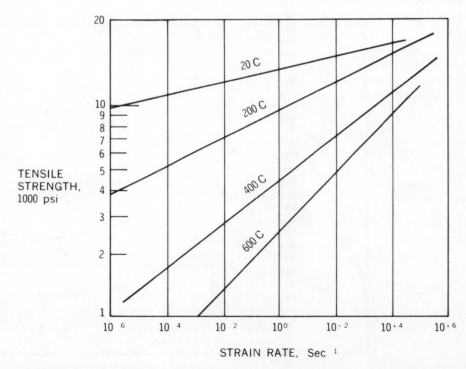

FIGURE 3-10. The relationship between tensile strength and strain rate for high-purity aluminum.

perature increase can be obtained from $\Delta T = \eta(\int \sigma d\epsilon)/\rho cJ$, in which $\sigma(\epsilon)$ is the stress-strain curve, ρ is the density, c is the specific heat, J is the mechanical equivalent of heat, and η is the fraction of work converted into heat that is retained in the metal (≈ 0.9 to 0.95 under adiabatic conditions).

In metals that have a strong dependence of flow stress on temperature (e.g., nickel-base and titanium-base metals), negative strain hardening or flow softening due to deformation heating occurs readily, particularly at hot-working temperatures. Moreover, for metals with high flow stresses at cold-working temperatures (and which therefore generate large amounts of deformation heat), the above heating effects may more than compensate for strain hardening in this temperature range and also lead to flow softening.

Strengthening Mechanisms in Metals

Pure metals may be strengthened by strain hardening (dislocation multiplication) and reduction of grain size.* Strengthening by reduction of grain size results from the fact that, as a grain deforms by slip, it must adjust to its neighboring grains in order to maintain coherency; the smaller the grain size, the more the resistance to this adjustment in shape and the higher the resistance to flow. Hardness is generally related to the grain size in a nonlinear fashion, but there is a linear relationship between hardness and the grain-boundary surface per unit volume. This illustrates the fact that increased hardness resulting from reduction of grain size is fundamentally related to grain boundaries.

The addition of alloying elements provides the basis for the following additional hardening mechanisms:

- Solid-solution hardening
- Second phase hardening
  Precipitation hardening
  Dispersion hardening
- Strain aging
- Transformation hardening

These mechanisms can have important influence on the forgeability of alloys, and plastic straining in turn can affect these hardening processes.

A solute atom may be completely soluble, partly soluble, or essentially insoluble in the lattice structure of the base metal. When the solute atom is present in amounts less than the solubility limit, it will be found in solid solution, i.e., dissolved in the lattice of the host atom.

When the solute atom is present in excess of its solubility, two structural components exist: namely, a saturated solid solution and a second phase composed of either the pure solute or a mixture of the solute and the solvent. The second phase may exist in a greater variety of shapes, sizes, and distributions.

The hardening of a metal by an alloying element, accordingly, occurs in two general ways.

- Formation of solid solutions, giving rise to solid-solution hardening
- Formation of a second phase, giving rise to hardening by interruption of the matrix by the second phase

Solid-Solution Hardening

Two basic types of solid solutions are formed in metals—namely, "interstitial" solid solutions in which the solute atom enters into interstitial sites in the lattice, and "substitutional" solid solutions in which the host atom is replaced by the solute atom. Interstitial solid solutions in general follow Hagg's Rule, which states that the ratio of the radius of the solute atom to the radius of the solvent atom must be less than 0.59. These solid solutions are found to be largely restricted to C, O, N, H, and B in transition metals that have unfilled inner shells of electrons. Substitutional solid solutions generally follow, as one condition to be met, the Hume-Rothery Rule, which states that the ratio of the radius of the solute atom to the solvent atom must lie in the range of 0.85 to 1.15 (15% rule).

Solid-solution hardening occurs fundamentally as a result of a decrease in the periodicity or regularity of the lattice. The effectiveness of hardening by a solute is found to be inversely proportional to its solubility. The solubility can be correlated with size effects, differences in electronegativity, and differences in valence. As the limit of solid solubility is approached, short-range ordering probably becomes important.

Second Phase Hardening

When the solubility limit for a solute is exceeded, a second phase appears in the microstructure (Figure 3-11). The second-phase particles act as barriers to slip and thereby harden and strengthen the alloy. The second phase may come out of solution "coherently" or "noncoherently," depending on whether the lattice of the second phase is continuous or discontinuous with the matrix. When a noncoherent precipitate is formed, it is called a dispersed second phase and the process is called dispersion hardening. The strength of the alloy is inversely proportional to the mean free path in the matrix (between precipitated particles) when the second phase is harder than the matrix.

The hardening process is sometimes called "precipitation hardening" or "age hardening" when the precipitate initially forms coherently. The forced fit of the precipitate on the matrix lattice creates large strains over moderate distances and results in effective hardening. The particles grow with time and reach a critical size at which they cannot sustain the strains required; they then break away from the matrix and become noncoherent. A typical age-hardening system is illustrated in Figure 3-12 and a typical aging curve in Figure 3-13.

Cold forging an age-hardenable alloy after solution quenching, but before aging, will accelerate aging, create favorable nucleation sites within the grain, and provide more general nucleation. With no deformation, the precipitation will nucleate at grain boundaries ahead of the sites in the interior of the grain. Plastic deformation thus makes precipitation occur more uniformly.

Strain Aging

The term "strain aging" is used in two senses. One refers to the return of the yield point after plastic straining of iron-base alloys. This is thought to be due to interstitial atoms locking dislocations by association with the dislocations. The other sense refers to the inducing of precipitation of, for example, iron nitride from iron by cold working. The conditions of supersaturation are attained through cold working. Perhaps these two connotations differ only in the degree of completion of the same process.

Hardening by Allotropic Transformation

In pure metals, allotropic transformations may lead to only small degrees of hardening. On the other hand, alloys of

---

*An exception to this rule occurs at hot-working temperatures and low deformation rates at which flow takes place at lower flow stresses in fine-grained materials because of grain-boundary sliding.

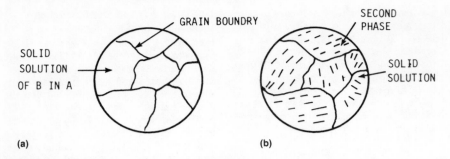

FIGURE 3-11. Change in structure upon exceeding the solubility limit of the solute. (a) Microstructure of a solid-solution alloy. (b) Structure of alloy after solubility lilmit has been exceeded.

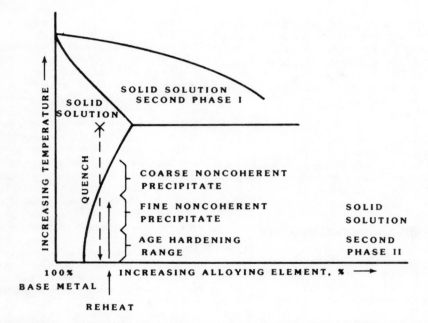

FIGURE 3-12. Steps in the age hardening of an alloy exhibiting a typical age hardening (precipitation hardening) system.

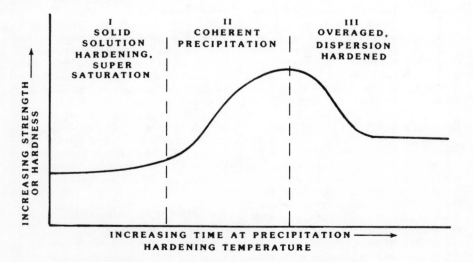

FIGURE 3-13. Typical precipitation hardening curve.

metals that undergo allotropic transformations may be hardened to a large extent by such transformations, because supersaturation in the low-temperature allotrope can be produced by cooling the high-temperature phase rapidly. This is true when the solubility of an element in the high-temperature allotrope is higher than in the low-temperature modification. If the quenched product is a supersaturated solid solution, it can be reheated to cause precipitation of a second phase either coherently or noncoherently. The system will relax to an equilibrium solid solution and a second phase. No new hardening mechanisms are involved; the lattice transformation serves only to achieve significant supersaturation. Plastic deformation aids in accelerating the decomposition of the high-temperature phase.

In steel (iron and carbon), the allotropic transformation results in the transition structure (martensite) upon rapid cooling of the high-temperature phase (austenite). This transformation is the basis for the heat treatment and hardening of steels and their outstanding industrial importance.

Stability of Strengthening Mechanisms at Elevated Temperatures

The stability of various elements of microstructure and hardening mechanisms at elevated temperatures is of interest both in determining the flow stress (and, therefore, the forging forces required) and the load-bearing abilities. Elevated temperatures bring about many changes in microstructure and hasten the attainment of the equilibrium structures. Some of the expected effects are:

- The rates of diffusion increase.
- Grain coarsening occurs.
- Solubilities usually increase, tending to dissolve second-phase particles.
- Aging and overaging are accelerated.
- Second-phase particles tend to agglomerate.
- The strength of the matrix continuously decreases with temperature, barring special metallurgical effects, and goes to zero at the melting point.
- The reaction with chemically reactive atmospheres increases, leading to contamination or loss of constituents to the atmosphere.
- Solute atoms can be absorbed at grain boundaries or in grains at elevated temperatures.

The strengths of hardened aluminum and aluminum alloys give some insight into the temperature regimes in which the various

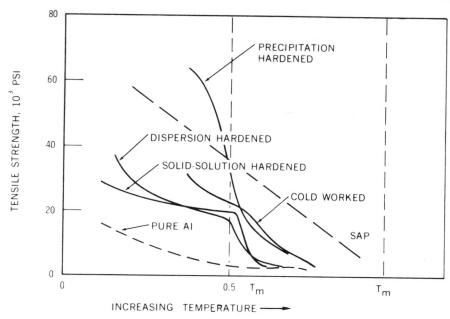

FIGURE 3-14. Effect of temperature on the stability of various hardening mechanisms.

strengthening mechanisms become unstable. Figure 3-14 shows that with the exception of sintered aluminum powder (SAP) all the variously hardened alloys lose their strength abruptly at about one half of the absolute melting point.

The solid-solution-hardened and dispersion-hardened alloys lose their strength first, whereas the cold-worked and precipitation-hardened alloys lose their strength more slowly with increasing temperature.

At about three quarters of the melting point, the strength of all the alloys becomes comparable to that of pure aluminum.

The case of SAP is somewhat special because it is a dispersion of Al_2O_3 in aluminum. Aluminum has a very small solubility for oxygen; therefore, Al_2O_3 has little tendency to dissolve, explaining the stability of this material. However, even at temperatures approaching T_M, SAP loses its strengthening capability.

deal on grain size and grain structure, factors that also influence the choice of workability tests. The variation is shown schematically in Figure 3-15. When the grain size is large relative to the overall size of the workpiece, as in conventionally cast ingot structures, cracks may initiate and propagate easily along the grain boundaries. Also, as is often the case in cast structures, impurities are segregated to the center and top of ingots (producing areas of low workability there), and chemical elements are not distributed uniformly either on a microscopic or macroscopic scale. Thus, the temperature range over which ingot structures can be forged is very limited. Typically, they must be hot worked.

The upper temperature limit in this range is usually slightly below the melting point of the metal. As cast, the melting point of alloys is usually below that in the finer grained, recrystallized state because of chemical inhomogeneities and the presence of low-melting-point compounds, which are often found at the grain boundaries. Forging too close to the melting point of these compounds may lead to grain-boundary cracking when heat developed during deformation raises the workpiece temperature and causes incipient melting of these compounds. This fracture mechanism is called hot shortness, and it is avoided sometimes by forging at deformation rates sufficiently slow to allow the

Forgeability and Workability Tests for Open Die Forging

Because forging is a relatively complex process, fracture or generation of undesirable defects may take a wide variety of forms, a number of which are summarized in Table 3-1. For this reason, it is not surprising that no single workability criterion can be relied on to determine forgeability. However, a number of testing techniques have been developed to gage forgeability depending on alloy type and microstructure, die geometry, and processing variables. In this section, typical forging defects and workability tests to characterize them are discussed.

In open die forging, metal flow patterns and stress states are not too complex, and forgeability is determined primarily by material structure and properties and process conditions and only secondarily by die geometry. Material structure variables include grain structure, crystal structure, and the presence of second phases or solid-solution elements. Material property considerations include temperatures at which melting, recrystallization, or phase changes occur, variation of flow stress with strain, strain rate, and temperature, and physical properties such as density, specific heat, and thermal conductivity. Of all the process variables, workpiece temperature is the most important.

Forgeability of Cast Structures

The variation of workability with workpiece preheat temperature depends a great

TABLE 3-1. Common Metallurgical Defects in Forging

Temperature regime	Cast grain structure	Wrought (recrystallized) grain structure
Cold working .... (a)		Free surface fracture, dead metal zones (shear bands, shear cracks), center bursts, galling
Warm working .... (b)		Triple-point cracks/fractures, grain-boundary cavitation/fracture
Hot working ...Hot shortness, center bursts, triple-point cracks/fractures, grain boundary shear bands/fractures		Shear bands/fractures, triple-point cracks/fractures, grain-boundary cavitation/fracture, hot shortness

(a) Cold working of cast structures performed typically only for very ductile metals (e.g., dental alloys), usually involves many stages of working with intermediate recrystallization anneals. (b) Warm working of cast structures is rare. Source: Ref. 12.

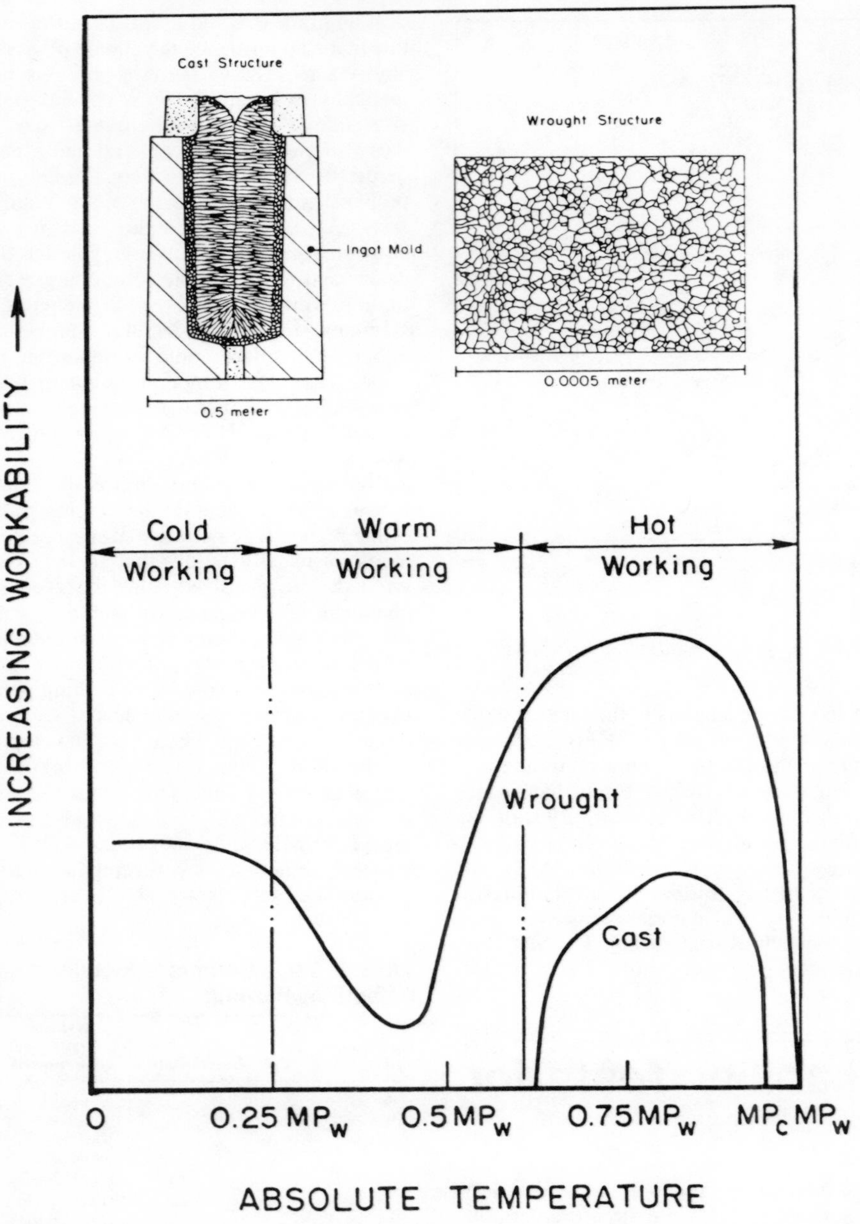

FIGURE 3-15. Schematic showing relative workability of cast and wrought and recrystallized metals at cold-, warm-, and hot-working temperatures. The melting point (or solidus temperature) is denoted as MP_C (cast metals) or MP_W (wrought and recrystallized metals).

in the lattice too much, and deformation may be forced to occur at the weak grain boundaries, leading to fracture at triple points (where three grain boundaries intersect) or to general cavitation (cavity formation) along the grain boundaries.

Low temperatures may also lead to the precipitation of alloying elements that remain in solution at higher temperatures, an occurrence that tends to pin matrix dislocations and to cause grain boundary failure. Precipitation at grain boundaries may also be harmful when it offers sites for grain-boundary cavitation.

From a process variable standpoint, the lower temperature limit for working of cast (and wrought) structure can be further restricted by the effects of die chilling. Because hot working is conventionally carried out with a hot workpiece and usually much cooler tool steel dies, heat transfer brought about by die chilling may lead to workpiece temperatures during deformation that are much lower than the preheat temperature.

In terms of die design, hydrostatic and other "secondary" tensile stresses are often set up in open die forging. These stresses may assist the grain-boundary fracture process, giving rise to gross center bursts, and should be evaluated in workability tests as well.

From the previous discussions, workability tests for open die forging of cast structures should be designed in order to determine realistic temperature regimes in which a recrystallized structure may be obtained and fracture avoided. Two such tests are the wedge-forging test and the sidepressing test.

The wedge-forging test is ideal for determining breakdown temperature ranges for ingot structures. In the test, a wedge-shaped piece of metal is machined from the ingot (Figure 3-16) and forged between flat, parallel dies. The overall dimensions of the wedge are selected based on the starting grain size. Large-grained material requires large specimens; fine-grained material requires small specimens. In either case, a specimen large enough to be representative of the entire ingot should be selected. Note that the wedge test is different from other workability tests (e.g., the tensile test and the uniaxial compression test) designed around small specimens of material with relatively fine wrought grain sizes that are subjected to nominally uniform deformation throughout the test section. In contrast, wedge specimens forged between flat, parallel dies suffer strains that vary with position.

heat developed by deformation to be dissipated into the dies, by using lower working temperatures, or by homogenization heat treatments prior to working. The last of these methods is especially useful for the breakdown of highly alloyed metals such as nickel-base superalloys.

The lower temperature limit for hot forging cast structures is the temperature at which there is some, but not a great deal of, dislocation multiplication. Dislocations are needed to build a reservoir of stored energy, which, in combination with thermal energy (supplied during hot deformation or during post-deformation heat treatment), is useful in breaking down the boundaries between the coarse grains and in generating a recrystallized grain structure. On the other hand, an excessive amount of dislocation multiplication, which occurs at temperatures below the hot-working range, hinders dislocation motion

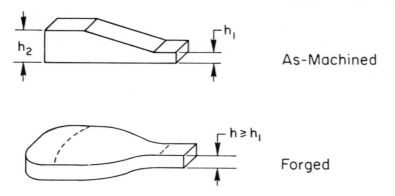

FIGURE 3-16. Specimens for the wedge test.

Machined wedge specimens are forged (or, equivalently, flat rolled) in the equipment in which the actual forging is to take place. This allows the effects of working speed, and hence die chilling due to heat transfer on workability, to be evaluated. Wedge preheat temperatures are selected based on previous experience with the alloy, or, for new alloys, on experience with similar alloys. If no such information is available, a temperature of approximately 0.8 to 0.9 times the measured solidus temperature is a good starting point. This choice can be modified based on examination of data for possible binary, ternary, etc. phases that may be formed from the alloy constituents and that are known to possess low melting points or that precipitate from solution at high temperatures. Even if no such phases are expected to be present, very high temperatures are often not advisable because recrystallization may be accompanied by grain growth and a yet less workable microstructure.

Following deformation, wedge specimens are quenched or put into annealing furnaces to determine dynamically and statically recrystallized grain sizes, respectively, as a function of working temperature, amount of deformation, and annealing temperature and time. In addition, they can be examined to determine temperatures at which the ingot structure can be broken down without cracking.

Sidepressing Test. Because specimen sizes can be varied, the sidepressing test is another test well suited for establishing the workability of large-grained cast structures. It is particularly useful for estimating the interaction of incipient grain boundary cracks, on the one hand, and secondary tensile and hydrostatic stresses often present in open die forging, on the other. In the test itself, round bars are laterally pressed between flat, parallel dies. The maximum tensile stresses in this in-

stance occur at the center of the bar at the beginning of the test (Figure 3-17).

As deformation proceeds, the bar assumes more of a rectangular cross section. As this occurs, the magnitude of the tensile stresses decreases, and the location of the maximum is shifted away from the center of the bar. Similar changes in magnitude and location of the maximum in tensile stress can be made to occur at early stages of deformation by changing from flat dies to dies that "encompass" more of the workpiece (Figure 3-17); secondary tensile stresses are minimized as less of an open die geometry is employed, and thus as more constraint is imposed. Moreover, friction and chilling may act in concert to

promote bulging and thus increase (or maintain) high levels of tensile stress during sidepressing above that due to geometry alone.

The application of sidepressing as a workability test for cast materials is illustrated in Figure 3-18 with the breakdown of a high-nitrogen stainless steel. When round ingots are forged "round-to-round" via sidepressing at 2100 °F (1150 °C), at which temperature dynamic recrystallization is found to occur, center bursting is observed. This must be attributed to the inability to recrystallize weak interfaces early enough in the deformation to avoid defect initiation and growth. Such an occurrence can be avoided by forging at 2100 or 2050 °F (1150 or 1125 °C), at which dynamic recrystallization is also favorable, and by forging/sidepressing in a sequence of round-to-square operations. In the latter instance, the decrease in tensile stresses can be assumed to be beneficial in eliminating center bursting and is in line with the typical press-forging practice for ingot breakdown.

Forgeability of Recrystallized Structures: Hot/Warm-Working Temperature Regime

Breakdown of coarse-grained cast structures to produce fine-grained wrought and recrystallized structures generally leads to improved workability in forging, as

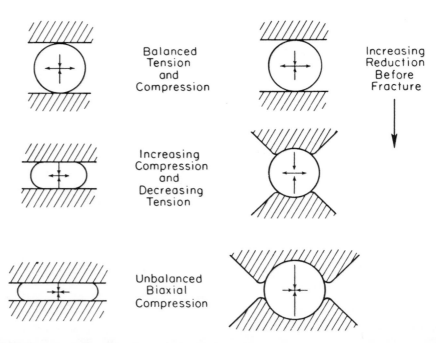

FIGURE 3-17. Schematic effect of billet shape and degree of enclosure on stress state in forging with good lubrication and no chilling.

FIGURE 3-18. Sidepressed bar of a high-nitrogen stainless steel cut from a cast ingot. Bar was forged round-to-round at 2100 °F (1150 °C) and developed a center burst. Courtesy of A. L. Hoffmanner, Battelle's Columbus Laboratories.

shown schematically in Figure 3-15. Even though most wrought metals can be worked over a wide range of temperatures and deformation rates, care must yet be exercised in selecting the forging temperature. Hot and warm forging require lower working loads than cold forging, but a number of defects can still arise in open die (and closed die) forging. These defects include many of those previously discussed for cast structures in Table 3-1.

Forgeability Involving Fracture-Controlled Defect Formation

Examination of the common defects in hot and warm working of wrought and re-crystallized structures reveals a convenient grouping that is helpful in devising different types of workability tests for open die forging. This grouping consists of workability problems related to (1) fracture-controlled failures and (2) flow-localization controlled failures. The former set includes such problems as hot shortness, triple-point cracking, and grain-boundary cavitation (Figure 3-19). The latter group includes shear-banding (which can lead to fracture), whose main source is heat transfer effects (due to chilling).

Fracture-related problems in forging involve the interaction of material aspects, process variables, and secondary tensile stresses. Material aspects include those mentioned before, such as crystal structure, phase changes, solidus, solutioning, and recrystallization temperatures, and mechanical and physical properties. Often, simple workability tests such as the hot compression test and hot tensile test are useful in making an initial selection of forging temperature and strain rate and in obtaining a gross estimate of forgeability. Usually, this estimate is based on a parameter such as upset test reduction at fracture or reduction in area in a tensile test. A useful summary of such measurements for different types of alloy systems is shown in Figures 3-20 and 3-21.

Generally, pure metals and single-phase alloys exhibit the best ductility/forgeability, except when grain growth occurs at high temperatures (Figure 3-20). On the other hand, alloys that contain low-melting-point phases (e.g., iron alloys with sulfur) or brittle phases (e.g., gamma-prime-strengthened nickel-base superalloys) tend to be difficult to forge and have limited forging-temperature ranges. The last behavior, that for alloys which contain brittle second phases, is further illustrated in Figure 3-22. This figure shows the relative forgeability of such an alloy system as a function of temperature and percentage of alloy content that forms second phases. The trends shown are very representative of gamma-prime-strengthened nickel-base superalloys, where a lean alloy might be Inconel 718 and an alloy that is highly strengthened by such precipitates might be Udimet 700. As the amount of alloying increases, the possibility of forming a low-melting-point phase increases in much the same way that the liquidus temperature in a binary eutectic system decreases as the eutectic composition is approached. Furthermore, the temperature for precipitation increases as the amount of second-phase elements, and hence the possibility for supersaturation, increases.

Workability Tests for Establishing Effects of Process Variables

Tests for establishing the effect of process variables on workability in forging

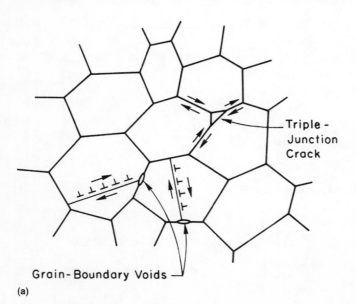

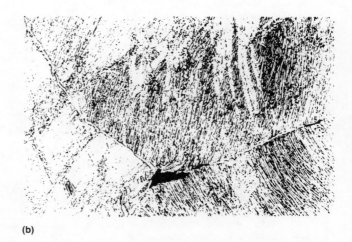

(a)   (b)

FIGURE 3-19. Formation of grain-boundary voids (cavitation) and triple-point cracks at warm- and hot-working temperatures. (a) Schematic showing how grain-boundary voids are formed under the action of matrix deformation and how grain-boundary sliding in the absence of grain-boundary migration and recrystallization may cause cracks to open at triple points. (b) Examples of grain-boundary voids and triple-point cracking at the prior beta grain boundaries in hot-forged Ti-6Al-2Sn-4Zr-2Mo-0.1Si with a Widmanstätten alpha starting microstructure.

include the high-speed isothermal tensile test (or compression test) and the "temperature-cycling" tensile test. The former test gives some estimate of the effect of strain rate and deformation heating on ductility. Performed under nominally isothermal conditions, the high-speed tensile test may be used to gage when deformation-induced heating may lead to temperature increases sufficient to cause incipient melting. The temperature rise is easily estimated if the specimen gage length is long enough and the tensile test is fast enough to give rise to adiabatic conditions at the center of the gage section.

As mentioned previously, under these circumstances, the temperature rise is derivable from the measured flow curve and the material's density and specific heat by assuming that a certain fraction (usually around 0.90 to 0.95) of the deformation work is transformed into heat. When the gage section of the tensile specimen is short or the deformation is slow such that a substantial amount of heat transfer occurs between the gage section and the undeformed shoulders, temperature changes and workability observations must be interpreted with the aid of a numerical thermal simulation.

The temperature-cycling tensile test method is useful in establishing the effects of die chilling in forging on workabillity. In such a technique, a series of nominally isothermal tensile tests is first performed over a range of temperatures by simply heating to the test temperature and running the tests. Then, a second set is run by heating to a variety of temperatures, holding at those temperatures for fixed or variable times, and then running the test after first cooling to a lower temperature. It is obvious that such a test is useful in determining the effects of chilling as a function of working speed and thermal properties on the forgeability of alloys that form second phases on cooling or that undergo grain growth during high-temperature preheating.

Results from tensile tests of this type are shown in Figure 3-23 for the high-nitrogen stainless steel discussed previously. However, the starting structure in this case is a wrought and recrystallized one. Isothermal hot tensile test data show a ductility maximum at approximately 1740 °F (950 °C). Isothermal transformation data for the wrought alloy show chromium nitride precipitation kinetics to be most rapid at 1740 °F (950 °C) also. Hence, at 1740 °F (950 °C), most of the solute was precipitated from solid solution prior to tension

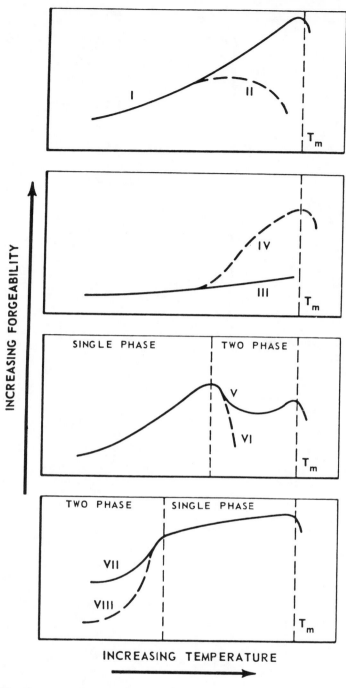

FIGURE 3-20. Typical forgeability behaviors exhibited by different alloy systems. I: Pure metals and single-phase alloys (aluminum alloys; tantalum alloys; columbium alloys). II: Pure metals and single-phase alloys exhibiting rapid grain growth (beryllium; magnesium alloys; tungsten alloys; all-beta titanium alloys). III: Alloys containing elements that form insoluble compounds (resulfurized steel; stainless steel containing selenium). IV: Alloys containing elements that form soluble compounds (molybdenum alloys containing oxides; stainless steel containing soluble carbides or nitrides). V: Alloys forming ductile second phase on heating (high-chromium stainless steels). VI: Alloys forming low-melting second phase on heating (iron containing sulfur; magnesium alloys containing zinc). VII: Alloys forming ductile second phase on cooling (carbon and low-alloy steels: alpha-beta and alpha titanium alloys). VIII: Alloys forming brittle second phase on cooling (superalloys; precipitation-hardenable stainless steels). T_m is melting temperature. Source: Ref. 13.

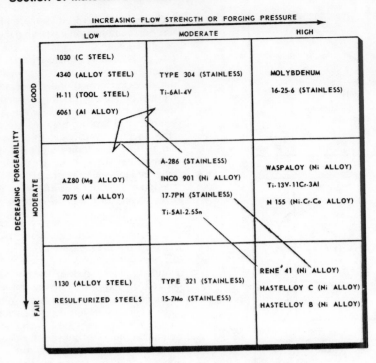

FIGURE 3-21. Interaction of workability, flow strength, and die-filling capacity in forging. Arrow denotes increasing ease of die filling. Source: Ref. 13.

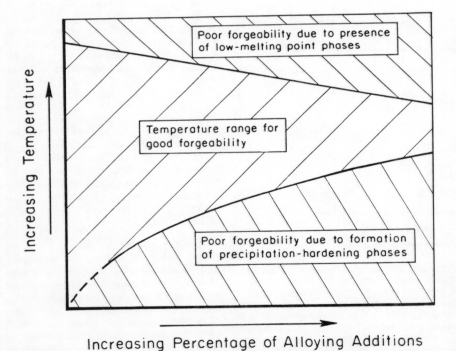

FIGURE 3-22. Influence of solute content that forms second phases on melting and solution temperatures and thus on forgeability.

testing. Because of this, little dynamic precipitation onto dislocations and at grain boundaries (which could nucleate grain boundary cavitation) probably occurred. Because such dynamic precipitation can limit workability, the correlation between transformation data and ductility in the present instance is reasonable not only for 1740 °F (950 °C) behavior but also for that above and below this temperature.

The temperature-cycling data can also be interpreted in terms of the interaction of precipitation and deformation. By preheating to high temperatures and then cooling to and testing at lower temperatures, more chromium and nitrogen had been retained in solution prior to testing, compared to isothermal tests. This solute was thus available for dynamic precipitation, which decreased the ductility relative to the isothermal tests.

The standard tensile test and the temperature-cycling tensile test may also be useful in evaluating heat treatments prior to deformation, which may improve forgeability. For example, a pretreatment to precipitate and overage (i.e., coarsen) second phases in various iron- and nickel-base alloys has been found to be very beneficial in subsequent working, an effect that can be evaluated most readily with simple workability tests.

Tests for Establishing the Effect of Secondary Tensile Stresses on Forgeability

Sometimes the uniaxial tension and compression tests are insufficient for determining forgeability in actual open die and closed die forging operations. This may arise because of secondary and hydrostatic tensile stresses, which can exacerbate a marginal workability problem. Several tests have been developed particularly to establish forging guidelines in these instances. These tests include the notched-bar, or U-notch, upset test, developed by Ladish Company.

The notched-bar upset test is similar to the conventional upset test, except that axial notches are machined into the test specimens prior to compression. The notch magnifies the effects of secondary tensile stresses that may arise during a conventional upset test due to bulging caused by chilling and/or poor lubrication. The higher levels of tensile stress in the test are supposedly more typical of those in actual forging operations.

In preparing specimens for the test, a forging billet is quartered longitudinally, exposing center material along one corner

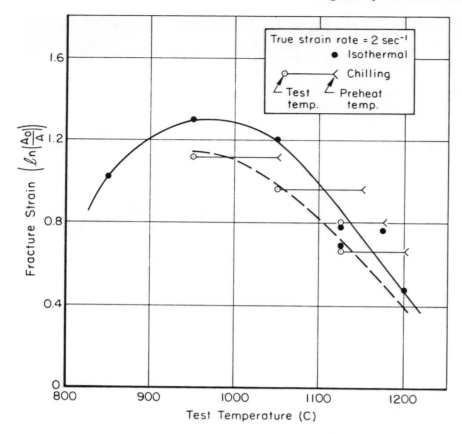

FIGURE 3-23. Tensile fracture strain dependence on test temperature for a wrought and recrystallized high-nitrogen stainless steel. Data from A. L. Hoffmanner, Battelle's Columbus Laboratories.

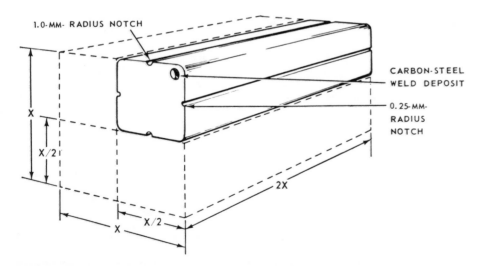

FIGURE 3-24. Method of preparing specimens for Ladish notched-bar upset forgeability test. Courtesy of R. Daykin, Ladish Co.

of each test specimen. After sectioning, notches with either 1.0-mm or 0.25-mm radii are machined into the faces, as shown in Figure 3-24; a weld button is often placed on one corner to identify the center and surface material in metals that exhibit forgeability problems because of segregation. Assuming a uniform bulge, the faces with the deeper 1.0-mm notches will be subjected to the highest tensile stresses and thus should be more likely to indicate a workability problem.

The notched-bar specimens are heated to one or a number of possible forging temperatures and upset forged to a reduction in height of approximately 75%. Because of the stress concentration effect, fractures in the form of ruptures are most likely to occur in the notched areas and may be assumed to have been initiated by hot shortness, triple-point cracking, or grain-boundary cavitation. These ruptures may be classified according to the rating system shown in Figure 3-25. A rating of 0 is applied if no ruptures are observed. If they are small, discontinuous, and scattered, the rating is 1; higher rating numbers indicate an increasing incidence of depth of rupture.

Notched-bar test results may be applied to forgeability predictions for a wide range of alloys and forging processes. In one example, rolled rings of 403 stainless steel with a notched-bar rating of 0 were found to be sound, whereas those with a rating of 4 ruptured extensively (Figure 3-26).

In a similar vein, the rupturing sensitivity of Inconel 718 has been demonstrated with the notched-bar upset test (Figure 3-27). The alloy was shown to be much more notch-sensitive as the test temperature was increased. The combined effects of interface frictional and deformation heating, in conjunction with the stress concentration, were able to produce very large ruptures when the alloy was forged at and above 2100 °F (1150 °C). Such an effect is not surprising in view of the fact that Inconel 718 starts to melt at approximately 2200 °F (1200 °C).

From these results, one may conclude that the notched-bar upset test is more sensitive than the simple upset test. In fact, it has been reported that unnotched billets from heats having a notched-bar rating of 3 are perfectly sound after similar reductions in simple upsetting. Thus, the simple upset test may indicate a deceptively higher degree of workability than can be realized in an actual forging operation, and the notched-bar upset test may be particularly useful for identifying materials having marginal forgeability.

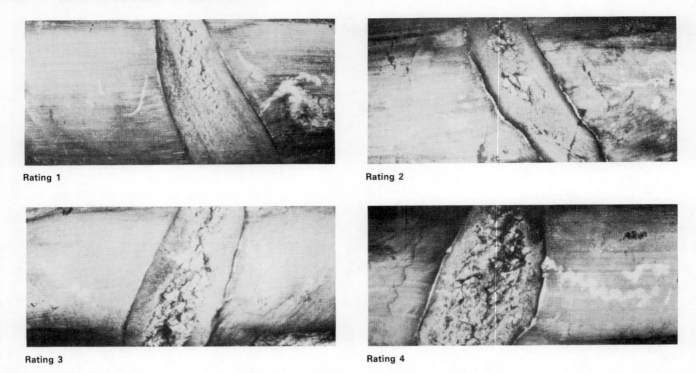

Rating 1

Rating 2

Rating 3

Rating 4

FIGURE 3-25. Suggested rating system for notched-bar upset specimens that exhibit progressively poorer forgeability. Courtesy of R. Daykin, Ladish Co.

Forgeability Involving Flow-Localization-Controlled Failure

Nonisothermal Upset Test. Workability problems in hot and warm open die forging that are related to flow localization are very common, and a variety of terms such as chill zones, dead metal zones, and locked metal have become synonymous with defects that result from the phenomenon. In open die forging where lateral constraint to metal flow is minimal, the occurrence of flow localization is a result of heat transfer between the hot workpiece and cold dies or excessive friction. As such, it is relatively easy to devise workability tests for investigating and resolving these problems when they occur.

The simplest workability test that can be employed to study the effects of heat transfer on flow localization is the nonisothermal upset test. Like the uniform, isothermal hot compression test, a cylindrical workpiece is upset between flat, parallel dies. Workpiece and die temperatures can be varied, as can working speed, lubrication, and dwell time on the dies prior to forging.

Following deformation, specimens are sectioned and metallographically prepared in order to determine the extent of chilling and the formation of shear bands (regions of intense localized deformation) between the chill zones and the remainder of the deforming bulk. As shown in Figure 3-28 for several Ti-6Al-2Sn-4Zr-2Mo-0.1Si (Ti-6242Si) specimens that had a starting equiaxed alpha microstructure, the chill zones are revealed usually as areas that etch differently from the rest of the specimen.

By varying the process variables in a series of nonisothermal upset tests, the extent of chill zones and severity of shear banding can be estimated. These results can be interpreted using a variety of mathematical techniques; they reveal that there are three important material/processing components that determine the magnitude and extent of strain rate (and thus strain) gradients: (1) the material's flow stress dependence on temperature, (2) the material's strain-rate sensitivity, and (3) the magnitude of the temperature gradient.

For a given material, it is obvious that the best way to minimize chilling effects is to forge at a temperature at which the flow stress is not too temperature-sensitive or to decrease the magnitude of the temperature gradient. This is accomplished by using forging equipment with faster rams and/or by employing glass lubricants with low thermal conductivity.

The nonisothermal sidepressing test is another workability technique that can be used to gage the interaction of material properties and process variables during flow localization in hot forging. As for the nonisothermal upset test, a number of test specimens are sidepressed between flat, parallel dies using a variety of workpiece temperatures, die temperatures, and working speeds, and the formation of defects is determined by metallography. Unlike the upset test with its axisymmetric chill zones, however, flow localization is primarily manifested by shear-band formation and propagation.

The absence of well-defined chill zones in the nonisothermal sidepressing test may be attributed to two main factors. First, the amount of contact area in sidepressing of round bars starts out at zero (i.e., line contact) and builds up rather slowly at the beginning of deformation. Second, because the operation is basically a plane strain one, surfaces of zero-extension, along which block shearing can initiate and propagate, are present in sidepressing. These surfaces are the natural ones along which shear strain can concentrate into shear bands. Formation of well-defined chill zones prior to shear-band formation would probably involve more deformation work and hinder the propagation of shear bands. This is not to say, however, that

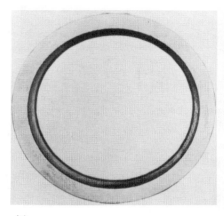

(a)

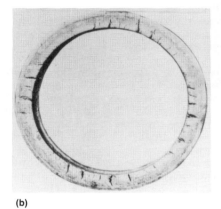

(b)

FIGURE 3-26. Rolled rings made from two heats of type 403 stainless steel exhibiting different forgeability ratings in notched-bar upsetting tests. (a) Forgeability rating of billet = 0. (b) Forgeability rating of billet = 4. Courtesy of Ladish Co.

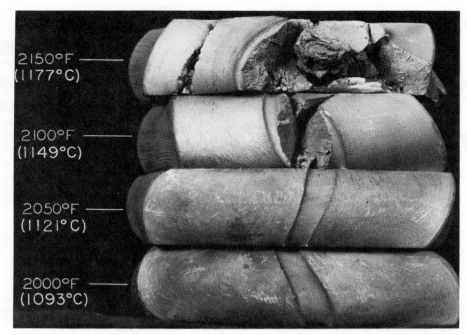

FIGURE 3-27. Inconel 718 notched-bar upset test specimens showing rupture sensitivity associated with forging temperature. Courtesy of R. Daykin and R. Koch, Ladish Co.

temperature gradients are not important in supplying a driving force or flow localization.

Sidepressing results for Ti-6242Si illustrates the kinds of localization behavior that can be observed. In Figure 3-29, cross sections of bars sidepressed in a mechanical press (strain rate $\sim 30\ s^{-1}$) using dies preheated to 375 °F (191 °C) are shown. When the bars are preheated to a temperature much below the transus temperature (1675 °F, or 913 °C), at which the flow stress is very sensitive to temperature (Figure 3-30), shear bands whose intensity increases with reduction are developed (Figure 3-29, top). In contrast to these results, no shear banding is observed (Figure 3-29, bottom) when the workpiece temperature is approximately 1800 °F (980 °C), which is close to the beta transus temperature for the alloy. At this temperature, the flow stress dependence on temperature is minimal. Hence, the high strain rates employed in this instance, in conjunction with the low temperature sensitivity of the flow stress, may be surmised to have minimized stress and strain gradients during sidepressing. It has been found that these results correlate well with observations of shear banding in conventional forging of Ti-6242Si jet-engine compressor blades (Figure 3-31).

The effect of lower working speeds, and therefore more heat transfer, on shear-band formation in sidepressing can also be determined by sidepressing tests; the results for Ti-6242Si illustrate extremes in behavior that can be investigated via this test. These experiments were conducted in a hydraulic press (strain rate $\approx 1\ s^{-1}$); for a given reduction, deformation thus lasted approximately 30 times as long as in the mechanical press trials.

At low reductions, shear bands, similar to those previously observed, were formed in the hydraulic press specimens (Figure 3-32). On the other hand, at high reductions, shear cracks were found to develop along these shear bands. The same effect was found regardless of whether the workpiece preheat temperature was much below the transus temperature or rather close to it, a trend that is considered to be significant for the mechanical press tests.

From these results, it can be concluded that (1) long deformation times drop the average workpiece temperature from one at which the flow stress is not too sensitive to temperature to one at which it is and at which thermal gradients may lead to shear bands, and (2) long deformation times drop the average workpiece temperature into a regime of low ductility. The latter conclusion explains why cracking was observed in hydraulic press sidepressing inasmuch as process simulation results suggested that, for the die temperatures used, the average workpiece temperature had indeed fallen to a temperature at which ductility is low. The shear bands in these instances probably contained microfis-

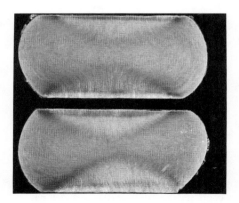

FIGURE 3-28. Axial cross sections of specimens of Ti-6Al-2Sn-4Zr-2Mo-0.1Si.

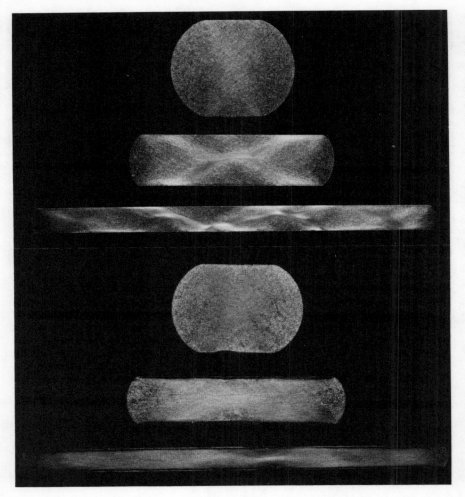

FIGURE 3-29. Transverse metallographic sections of specimens of Ti-6Al-2Sn-4Zr-2Mo-0.1Si. See text for discussion.

sures, which could serve as fracture initiation sites.

Forgeability of Recrystallized Structures: Cold-Working Temperature Regime

As the forging temperature is lowered into the cold-working regime, forgeability problems change from failure related largely to intergranular cracking and cavitation to a mode involving transgranular ductile fracture. Ductile fracture consists of the nucleation of voids at second-phase particles and inclusions, and the growth and eventual coalescence of these voids under the action of tensile stresses to cause fracture. In cold forging, ductile fracture occurs most readily at free surfaces where barreling has set up secondary tensile stresses.

A simple workability test for analyzing these kinds of failures is the upset test, in which various combinations of tensile and compressive stress fields are developed by using various lubrication conditions and initial specimen height-to-diameter ratios. From tests of the sort, it has been found that surface fracture follows a critical strain criterion, $\epsilon_1 + 2\epsilon_2 = C$, where ϵ_1 and ϵ_2 are tensile and compressive surface principal strains at fracture, and C is a material constant (Figure 3-33). The determination of the value of the constant in this fracture criterion is very useful in ranking the resistance of various materials and different lots of a given material to ductile fracture.

Other tests besides that employing simple upsetting of cylindrical specimens have been devised to obtain a measure of workability in cold forging. These include the grooved compression test and an indentation-type method utilizing a truncated cone. Both are useful for determining the deformation response of metals such as low-carbon steel that do not readily fracture in ordinary upsetting.

In the grooved compression test, cylindrical samples of various initial aspect ratios and with axial grooves are upset under various conditions of lubrication in much the same way that standard upset testing for workability analysis is conducted. Fracture is initiated in the groove at the sample mid-height (at which tensile stresses are greatest), and the average circumferential fracture strain here is taken as a measure of workability.

Available data also show that, for a given material, the value of this circumferential fracture strain is independent of the initial groove depth; however, as the groove depth is increased, the axial reduction at which this strain is reached is diminished. For this reason, the grooved compression test appears to be useful for establishing the effects of surface flaws on forgeability and in ranking the quality of different lots of the same material. In this regard, it is similar to the previously discussed notched-upset test used for hot workability and has been suggested as a good test for obtaining an index of surface quality of steel wire used in cold heating, which is a common forging operation involving upsetting-type deformation.

The truncated cone test is a test that was developed in an attempt to minimize the effect of surface flaws and the random errors they produce on workability data. Unlike the simple upset and notched-upset techniques, the truncated cone method involves indentation of a conical tool into a cylindrical specimen whose initial diameter is greater than that of the truncation (Figure 3-34). By this means, cracking is made to occur on the inside of the workpiece at the tool-material interface, thereby eliminating the influence of free-surface flaws and surface finish on workability limits. The reduction (measured at the specimen axis) at which cracking occurs may be used to compare the workability of different materials. Alternatively, the reduction at which a fixed crack width is obtained or the width of the crack at a given reduction has been suggested as a parameter to provide an index of workability.

Ductile fracture may also occur in the workpiece interior during cold forging. As with free-surface fracture and center-bursting at hot-working temperatures, secondary tensile stresses play a large role in such failures. In addition, the voids generated by these stresses may coalesce to form gross center bursts as in hot forging.

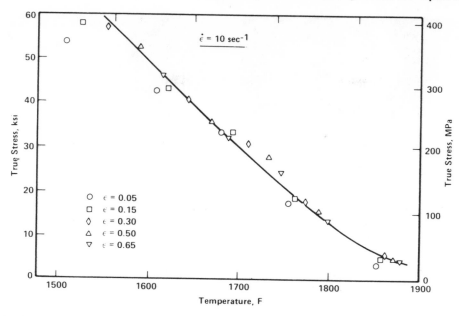

FIGURE 3-30. Flow stress dependence on temperature for Ti-6242Si with an equiaxed alpha starting microstructure.

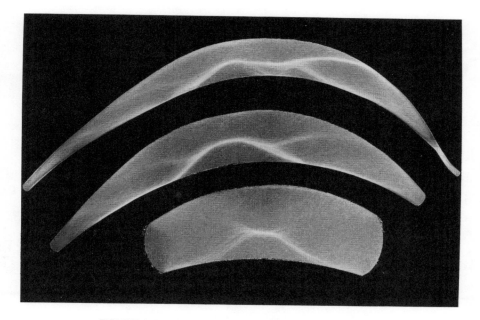

FIGURE 3-31. Shear bands in blade forging of Ti-6242Si.

FIGURE 3-32. Ti-6242-Si specimens with an equiaxed alpha starting microstructure that were nonisothermally sidepressed in a hydraulic press ($\dot{\epsilon} \approx 1\ s^{-1}$). (a, b) Transverse metallographic sections. (c) Micrograph of region with shear band and crack from section shown in (b). Specimen preheat temperature, die temperature, and dwell time were 913 °C, 191 °C, and 14 s, respectively.

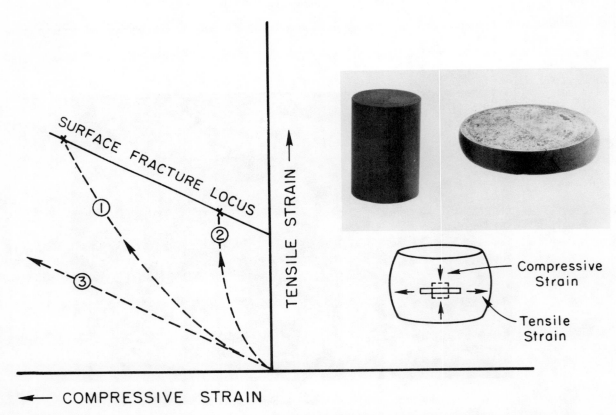

FIGURE 3-33. Ductile fracture criterion for cold forging. 1: Good lubrication, large h/d. 2: Poor lubrication, small h/d. 3: Frictionless compression, large or small h/d.

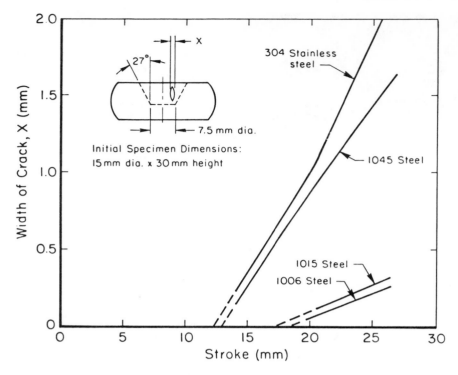

FIGURE 3-34. Relationship between crack width and stroke in truncated cone indentation test for workability of various steels at cold-forging temperatures.

Forgeability and Metal Flow in Closed Die Forging

In closed die forging, workability problems such as those previously described may be encountered depending on alloy type, forging temperature, die temperature, and press speed. Thus, these problems can often be investigated and solved based on consideration of material properties (e.g., flow curves) and results from simple workability tests. At other times, forging difficulties in closed die situations may not be related to fracture occurrence but rather to poor grain-flow patterns, lack of die fill, or related metal-flow defects. These problems and means of alleviating them are discussed next.

Control of Grain Flow in Closed Die Forging

During initial breakdown of cast ingots or subsequent working—for example, by forging—nonuniformities in alloy chemistry, second-phase particles, inclusions, and the crystalline grains themselves are aligned in the directions of the greatest

metal flow, as illustrated schematically in Figure 3-35. The directional pattern of the crystals following working is known as the grain-flow pattern. An example of grain

flow in an actual forging is shown in Figure 3-36.

To a degree, grain flow (or "fibering," as it is sometimes known) produces directional characteristics in properties such as strength, ductility, and resistance to impact and fatigue. The forging process uses this directionality to provide a unique and important advantage by orienting grain flow within the component so that it lies in the direction requiring maximum strength. The variation in yield strength and tensile strength is not particularly important.

In the case of heat-treated wrought steels, these strength properties are independent of the direction of testing. The important anisotropies are in ductility, notched-bar impact strength, and fatigue (Figure 3-37). The resistance to stress corrosion cracking can be highly directional, for example, in high-strength aluminum alloys. The maximum load-bearing capacity of a forging is realized when the component is loaded along the fiber, or grain-flow direction. It should be noted that deformation of a cast alloy leads to a significant improvement in ductility and toughness, which is maintained in the direction of the grain flow, or the longitudinal direction.

Properly developed grain flow in forgings closely follows the outline of the component. In contrast, bar stock and plate have grain flow in only one direction, and changes in contour require that flow lines be cut, exposing grain ends and rendering the material more liable to fatigue and more sensitive to stress corrosion (Figure 3-38). Also, forgings designed to the approximate final shape of the component better utilize material than parts machined from bar stock or plate.

Location of the parting line has a critical bearing on grain flow and the direc-

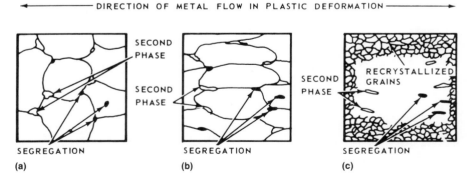

FIGURE 3-35. Typical development of grain flow by plastic deformation of metals. (a) As-consolidated; grains, segregation, and second phase are randomly oriented. (b) As-wrought; grains, segregation, and second phase are elongated in the direction of plastic deformation. (c) Wrought and recrystallized; grains are randomly oriented; segregation and second phase are elongated in the direction of plastic deformation. Source: Ref. 13.

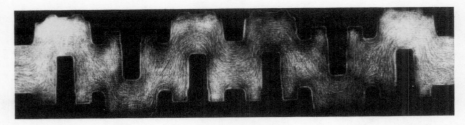

FIGURE 3-36. Polished and etched surface of this sectioned crankshaft forging offers an example of controlled grain flow.

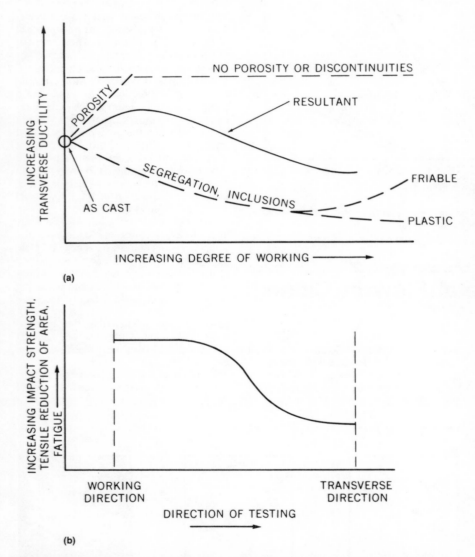

FIGURE 3-37. Anisotropy in wrought alloys. (a) Development of ductility anisotropy. (b) Nature of anisotropy.

tional properties of the forged piece. In the forging process, excess metal flows out of the cavity into the gutter as the dies are forced together. In this flow toward the parting line, objectionable flow patterns may be created if the flow path is not smooth (Figure 3-39). A parting line on the outer side of a rib should be placed either adjacent to the web section or at the end of the rib opposite the adjoining web section. The placement of the line at an intermediate step can result in irregularities.

In the forging of some parts, the die sequence as well as the designs of the dies themselves may be altered in order to control the grain-flow pattern. The forging of steel hemispherical parts, for example, often includes a preliminary side-forging step instead of conventional upsetting common to most other circular parts. Figure 3-40 compares two grain-flow patterns developed in hemispherical parts. The lower sketch shows how smooth, unbroken flow lines results from a preliminary side-forging step. The upper sketch shows how the weaker center material is concentrated in the polar region by normal forging practices.

Most methods for altering grain flow are not this simple. As stated earlier, inclusions ordinarily align with grain flow; if the grain-flow pattern is straight, so are the inclusions. Because cylindrical parts forged from wrought bar will retain this basic pattern, it is sometimes desirable to alter the forging practice to break up the straight-line pattern in such parts. This reduces the probability of long inclusions appearing on the surface of finished parts and causing rejections. Forge shops sometimes use a corrugated die design to provide a wavy grain-flow pattern for this purpose. This practice has been known to reduce rejections from nearly 30% to less than 5% on parts of cylindrical shape.

Forging and lubrication techniques can also be used for controlling grain flow. If, for example, a disk is hammer forged between contoured dies, the top surface of the forging is generally better lubricated than the lower surface. This can cause a nonuniform grain flow. Essentially, the metal flows readily against the upper die surface, but is virtually undeformed near the lower die surface.

In addition to inadequate lubrication, die-chilling effects contribute to this type of flow. The out-bent fibers formed in this way result in poorer toughness and surface finish. The condition can be corrected by turning the disk over during forging or by providing better lubrication on the lower die surface.

A third alternative consists of upsetting between flat dies to a disk slightly smaller than the final die, turning the disk over, then finishing the impression die. This latter method is the one most often used in normal forging practice, because the undersized flat disk is easier to locate in the die and the grain-flow patterns are gen-

GRAIN

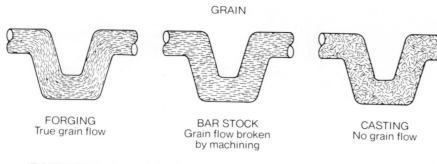

FORGING
True grain flow

BAR STOCK
Grain flow broken
by machining

CASTING
No grain flow

FIGURE 3-38. Comparison of grain structures of forging, bar stock, and casting.

erally more uniform. The flow patterns can be adjusted in other shapes by similar lubricating and forging techniques.

Factors Affecting Die Filling

Many variables influence the ability of metals to flow during forging and to fill the cavities of impression dies. Some of the more important factors are:

- Forging temperature
- Forgeability and flow strength
- Friction and lubrication

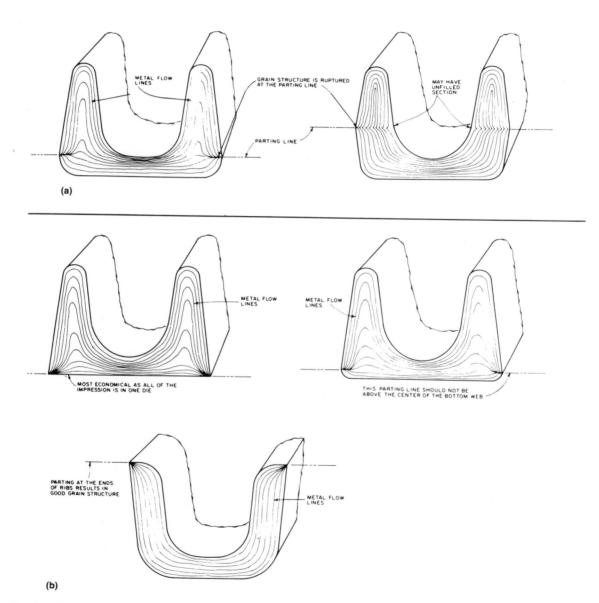

FIGURE 3-39. Various parting line locations on a channel section have differing effects. (a) Undesirable: These parting lines result in metal flow patterns that cause forging defects. (b) Recommended: The flow lines are smooth at stressed sections with these parting lines. Source: Ref. 14.

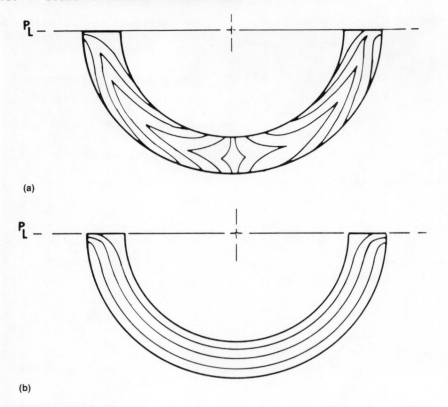

(a)

(b)

FIGURE 3-40. Comparison of flow lines developed in hemispherical forgings with different preliminary forging techniques. (a) Upset before die forging, showing weakness in polar region. (b) Side forged before die forging, showing no weakness in polar region. Source: Ref. 13.

- Die temperature
- Shape and size factors

Forging Temperature. In general, increasing forging temperature increases ductility and decreases flow stress of a metal or alloy and therefore promotes or improves die fill. These improvements continue until a second phase appears, melting begins, or, in some alloys, grain growth becomes excessive. In many materials, metallurgical considerations restrict the forging temperature to levels well below the optimum for die filling in order to produce a forging with desirable structure, strength, and toughness properties.

Forgeability and Flow Strength. Difficulties in filling die cavities are encountered in forging metals characterized by either poor forgeability or high flow stress. Poor forgeability causes rupture before the die is filled; high flow strengths may cause metals to flow past recesses without filling them or may result in underfilling with the maximum loads available. Figure 3-21 suggests, in terms of forgeability and flow strength, the ease with which different metals and alloys can be deformed to fill intricate die cavities. Some of the difficulties attributed to these characteristic differences among materials can be illustrated by considering two separate types of forgings—connecting rods and hubs.

Table 3-2 provides data obtained in experimental hammer forging of several alloys in a set of dies ordinarily used for 1030 steel connecting rods. As would be expected, the stronger materials required more reheating operations and a larger number of blows. Even so, the drawing and fullering operations, ordinarily employed to achieve the proper stock distribution for impression die forging, were unsuccessful with the two nickel-base superalloys. Figure 3-41 shows four of the specimens after forging. The forgeability of the higher alloy stainless and the iron-base superalloy was good enough to avoid cracking, but the hammer was too small to fill the dies completely.

Table 3-3 gives data from a similar study on seven metals forged in dies ordinarily employed for forging hubs from 4340 steel. Both ruptures and pronounced underfilling occurred in the stronger nickel-base superalloys with poor forgeability. Sound hubs were made from the other four alloys, but the dies were not completely filled. The degree of underfilling appears to depend on the flow strengths of the alloys at the forging temperatures used. In both of the above cases, sound and completely filled parts could have been produced from any of the materials involved by the use of dies and operations designed specifically for each of the materials.

Friction and Lubrication. Die lubricants are used principally to reduce forging loads, reduce die wear, and control or improve the uniformity of metal flow. They are also useful as parting compounds when forging metals with tightly adhering oxides. In general, the selection of forging lubricants usually represents a compromise among several desirable attributes: (1) insulating qualities, (2) low coefficient of friction, (3) ease of removal, and (4) low corrosiveness and toxicity. Graphite,

TABLE 3-2. Forging Data on Connecting Rods Forged from Eight Different Materials Representing Increasing Forging Difficulty

Material	Forging temperature °F	Forging temperature °C	No. of times heated	Total No. of blows to complete forging	Remarks
1030 steel	2300	1260	1	<20	Completely filled
Type 304 stainless steel	2200	1200	1	40	Completely filled
Type 347 stainless steel	2200	1200	1	50	Completely filled
Monel	2050	1120	2	51	Completely filled
16-25-6	2000	1100	5	76	Incomplete fill on shaft end
N-155	2000	1100	5	89	Incomplete fill on both ends and ribs
Hastelloy B	2200	1200	3	. . .	Extreme difficulty in drawing and fullering shape, experiment discontinued
Hastelloy C	2150	1180	4	. . .	Same as for Hastelloy B

Source: Ref. 13.

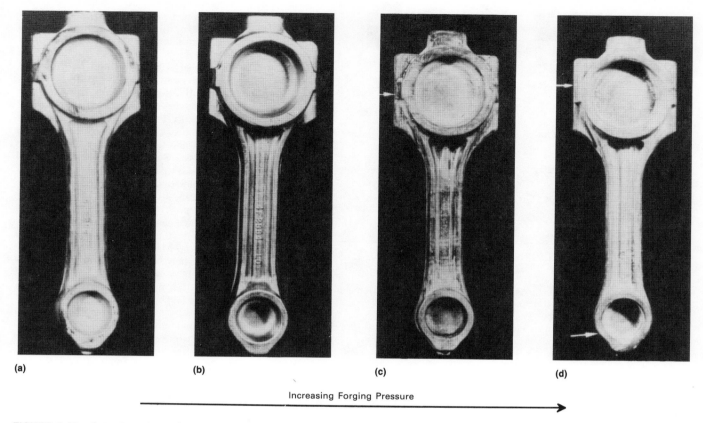

Increasing Forging Pressure

FIGURE 3-41. Parts forged experimentally in connecting-rod dies from four alloys representing increasing forging difficulty. (a) Type 304 stainless steel: completely filled. (b) Monel: completely filled. (c) 16-25-6 stainless steel: arrow indicates incomplete fill on large end. (d) N-155 superalloy: arrows indicate incomplete fill on ends and ribs. Source: Ref. 13.

TABLE 3-3. Forging Data on Hubs Forged from Seven Different Materials Representing Increasing Forging Difficulty

Material	Forging temperature °F	°C	No. of times heated	No. of hammer blows required to complete forging Upset	Block	Finish	Total	Remarks(a)
4340	2300	1260	1	2	4	5	11	Sound, completely filled
Monel	2100	1150	1	2	5	7	14	Sound, shaft underfilled by $9/16$ in.
Type 304	2100	1150	1	3	8	10	21	Sound, shaft underfilled by $5/8$ in.
A-286	2100	1150	2	5	8	9	22	Sound, shaft underfilled by $7/8$ in.
16-25-6	2000	1100	2	5	7	12	24	Sound, shaft underfilled by $7/8$ in.
Hastelloy C	2200	1200	3	5	10	18	33	Ruptures, shaft underfilled by 1 in.
Hastelloy B	2200	1200	2	4	18	9	31	Ruptures, shaft underfilled by $1 1/4$ in., forging stopped when rupture noticed

(a) Sound refers to forgings that did not rupture. Underfill dimensions are given as average amounts based on measured heights of hubs.

in various dispersions and preparations, appears to be the most widely used lubricant for forging.

Lubricants suitable for forgings in flat dies do not always give good results in impression dies, because the requirements are dissimilar. Low friction coefficients favor lateral metal flow, the main requirement in flat dies. In most impression dies, however, lateral flow must be selectively impeded in order to ensure enough vertical flow to fill the cavities. This is one of the principal functions of the narrow, annular flash openings. Consequently, an ideal lubricant for closed dies would provide minimum friction on vertical surfaces of cavities and maximum friction near the flash openings. These contradictory requirements can sometimes be met by using different lubricants at critical locations.

Die Temperature. The use of heated dies improves die filling and reduces forging pressures. The benefits result from the

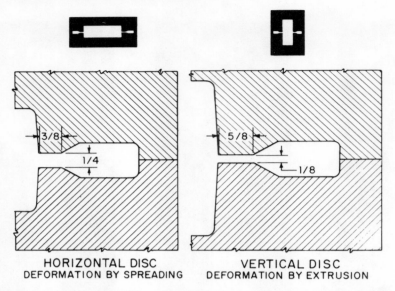

FIGURE 3-42. Schematic comparison of flash designs for 10-in.-diameter × 2-in.-thick disk forged in horizontal and vertical positions. Source: Ref. 13.

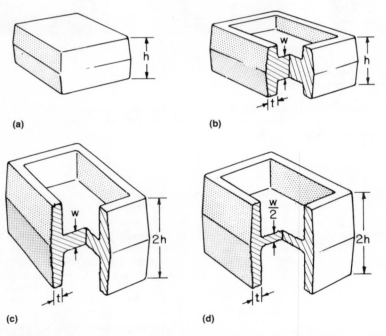

FIGURE 3-43. Basic rectangular shape (a) and examples of three modifications illustrating increasing forging difficulty with increasing rib height and decreasing web thickness. (b) Rib-and-web part. (c) Part with higher rib. (d) Part with higher rib and thinner web. Source: Ref. 13.

ferential between the die and the billet. Therefore, heated dies offer more advantages when forging with hydraulic presses than with hammers and high-velocity machines.

The use of hot dies limits the choice of lubricants and die materials. Most water-base and oil-base forging lubricants are unsatisfactory for die temperatures above 500 to 800 °F (260 to 425 °C). High die temperatures are difficult to maintain and control, and the unusual die materials are expensive. Steel dies are rarely heated above 800 to 1000 °F (425 to 535 °C) for production operations, principally for economic reasons.

The specific heat and thermal conductivity of the workpiece influence the effects of die temperature on metal flow and die filling. Materials with a relatively low specific heat but good thermal conductivity, such as molybdenum, will be chilled rapidly, but the variations in temperature from the surface to the center will be small. Conversely, metals such as titanium, with poor conductivity, can develop appreciable temperature differentials.

Some materials, such as nickel-base superalloys, exhibit satisfactory forgeability in only a relatively narrow range of temperature. It is possible, therefore, for chilling of the billet by cold dies to cause cracking as well as nonuniform metal flow.

Shape and Size Factors. Spherical and block-like shapes are the easiest to forge in impression dies. Parts with thin and long sections or projections (webs and ribs) are more difficult because they have more surface area per unit volume. Such variations in shape maximize the effects of friction and temperature changes, and hence influence the final pressure required to fill die cavities. Neglecting the problems of forging defects, there is a direct relationship between the surface-to-volume ratio of a forging and the difficulty of producing it. Sometimes the difficulties of controlling the direction of metal flow in the dies also have substantial effects on forging loads.

Vertical projections such as ribs and bosses are appreciably harder to forge than lateral projections of comparable dimensions. Forging loads are higher because additional lateral constraint is needed to ensure complete die filling. Restraint must be provided during the entire time required to form a vertical projection; in a die filled by spreading, this is necessary only when the metal reaches the periphery. Figure 3-42 compares the flash designs required for forging disks in horizontal and in vertical positions. The thinner and wider flash design suggested for the

fact that hot dies do not chill the surface layers of the workpiece so drastically, a factor of particular concern in metals such as titanium and nickel-base alloys, the flow stresses of which are very temperature-sensitive. The effects of die temperature and chilling are most noticeable when the

dies are heated close to the workpiece temperature. This is feasible with metals such as aluminum and magnesium, but it is extremely difficult with steels, titanium alloys, and refractory metals. Temperature changes in the workpiece increase with time as well as with the temperature dif-

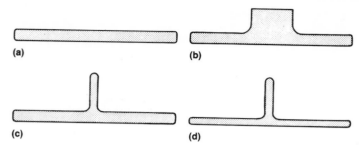

FIGURE 3-44. Basic disk shape (a) and examples of three modifications that affect forging difficulty. (b) Forging difficulty of basic disk predominates. (c) Forging difficulty of tall, slender projection predominates. (d) Forging difficulty of disk is increased to equal that of tall, slender projection. Source: Ref. 13.

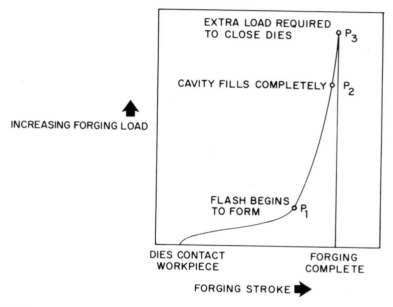

FIGURE 3-45. Three distinct stages of a representative forging load curve for impression die forging.

As the dies continue to close, the loads increase sharply to point P_2, the stage where the die cavity is filled completely. This would correspond to the maximum forging load if designs were figured that closely and no flash was formed. In that respect, the most difficult detail determines the minimum load for producing a satisfactory forging. Usually, however, flash is necessary and the work done in extruding metal through the narrow flash opening causes the load to increase from P_2 to P_3. Therefore, the dimensions of the flash determine the load required for closing the die or the maximum load developed during forging.

Effect of Metal Flow on Grain Structure Development

The development of grain-flow patterns and die filling in closed die forgings has been described previously. Implicit in these discussions has been the fact that the deformation in a given forging may vary greatly from one location to another. An extreme example of such variations might

disk forged vertically produces more lateral confinement during the critical period of deformation. Roughly speaking, vertical configurations are characterized by one third the die life and three times the flash-metal loss of horizontal configurations.

The ease of forging more complex shapes depends on the relative proportions of vertical and horizontal projections on the part. Figure 3-43 illustrates the fact that relative dimensions of ribs and webs affect the forging operation. Bulky projections on thinner plates have little or no effect. Shapes with tall, slender projections are more difficult to forge than plates or disks. Figure 3-44 is another schematic representation of the effects of shape on forging difficulties. Figure 3-44(c) and (d) would require not only higher forging loads but

also at least one more forging operation than the other to ensure die filling.

The increase in load during forging in impression dies is shown schematically in Figure 3-45. Loads are comparatively low until the more difficult details are partly filled and the metal reaches the flash opening. This stage corresponds to point P_1 in Figure 3-45. For successful forging, two conditions must be fulfilled when this point is reached:

- A sufficient volume of metal must be trapped within the confines of the die to fill the remaining cavities.
- The difficulty of extruding metal through the narrowing gap of the flash opening must exceed that of filling the most difficult detail in the die.

TABLE 3-4. Importance of Controlled Forging Deformation When Forging Several Alloys

Alloy group and examples	Type of forging cycle	Need for controlled deformation
Carbon and alloy steels (1035, 4340)	Hot work	Level of deformation not important for final properties
Superalloys (A-286, Incoloy 901)	Hot work	Level of deformation must be large enough to avoid abnormal grain growth; above certain minimum levels, the amount is not essential
Austenitic stainless steels (16-25-6, 19-9DL)	Hot/cold work	Level of deformation in final die requires control and uniformity
Refractory metals (unalloyed molybdenum)	Cold work	Level of deformation accumulated during all forging steps requires control and uniformity

Source: Ref. 13.

be in forgings with chill or dead-metal zones, a common occurrence in metals whose flow stress is very temperature-sensitive. Sometimes, when there are large deformation variations, heat treatments that involve a phase change can alleviate consequences such as variations in strength, ductility, or fracture toughness. At other times, such variations are very important.

Critical grain growth during heat treatment in a variety of metals such as nickel-base alloys occurs in regions that have suffered only relatively small deformations. Table 3-4 lists several alloy systems

and the importance of controlling variations in deformation. Often, when problems associated with deformation variations cannot be resolved through heat treatment or other changes in processing parameters, changes in die design (e.g., changes in forging sequence or flash design) may be necessary.

Metal Flow Defects

In extreme cases, the metal flow during die fill in closed die forging may be extremely nonuniform and result in defects

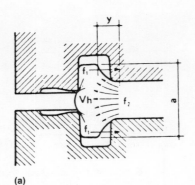

(a)

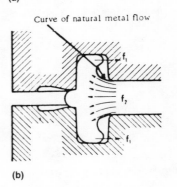

(b)

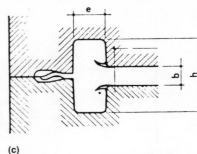

(c)

FIGURE 3-47. Lap formation in the rib of a rib-web part due to improper preform geometry. (a) Formation of flash. (b) Reverse flow forming a fold. (c) Formed forging defect. Source: Ref. 13.

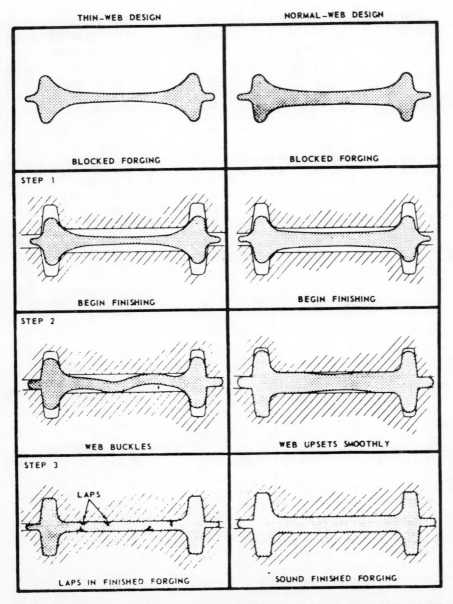

FIGURE 3-46. Typical deformation sequence in closed die forging of a rib-web part, showing how laps can be generated if preform geometry is selected improperly. Source: Ref. 13.

that do not necessarily involve fracture. These defects may result from improper choice of the preform shape, poor die design, and poor choice of lubricant or process variables. All of these may contribute to formation of flaws such as laps, flow-through defects, extrusion defects, and cold shuts.

Laps are defects that form when metal folds over itself during forging. This may occur in rib-web forgings at a variety of locations. One location is in the web of a forging in which the preform web is too thin. During finish forging, such a web may buckle, causing a lap to form (Figure 3-46). Another location is in a rib in which metal is made to flow nonuniformly (Fig-

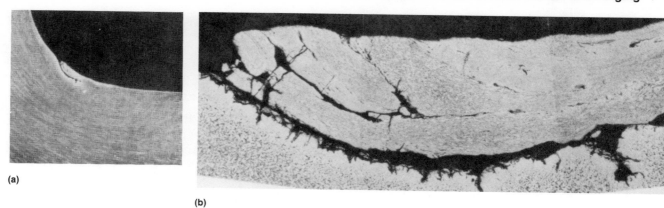

FIGURE 3-48. Lap defect in Ti-6Al-4V bulkhead forging. (a) 3¹/₃×. (b) 50×. Courtesy of F. Lake and D. Moracz, TRW, Inc.

ures 3-47 and 3-48). Most often, the lap at this location is a result of too sharp a fillet radius in the forging die.

Flow-through defects are flaws that form when metal is forced to flow past a recess after it has filled or regions that have stopped to deform because of chilling (Figures 3-49 and 3-50). Similar to laps in appearance, flow-through defects may be shallow but indicative of an undesirable grain flow pattern or shear band that extends much farther into the forging. They may also occur when trapped lubricant forces metal to flow past an impression.

Extrusion-type defects are formed when centrally located ribs formed by an extrusion-type flow draw too much metal from the main body or web of the forging. A defect similar to a pipe cavity is thus formed (Figure 3-51). Means of minimizing the occurrence of these defects include increasing the thickness of the web or de-

signing into the forging a small rib opposite to the larger rib.

Most of the defects summarized above are found in hot forging, which is most common for impression die forging. Because of this, defect formation may also involve the entrapment of oxides (and lubricant). When this occurs, the metal is incapable of rewelding itself back together under the high forging pressures, and the term cold shut is often applied, in conjunction with laps and flow-through defects, in describing the flaws that are generated.

Three techniques under investigation to correct metal flow defects involve the application of: (1) empirical guidelines in designing preforms and finish forging dies, (2) physical modeling studies, and (3) computer-aided design (CAD) process modeling techniques. Computer-aided design process modeling techniques are use-

ful in calculating the required volume distribution in forging preforms and in determining neutral planes and general metal-flow patterns. With recently developed, more advanced FEM techniques, it is becoming possible to simulate the metal-flow pattern in complex forgings and thus determine preform geometries that are likely to cause defects. Some of those developments are summarized in Section 2 of this book. Others are discussed below.

Empirical Guidelines for Die and Preform Design

A number of simple guidelines for designing forging dies and preforms to avoid defect formation have been developed through years of experience. These guidelines are for conventional hot forging in which the dies are much cooler than the workpiece. The most important design parameters for dies are draft angles, corner and fillet radii, and rib and web dimensions.

Draft angles are needed to allow part removal following forging; they are generally in the range of 1 to 7°, with lower values used when part knockout mechanisms are employed or when surface area must be minimized because of frictional or chilling effects. In addition, easily forged aluminum alloys tend to require small draft angles (~1°), carbon and alloy steels require moderate draft angles (3 to 5°), and hard-to-work alloys such as nickel-base superalloys need large draft angles (7° or greater) because of higher forging pressures and greater tendency for sticking in the dies.

Radii are also important in die design, because too small a radius may lead to lap formation or incomplete die fill. As with drafts, required radii increase with increasing forging difficulty of the work-

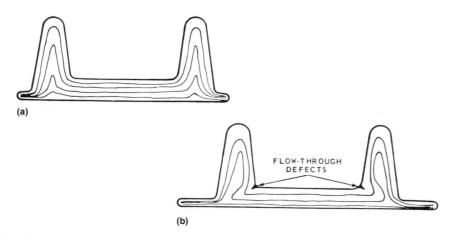

FIGURE 3-49. Metal flow patterns in forging without and with flow-through defects. (a) Normal forging: Smooth flow lines. (b) Excess metal flow in flash region: disrupted flow lines. Source: Ref. 13.

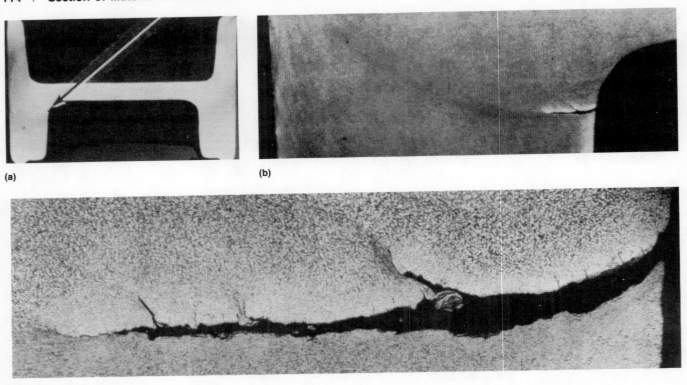

FIGURE 3-50. Flow-through defect in Ti-6Al-4V rib-web structural part. (a) $^3/_5\times$. (b) 4.75$\times$. (c) 40$\times$. Courtesy of F. Lake and D. Moracz, TRW, Inc.

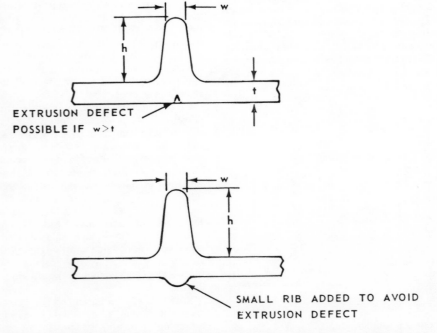

FIGURE 3-51. Extrusion-type defect in centrally located rib and die design modification used to avoid defect. Source: Ref. 13.

piece alloy. Typical values when making 2.5-cm-high ribs are given in Table 3-5. Guidelines for minimum dimensions of ribs and webs in conventional hot forgings may also be suggested based on experience (Table 3-6). The important or deciding factors here are the cooling conditions of the forging, which depend on the work-piece alloy and deformation speed, among other considerations, and the minimum web dimension, which determines the amount of lateral constraint.

Because the formation of metal flow defects depends a great deal on the shape of preforms, it is not surprising that guidelines for their design have been devised. These guidelines, primarily for conventional hot forging, give a basic idea of how much metal flow can be imposed under typical forging conditions. In this regard, useful relations between preform and finish forging dimensions for steel, aluminum, and titanium rib/web structural forgings have been found from actual forging trials. Those for the latter two alloy types are summarized in Table 3-7. Draft angles which are selected to be the same in the preform as in the finish forging are not listed.

Physical Modeling

An alternative to applying the empirical guidelines for the design of preform and finish forging dies is to apply the principles of similarity, commonly known as physical modeling. These techniques are often applied to design the dies for new parts when previous experience may not be sufficient. In these situations, an improper design may be quite costly in terms of expensive die materials and labor needed to manufacture the dies. Hence, it may be wise to use small-scale model dies and model materials to determine the forging sequences and the possibility for metal-flow defects. For the model system to have application, it is best that the laws of similarity be adhered to.

The laws of similarity consist of rules governing geometrical, physical, and boundary condition similarity. From a geometric standpoint, model dies must be designed, as nearly as possible, to the exact scale model of the envisioned forging dies. That is to say, all linear dimensions, radii, angles, and so forth of the model dies must be in fixed proportion to those of the real tooling. A similar relation between the model and actual workpiece dimensions should be followed. Physical similarity pertains largely to selection of model materials whose deformation and thermal response is like that of the intended workpiece material. Boundary condition similarity is related to interface friction and heat transfer considerations.

In forging, physical modeling has been applied most often in the simulation of conventional and isothermal hot forging. These studies are usually performed at room temperature with model materials that are rate-sensitive and non-strain-hardening, as are typical forging alloys at hot-working temperatures.

Lead, plasticene, and wax are the most common choices for materials. With these materials, model dies need not be too strong and can be made from inexpensive steels,

aluminum alloys, or even hard plastics. Because the tests are usually conducted at room temperature, boundary condition similarity with regard to heat transfer is satisfied for isothermal but not conventional hot forging. Therefore, information on the possible effects of chilling on metal flow may not be obtainable from simple modeling studies. On the other hand, it may be possible to adjust the frictional boundary conditions to compensate for this deficiency.

In physical modeling of forging, model specimens or workpieces are usually sectioned and glued back together after a grid pattern is inscribed on one of the flat faces. By this means, the flow pattern may be visualized, and regions where defects are likely to form in an actual forging are determined. Such a technique is similar to the grid technique drawn upon for diagnosing formability problems in sheet metal shops.

Figure 3-52 gives an example of how a lap defect is visualized in a closed die forging. Here, the grid lines give valuable information not only on the development of the lap in the forging, but also on the occurrence of a dead-metal zone in contact with the bottom die.

Process Modeling

In recent years, the need to conserve raw materials and to make manufacturing production more economical has led to the development of a scientific approach to the design and understanding of metalworking processes such as forging. This approach, being scientific in nature, depends not so much on the design rules and observations of experienced forging engineers, but rather on the development of detailed models of deformation processes themselves.

Such process models include a synthesis of information describing (1) the workpiece material and its properties, (2) the characteristics of the tooling-workpiece

interface, and (3) the mechanics of plastic deformation in the particular process of interest.

The workpiece material model includes descriptions of the flow stress (as a function of strain, strain rate, and temperature), the workability or forgeability, and the effects of processing conditions on final microstructure and service properties. Frictional and heat transfer properties are the components of the interface model. The deformation model allows metal flow and forming loads to be predicted using mathematical techniques such as the upper-bound or finite-element methods. These methods when applied to actual metalworking processes often rely on high-speed digital computers and peripheral graphics capabilities to perform actual process simulations and to present the final results.

Integration of the material model, the interface model, and the deformation model results in a system from which information on metal flow, defect development, and die loading can be derived. These data, therefore, are useful tools in the design of

TABLE 3-6. Suggested Minimum Section Thickness for Rib-and-Web Forgings of Several Alloys

Values listed are for rib-and-web forgings with 7° draft and standard fillet and corner radii.

Alloy	Minimum rib thickness, mm, for forgings of plan area, cm²:			Minimum web thickness, mm, for forgings of plan area, cm²:		
	Up to 25	25–250	250–750	Up to 25	25–250	250–750
2014 (aluminum)	3	5	. . .	3	6	8
AISI 4340 (alloy steel)	3	5	. . .	5	8	13
H-11 (hot-work steel)	3	5	. . .	5	8	13
17-7PH (stainless steel)	. . .	6	. . .	. . .	10	. . .
A-286 (superalloy)	. . .	6	. . .	. . .	10	. . .
Ti-6Al-4V	. . .	6	. . .	. . .	8	. . .
Unalloyed molybdenum	. . .	10	. . .	. . .	10	. . .

Source: Ref. 13.

TABLE 3-5. Suggested Fillet and Corner Radii(a) for Forgings of Several Alloys with 2.5-cm-High Ribs

Alloy	Fillet radius, mm	Corner radius, mm
2014 (aluminum)	6	3
AISI 4340 (alloy steel)	10–13	3
H11 (hot-work steel)	10–13	5
17-7PH (stainless steel)	6–13	5
A286 (superalloy)	13–19	6
Ti-6Al-4V	13–16	6
Unalloyed molybdenum	13	13

(a) From the viewpoint of the forging or workpiece. Source: Ref. 13.

TABLE 3-7. Relationship Between Preform and Finish Forging Dimensions for Rib-Web Structural Parts

Dimensions in finish forgings(a)	Dimensions in preform(a)	
	Aluminum alloys	Titanium alloys
Web thickness, t_F	$t_p \approx (1-1.5) t_F$	$t_p \approx (1.5-2.2) t_F$
Fillet radii, R_{FF}	$R_{PF} \approx (1.2-2) R_{FF}$	$R_{PF} \approx (2-3) R_{FF}$
Corner radii, R_{FC}	$R_{PC} \approx (1.2-2) R_{FC}$	$R_{PC} \approx (2) R_{FC}$
Draft angle, δ_F	$\delta_p \approx \delta_F (2-5°)$	$\delta_p \approx \delta_F (3-5°)$
Width of rib, W_F	$W_p \approx W_F - 0.8$ mm	$W_p \approx W_F - (1.6-3.2$ mm$)$

(a) The first subscript letter on each dimension indicates finish forging (F) or preform (P). Source: Ref. 21.

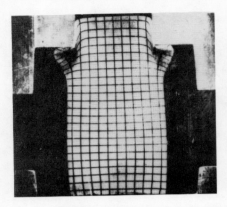

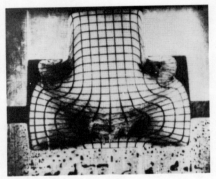

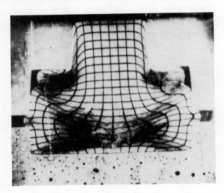

FIGURE 3-52. Forging sequence of gridded plasticene billet used in physical modeling studies to establish the effect of die and preform design on the occurrence of metal flow defects such as laps. Courtesy of S. Kobayashi, University of California—Berkeley.

Metal Flow and Microstructural Prediction Using Process Modeling. One of the first problems analyzed using process-modeling techniques was how to achieve optimal part properties in a jet engine compressor disk by controlling the microstructure. This was touched on briefly in Section 2. Low-cycle fatigue response is critical at the base of an alpha-beta titanium alloy jet engine compressor disk, whereas creep resistance is the most important design feature at its rim. The fatigue and creep behavior is optimized in alpha-beta titanium alloys with equiaxed alpha and acicular alpha microstructures, respectively. The former microstructure is stable during substransus deformation and heat treatment. In contrast, the acicular alpha microstructure is known to revert to an equiaxed one after substransus deformation and heat treatment. This phenomenon served as the basis for development of a so-called dual microstructure—a dual-property disk that was forged from a preform having an acicular alpha microstructure.

The material characteristics that were included in the process were the following:

- The flow stress of the alpha-beta titanium alloy of interest, Ti-6242Si. Although this was obtained for both an equiaxed and an acicular microstructure (Figure 3-53), only that for the acicular microstructure was used.
- The critical strains and temperatures for which the acicular alpha microstructure transformed to an equiaxed microstructure. It is found that deformation to strains of the order of 1.0 at a temperature of 1650 °F (900 °C), followed by heat treatment at 1750 °F (955 °C), produced the desired transformation.

The data were obtained in simple, homogeneous upset tests. Because the forging was to be hot forged isothermally (dies and workpiece at the same temperature), the interface model consisted solely of its frictional characteristics. For this purpose, ring tests were done at the expected forging temperatures using the glass-based forging lubricants, and friction factors were established. The deformation analysis part

dies and preforms (computer-aided design), the prediction of final microstructure and properties in finished parts, and the economics of tentative processing sequences. All of this can be done without actually constructing tooling, which may not be satisfactory in production of the desired part, and the large expenses associated with it. The application of the process-modeling techniques will be illustrated in various forging situations that follow.

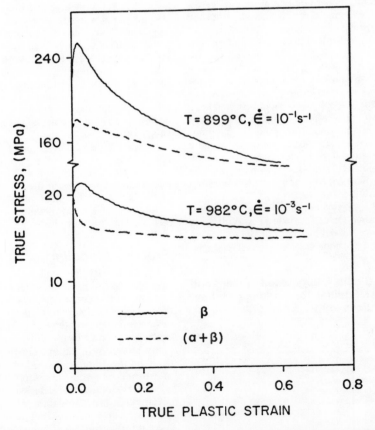

FIGURE 3-53. Flow stress data for Ti-6242 at various test conditions. Data from P. Dadras and J. F. Thomas, Jr., Wright State University.

of the process model was based on a rigid viscoplastic formulation suitable for large plastic flow problems typical of bulk forming. The formulation was implemented with the FEM code ALPID (analysis of large plastic incremental deformation).

Figures 3-54 and 3-55 illustrate the results of the process-model synthesis of the individual material, interface, and deformation models. These are for forging of a circular cylindrical preform in one forging operation to form the disk at 1650 °F (900 °C). The predicted strain contours (Figure 3-54, in which only one fourth of the disk is shown because of symmetry) show very high strains at the center of the disk and much smaller ones at the rim. From the data obtained in the upset tests, it was surmised that an equiaxed microstructure would be developed here during heat treatment. This prediction was validated by actual forging trials (Figure 3-55).

These results show that the simulation of metal flow using process-modeling techniques can be used to determine the effects of material properties, die and perform design, and interface conditions on critical grain growth or the development of heterogeneous microstructures and on grain-flow patterns. In regard to the last of these phenomena, process-modeling techniques have been applied successfully to predict the localization of metal flow into shear bands (Figure 3-56). These data are of particular use in the forging of ti-

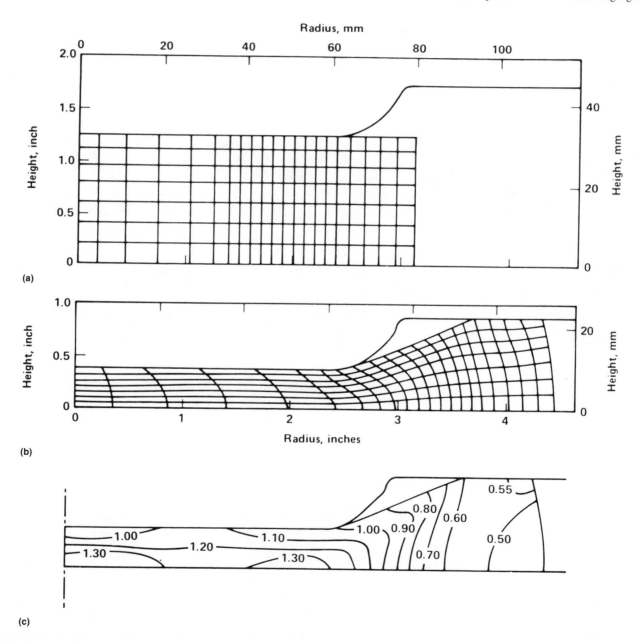

FIGURE 3-54. Compressor disk forging simulation results. (a) Undeformed grid. (b) Grid distortion at 70% reduction. (c) Predicted effective strain distribution at 70% reduction. Results courtesy of S.I. Oh, Battelle's Columbus Laboratories.

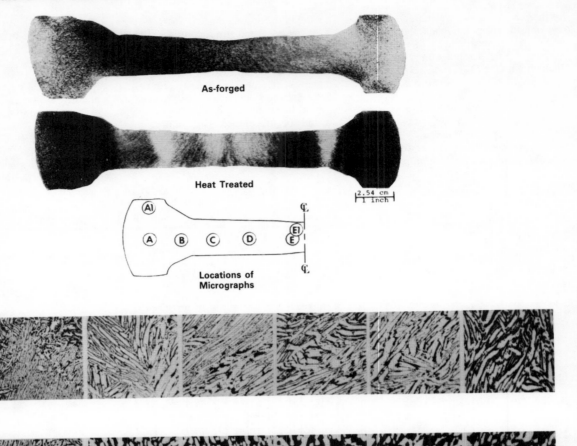

FIGURE 3-55. Structural features of disks produced by selected forging and heat treating combinations. Courtesy of C.C. Chen, Chen Tech Industries.

tanium and nickel-base alloys, both of which are expensive and whose properties are very sensitive to the microstructure developed as a result of forging.

Defect Formation Prediction Using Process Modeling. Process-modeling techniques may also be used to predict the occurrence of defects in forging. This is done by including a suitable workability criterion in the material model. One of the first applications of process modeling in this area involved the prediction of free-surface fracture during cold forging. It will be recalled that fracture in these cases follows a relatively simple criterion (Figure 3-33). Because fracture follows a critical

strain criterion, the process model is used to predict the free surface strains during the actual forging operation. These strain fields are examined to determine if the fracture criterion has been satisfied. If it has, the process modeling procedure can be used to determine whether changes involving preform design or lubrication, for instance, can be used to avoid surface fracture.

Process-modeling techniques can also be applied to predict the occurrence of defects such as voids or center bursts in which stress-based criteria are applicable. One such application is shown in Figures 3-57 and 3-58 for a tapered disk of Ti-6A1-2Sn-4Zr-2Mo-0.1Si with an acicular alpha

starting microstructure forged isothermally at approximately 1725 °F (940 °C) and a strain rate of approximately 2.5 s^{-1}. The computer solution was done with ALPID, and material property data were measured in isothermal hot compression tests.

Comparison of the results of this solution with test specimens showed that the magnitude of the circumferential tensile stress correlated best with the observed occurrence of cavitation and triple-point cracks. The areas of circumferential stresses are shown (as shaded areas) in Figure 3-58; the data are summarized as a function of reduction relative to the initial height of the specimen, measured along its axis.

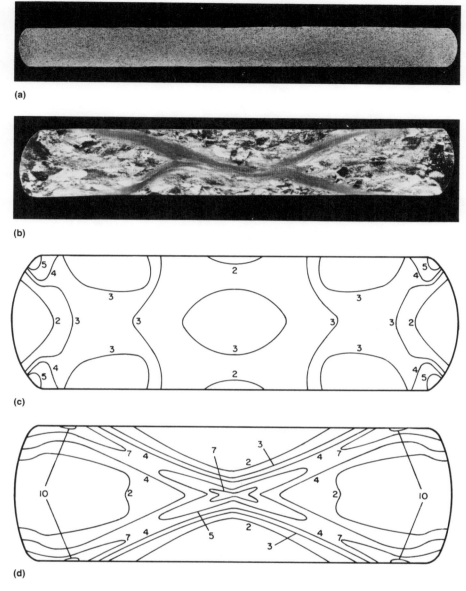

FIGURE 3-56. Specimen cross sections and simulated metal flow for isothermally side-pressed bars of Ti-6Al-2Sn-4Zr-2Mo-0.1Si. (a, c) Equiaxed alpha starting microstructure, (b, d) Widmanstätten alpha starting microstructure. T = 1675 °F (913 °C); strain rate ≈ 10 s^{-1}.

FIGURE 3-57. Cavity distribution revealed by dye penetrant in isothermal tapered disk forging of Ti-6Al-2Sn-4Zr-2Mo-0.1Si with a Widmanstätten alpha starting microstructure. Courtesy of J.C. Malas and H.L. Gegel, Air Force Wright Aeronautical Laboratories.

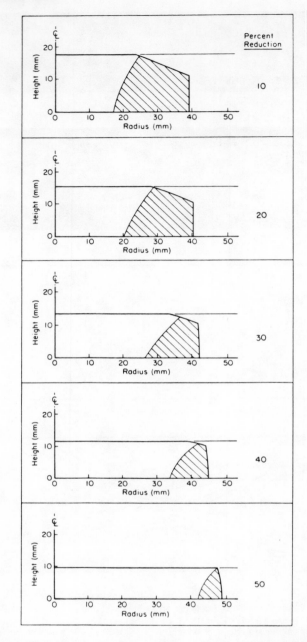

FIGURE 3-58. Regions of circumferential tensile stress (shaded areas) in isothermal tapered disk forgings of Ti-6Al-2Sn-4Zr-2Mo-0.1Si with a Widmanstätten alpha starting microstructure. Courtesy of S.I. Oh, Battelle's Columbus Laboratories.

Because of symmetry, the simulation used only one fourth of the disk, and it is for this section that results are shown.

To reveal the general distribution of voids in experimental tapered disk forgings, a fluorescent dye penetrant was used. Due to "bleeding" of the penetrant, a somewhat exaggerated idea of the void sizes is obtained. Nevertheless, the usefulness of the technique may be seen by comparing

the observation in Figure 3-57 for a disk forged to 50% reduction to the corresponding simulation result in Figure 3-58. Agreement between the area with voids revealed by dye penetrant and the predicted circumferential tensile stress area is very good.

Furthermore, the analysis predicts that these defects originate at low strains at which the tensile stresses are relatively large

and that the tensile stresses diminish as the amount of compression increases, giving rise to less of a taper on the forging. In effect, the simulation suggests that the defects move toward the outside of the disk as the reduction increases, leaving behind a sound structure—a trend verified by comparison to experimental observations. Therefore, voids may be formed and healed by further deformation.

Forging Ferrous Alloys

Carbon and Alloy Steels

Forging Behavior. The extent to which carbon and alloy steels can be forged into intricate shapes is seldom limited by forgeability problems, except when the grades contain sulfides, bismuth, or other intentional additions for other purposes such as machinability. Figure 3-59 shows the relative forgeability of selected steels. Probably the most important factor limiting section thickness, shape complexity, and forging size is the cooling that occurs when the heated workpiece comes in contact with colder dies.

In general, the forgeability of steels improves as deformation rate increases. It is believed that the improvement in workability is due primarily to heating effects that occur with the increasing deformation rates.

Selection of forging temperatures for carbon and alloy steels is based on (1) the carbon content, (2) the alloy composition, (3) the temperature range for optimum plasticity, and (4) the amount of reduction. Generally, the maximum temperature allowable by these factors is selected, because it ensures the lowest possible forging pressures.

The upper limiting forging temperatures for steels are influenced most noticeably by carbon content. The maximum forging temperatures for carbon steels with increasing carbon contents are approximately:

AISI steel designation	Carbon content, %	Maximum forging temperature °F	Maximum forging temperature °C
C1010 . . . . 0.1		2400	1315
C1030 . . . . 0.3		2350	1290
C1050 . . . . 0.5		2300	1260
C1080 . . . . 0.8		2200	1200
C1095 . . . . 1.0		2150	1180

These forging temperatures are approximately 300 °F (165 °C) below the solidus temperature of each composition. Above these temperatures, the steels are subject to possible damage by overheating or incipient melting.

Alloy steels have solidus temperatures lower than plain carbon steels of comparable carbon contents. Thus, the maximum forging temperatures for the alloy steels at each carbon level are generally 50 to 100 °F (30 to 55 °C) lower. In addition, the forging temperatures are usually reduced another 50 to 100 °F (30 to 55 °C)

for parts requiring only small reductions. This provides further assurance that the alloys do not overheat and helps to reduce unnecessary iron oxide scaling.

There are hundreds of steels that range in composition from about 0.06 to 1.0% carbon and from trace amounts to about 5% each of several metallic alloying elements. The carbon content affects the hardness and strength levels obtainable, while the alloying serves to increase hardenability and strength, improves resistance to thermal and mechanical shock, and provides grain refinement. The steels in this group range from carbon steels and standard alloy steels to specially alloyed super-strength steels such as 300M, D6ac, and H-11.

Carbon Steels. By definition, plain carbon steel is iron combined with amounts of carbon varying between approximately 0.06 and 2.0%. Carbon steels may also contain limited amounts of manganese (1.65% maximum), silicon (0.60% maximum), and copper (0.60% maximum). The more common grades are the standard types between C1006, with approximately 0.06% carbon, and C1095, with approximately 0.95% carbon.

Alloy steels are generally considered to be steels to which one or more alloying elements have been added to give them special properties that cannot be obtained in carbon steels. The standard alloy steels are commonly regarded as those found in the AISI 1300 through AISI 9800 series,

although modifications of these types are also used in special applications. From the standpoint of composition, steel is considered to be alloy steel when amounts of manganese, silicon, and copper exceed the maximum limits for carbon steel, or when definite minimum quantities of alloying elements such as aluminum, chromium, cobalt, columbium, molybdenum, nickel, titanium, tungsten, vanadium, and zirconium are specified.

Effects of Alloying. Alloying elements are incorporated in steel for one or more of the following reasons, of which the first two are more important:

- To improve mechanical properties through control of the factors that influence hardenability and to permit higher tempering temperature while maintaining high strength and improved ductility
- To improve mechanical properties at elevated or low temperatures
- To increase resistance to chemical attack and to elevate temperature oxidation
- To influence other special properties such as magnetic permeability and neutron absorption

As a general rule, alloy steels are specified when more strength, ductility, and toughness are required than can be obtained in carbon grades. Alloy steels are also used when specific properties such as wear resistance, corrosion resistance, heat resistance, and special low-temperature impact properties are desired.

Effects of Forging Variables on Mechanical Properties. Forging usually affects important mechanical properties of steel such as ductility, impact strength, and

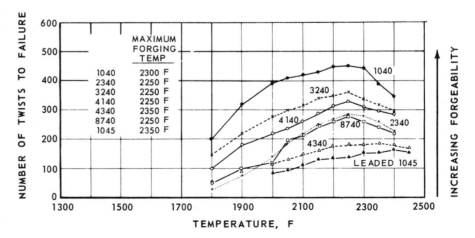

FIGURE 3-59. Indications of forgeability for seven selected ferrous materials as determined by hot-twist tests.

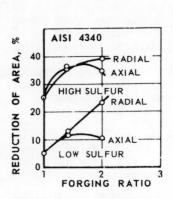

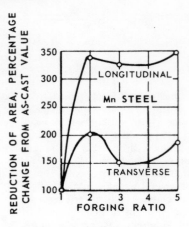

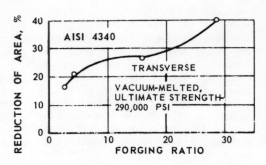

FIGURE 3-60. Effect of forging ratio on reduction in area of heat-treated steels. Forging ratio equals the final cross-sectional area divided by the original cross-sectional area.

fatigue strength, while hardness and strength are controlled by composition and heat treatment. The reasons generally held for these improvements are:

- Forging breaks up segregation, heals porosity, and aids homogenization.
- Forging produces a fibrous grain structure, which enhances mechanical properties parallel to the grain flow.
- Forging reduces grain size.

Figures 3-60 and 3-61 show the typical improvement in ductility and impact strengths of heat-treated steels obtained from increasing reductions. It is apparent that maximum improvement, in each case, occurs in the direction of maximum extension. After a certain amount of hot reduction on these steels, the toughness and ductility properties reach levels beyond

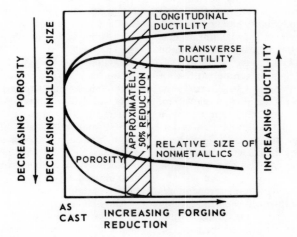

FIGURE 3-62. Diagram indicating the improvement in ductility of steels as a result of reducing both porosity and the size of inclusions during forging.

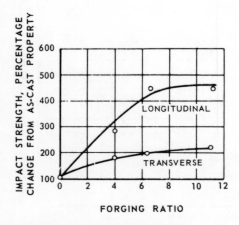

FIGURE 3-61. Effect of hot-working reduction on impact strength of heat-treated steels. Forging ratio equals final cross-sectional area divided by original cross-sectional area.

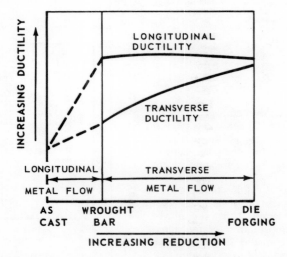

FIGURE 3-63. Typical influence of lateral reduction during forging on longitudinal and transverse ductility of steels.

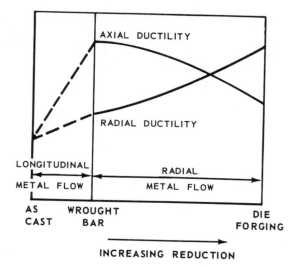

FIGURE 3-64. Typical influence of upset reduction during forging on axial and radial ductility of steels.

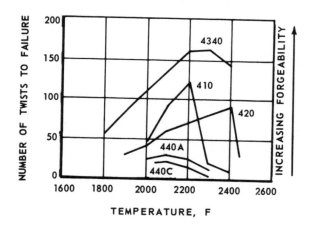

FIGURE 3-65. Hot-twist test results showing forging characteristics of several martensitic stainless steels compared to 4340 steels.

Martensitic Stainless Steels

Forging Behavior. The high-chromium, martensitic stainless steels have forging characteristics similar to those of alloy steels. Because of the higher chromium contents, however, forging load requirements are about 30 to 50% higher. At temperatures above 2000 °F (1090 °C), certain alloys transform partly to delta ferrite, which reduces forgeability. For this reason, maximum forging temperatures are generally 100 to 300 °F (55 to 165 °C) lower than those for alloy steels.

Figure 3-65 compares workability data for several martensitic stainless steels with those for AISI 4340 steel. The hot-twist ductility values for the AISI 410 and 420 grades drop sharply above 2200 and 2400 °F (1200 and 1300 °C), respectively. The 440A and 440C grades do not exhibit such drastic changes with temperature, but these grades also do not exhibit high levels of ductility due to higher carbon content. Such comparisons with more familiar alloy steels are useful for estimating the forces required to forge these alloys.

Control of Delta Ferrite. The most serious problem in the forging of martensitic stainless steels is the presence of delta ferrite. This phase significantly reduces the transverse ductility of these steels, particularly in the short transverse direction. In some heavy forgings, the ductility of martensitic stainless steels has been of the order of 1 to 2% in this direction, while in the longitudinal direction the ductility may be as high as 15%. The delta ferrite problem can be largely eliminated, however, by proper use of the "chrome equivalent" calculations to properly balance the alloy heat during melting.

Forging Temperatures. In general, maximum forging temperatures are determined by the first appearance of the delta ferrite phase on heating. In practice, these temperatures are readily determined by conducting upset forging trials at temperatures between 2000 and 2300 °F (1090 and 1260 °C). The first appearance of the phase on heating correlates with the onset of edge cracking.

Forging sequences for the martensitic stainless steels are essentially the same as those used for carbon and alloy steel. Because greater pressures are required, larger forging equipment is necessary for otherwise similar parts.

Controlled Cooling. Because the martensitic stainless steels are characterized by high hardenability, special precautions are taken in cooling them from forging temperatures. Normal practice is to place hot forgings in insulating materials for slow

which further hot reduction is of little significant value.

Figure 3-62 shows how increasing the forging reduction improves ductility by reducing both porosity and the size of inclusions present in the original as-cast ingot. Porosity is eliminated and inclusion size is markedly reduced. Some types of nonmetallic inclusions continue to break up with increasing reduction. Large reductions develop a highly oriented, fibrous structure. As a result, longitudinal ductility continues to improve. Transverse ductility shows little change after porosity is healed.

Because impression die forgings are usually made from wrought bars and billets that have received considerable prior

reduction, porosity is rarely a problem. It is also important to remember that metal flows in a number of directions during die forging. For example, the metal flow is virtually all in the transverse direction when die forging a rib-and-web shape. Such flow improves transverse ductility with little or no loss of longitudinal ductility (Figure 3-63).

Similar effects are observed when upsetting wrought material. In this case, however, the original longitudinal axis is shortened by upsetting and the lateral displacement of metal is in the radial direction. When upset reductions exceed about 50%, the ductility in the radial direction generally surpasses that in the axial direction, as shown in Figure 3-64.

cooling. For parts that have either heavy sections or large variations in section, it is often desirable to charge the forged parts into an annealing furnace immediately after forging.

In particular, the higher carbon grades, 440A, 440B, and 440C, and the modified 420 types such as Greek Ascoloy must be slow cooled carefully after forging. These steels often require furnace-controlled interrupted cooling cycles to ensure against cracks. Suitable cycles consist of air cooling the forgings to temperatures where the martensite transformation is partially complete (between 300 and 500 °F, or 150 and 250 °C), then reheating the forgings in a furnace at a temperature of about 1200 °F (650 °C) before finally cooling them to room temperature. This procedure also prevents the formation of excessive grain-boundary carbides that sometimes develop during continuous slow cooling.

Martensitic stainless steel forgings may be specified either for high-strength, wear-resistant applications or for long-time service at elevated temperatures up to 1100 °F (600 °C). For high-strength applications, full heat treatment is required. This entails austenitizing at 1750 to 1850 °F (950 to 1000 °C), depending upon the type of steel, cooling either in air or oil to room temperature, and subsequently tempering at 600 to 800 ° F (300 to 425 °C).

For high-temperature applications, air cooling from the austenitizing temperature followed by tempering at 1050 to 1250 °F (560 to 675 °C) is common practice. Often, double tempering is employed to reduce the likelihood of subsequent transformation of retained austenite, as well as secondary hardening effects through precipitation of metallic carbides.

Austenitic Stainless Steels

By comparison with carbon and alloy steels, the austenitic stainless steels require greater forging pressures. They are generally more difficult to forge, they exhibit work-hardening behavior at higher temperatures, and their forging-temperature ranges are narrower. For example, AISI 303 stainless steel contains sulfur for machinability purposes and therefore contains sulfide inclusions. The presence of these inclusions greatly impairs the hot ductility and forging characteristics. Grades AISI 321 and 347, which contain titanium and columbium, respectively, are prone to carbonitride stringers, which can also lead to forging difficulties, although these two steels are more readily forgeable than AISI 303. The 347 grade is much less prone to stringers than 321 and therefore is considered the preferable forging grade of the two.

Austenitic stainless steels can require increasing forging pressures with increasing reductions even at temperatures thought to be true hot-working temperatures. For example, at 1800 °F (980 °C), 1020 carbon steel requires pressures from 21 to 25 ksi over a range of upset reductions from 5 to 50%. The forging pressure for AISI 304 stainless steel, however, increases from 33 to 57 ksi under the same conditions.

Forging Behavior. Austenitic stainless steels possess higher hot strengths than carbon, alloy, or even martensitic stainless steels and, in fact, exhibit work hardening at relatively high temperatures. Maximum forging temperatures up to about 2300 °F (1250 °C) are possible for the 18-8 types, but this limit is considerably lower for those grades that tend to form delta ferrite. Minimum forging temperatures must generally be above about 1700 °F (925 °C). Special care is taken during forging with regard to the intensity of hammer blows, the amount of reduction between reheatings, and the amount of final reduction to control grain size.

Forging Temperature. In the selection of optimum forging temperatures, the final properties of the forging are carefully considered. Parts receiving small reductions are subject to excessive grain growth if forged at the maximum temperatures. It is common practice, therefore, to adjust forging temperatures downward for forging operations requiring small amounts of reduction. For example, the typical forging temperatures for AISI 304 stainless steel for various operations would be:

Operation	Forging temperature, °F (°C)
Severe reductions (ingot breakdown, roll forging, drawing, blocking, and backward extrusion)	2300 (1260)
Moderate reductions (finish forging and upsetting)	2200 (1200)
Slight reductions (coining, restriking, and end upsetting)	2050 (1120)

Similar adjustments are made for the other austenitic grades. The temperatures used for slight reductions rarely exceed 2100 °F (1150 °C), regardless of the alloy being forged.

Neither forging pressure nor the forgeability of austenitic stainless steels is influenced significantly by deformation rates. Hammers and other fast-acting equipment are preferred, but only because they minimize the problems associated with heat transfer. Two to three times as much energy is required for forging the 300 series stainless steels than for forging carbon and alloy steels. In hammer forging, this means either larger hammer capacity or more hammer blows per forging. In most cases, forging companies provide about 50% greater hammer capacity and 50 to 100% more hammer blows. Press forgers figure the maximum forging plan areas at about half that for the carbon and alloy steels for any given press size. For the hot/cold-working alloys, equipment and requirements vary depending on the strength desired.

The austenitic stainless steels are noted for their strength at elevated temperatures and for their resistance to corrosion. Because a large proportion of these steels are employed in applications where corrosion resistance is required, special care is taken to ensure that forgings of these materials are supplied in a condition best suited for resistance to corrosive attack.

When slowly cooled from the forging temperature, those steels that are not stabilized (i.e., do not contain columbium or titanium) will contain intergranular chromium carbides and will be subject to intergranular corrosion in certain media. To prevent this, the forged parts are annealed at about 1950 °F (1065 °C) and rapidly cooled through the temperature range from 1500 to 900 °F (815 to 480 °C), usually by water quenching. This retains the undesirable chromium carbides in solid solution and reduces subsequent intergranular attack.

The stabilized grades (Types 321 and 347) are also annealed after forging to ensure that all of the carbon is tied up as titanium or columbium carbides. This is accomplished by heating these stabilized grades at 1600 to 1650 °F (870 to 900 °C), a temperature at which the chromium carbides are dissolved but the titanium or columbium carbides are precipitated. After such annealing of AISI 321 and 347, slow cooling to room temperature will not impair the corrosion resistance of the steels.

Precipitation Hardening Steels

The composition of these alloys is such that some delta ferrite is usually formed during solidification of the ingot. During hot breakdown forging, the amount of delta ferrite may either decrease or increase, depending on the temperatures employed. In the finished product, the semiaustenitic, precipitation-hardenable stainless steels may contain as much as 10 to 20% delta ferrite. Under proper forging conditions, martensitic alloys will not contain delta ferrite. This difference is a matter of chemistry balance inherent in the composition of the alloys.

Forging Behavior. The relative forging behavior of the precipitation-hardenable stainless steels can be illustrated by the following comparison of forging characteristics between AISI 4340 steel and the 17-7PH, AM 355, and 17-4PH steels:

	4340	17-7PH	AM 355	17-4PH
Forging temperature, °F (°C)	2300 (1260)	2150 (1180)	2150 (1180)	2150 (1180)
Decarburization	High	Low	Low	Low
Scale	High	Low	Low	Low
Grain size control	Excellent	Fair	Fair	Good
Forgeability	Excellent	Fair(a)	Good	Good
Forging pressure (relative)	1.0	1.4	1.4	1.4
Thermal cracking	Low	None	Low	Medium
Die wear	Low	Medium	Medium	Medium
Delta ferrite content, %	None	10–20	1.0–5.0	None

(a) Forgeability of 17-7PH is good in drawing operations, but it is poor in upset operations.

Because of the combination of lower forging temperature and greater stiffness, 30 to 50% higher forging loads are required for the precipitation-hardenable stainless steels than for 4340 and, accordingly, heavier equipment is needed. On the other hand, the precipitation-hardenable grades are much less sensitive to decarburization than are higher carbon alloy steels. Also, they do not scale as much. Thus, it is possible to design some precipitation-hardenable stainless steel forgings for use with as-forged surfaces.

Forging practices play an important role in achieving the optimum mechanical properties in precipitation-hardenable stainless steels, particularly in the semi-austenitic grades in which the grain size is primarily controlled by prior processing. The quality of semiaustenitic-alloy forgings is also quite sensitive to the amount of forging reduction, as well as to the forging temperature.

In addition to improving forgeability, it is important to control the amount of delta ferrite present to retain satisfactory short transverse ductility levels. The typical effect of delta ferrite on transverse tensile ductility of semiaustenitic alloys is illustrated by the following:

Alloy	Ferrite content, %	Ultimate strength, ksi	Elongation, %
AM 355	0	178	22.0
17-7PH	0	184	19.0
17-7PH	0–2	180	5.0–8.0
17-7PH	More than 2	180	0–3
15-7PH Mo	About 2	205	1–4
15-7PH Mo	About 2	195	4–9

Because only small amounts of delta ferrite in the structure have a serious effect on transverse ductility, close control of the forging temperature and strain rate for a given reduction is necessary to avoid overheating. Delta ferrite can also be formed by overheating the martensitic alloys (although it will generally revert to austenite on cooling), so this holds true for the precipitation-hardenable stainless steels in general.

Because the basic alloy contents of these steels are quite similar to those of the 18-8 variety of austenitic stainless steels, forging pressure requirements are quite similar. At temperatures above about 1700 °F (925 °C), the AM 355 grade exhibits forgeability comparable to the 18-8 stainless. Below 1700 °F (925 °C), however, the forgeability drops rapidly due to carbide precipitation.

The forging characteristics of the martensitic grades are about the same as the 12 to 14% chromium stainless steels (such as Type 410). Care must be exercised in forging 17-4PH to avoid overheating at the center of heavy sections from too heavy or too rapid reductions to prevent thermal cracking. Excellent reproducibility of both forgeability and mechanical properties, however, is attainable with this material. A 2050 °F (1120 °C) anneal helps to recrystallize the grain structure of 17-4PH and greatly improves short transverse tensile ductility.

All of the semiaustenitic grades have a tendency to retain the austenite structure at room temperature after forging at temperatures higher than 2000 °F (1100 °C) and heat treating. To achieve uniform grain refinement it is frequently necessary to promote the austenite transformation to martensite by heating forgings in the vicinity of 1400 °F (750 °C). This treatment causes carbide to precipitate, renders the austenite less stable, and promotes complete transformation to martensite on cooling back to room temperature. Forgings heat treated this way are in a favorable condition for subsequent heat treatment and grain refinement.

The semiaustenitic stainless steels usually permit a certain amount of cold forging or coining to close dimensional tolerances. Reductions in the neighborhood of 20% are considered safe for coining. During the hardening treatment, however, a growth of about 0.004 in./in. occurs. In cold coining to final dimensions, this growth factor is taken into account in the process design. The martensitic precipitation-hardenable stainless grades are not amenable to cold coining.

Maraging Steels

The 18% nickel type of maraging steel has received great attention because of its outstanding fracture toughness and because it is hardened by a simple single-step schedule carried out at a moderate temperature. The 18Ni maraging steels are readily hot worked by conventional methods and standard equipment. These steels are heated at 2200 to 2300 °F (1200 to 1260 °C), and forging operations may be started at temperatures as high as 2300 °F (1260 °C).

The martensite formed in the 18Ni maraging steels is relatively soft and ductile before being aged. For bar stock, the yield strength is on the order of 100 ksi and the elongation in 1 in. is about 17%. Likewise, the tendency for this material to work harden is extremely low. Because of these factors, the 18Ni maraging steels are readily amenable to cold forging operations such as coining.

Effect of Forging Variables on Mechanical Properties. The effect that hot forging has on the microstructure and mechanical properties of the 18Ni maraging steels depends on the type of forging, the forging temperatures, and the extent to which the metal is worked. The effects are similar to those produced in the austenitic stainless steels and other malleable single-phase metals and alloys, but are probably of greater importance with the maraging steels.

As their name implies, the maraging steels develop their ultrahigh strength by the combination of transformation to martensite followed by age hardening. The strength of material aged directly after forging increases significantly for forging temperatures below 1700 °F (925 °C); whereas, when the steel is annealed before aging, the strength is essentially independent of forging temperature. As a result, for material forged at temperatures about 1700 °F (925 °C), annealing at 1500 °F (815 °C) before aging improves properties to some extent. On the other hand, for material forged at temperatures below 1700 °F (925 °C), this annealing treatment reduces strength.

Forging Practice. The selection of forging temperatures for the maraging steels is dominated by the type of forging operation intended. If the part is designed in such a way that it will receive sizeable reductions, forging temperatures are similar to those used for alloy steels. If the part contains portions that receive little reduction during forging, however, temperatures are adjusted downward to below 2000 °F (1100 °C). This is done to prevent excessive grain growth. In any case, the final forging steps are completed below 2000

°F (1100 °C).

Lower temperatures (1500 to 1700 °F, or 815 to 925 °C) are often required during final forging operations to control grain size and ductility. As a result, forging pressures higher than those normally used for other steels are required for the 18Ni maraging steels.

Because of their low carbon content, the maraging steels as a class are not troubled by decarburization problems; they are readily fabricated and reasonably weldable. They are not sensitive to heating or cooling rate either in annealing or in aging, and they are subject only to minor dimensional changes upon hardening.

Mechanical properties of aluminum alloy forgings are developed primarily through a duplex heat treatment. For most alloys, this consists of solution heat treating at a temperature just below the solidus temperature, water quenching, then aging between 250 and 350° F (120 and 180° C) for several hours.

Tensile strengths and elongations of several aluminum alloys at various temperatures are presented in Figures 3-66 and 3-67, respectively. Table 3-8 presents the elevated temperature yield strengths of two alloys when tested at two strain rates and shows that the yield strength increases with increasing strain rate.

None of the data in Figure 3-66 indicates the actual forging pressures, because the values were determined at strain rates considerably lower than those of conventional forging equipment. However, the data provide useful information about relative forging pressure and forgeability when comparing the alloys with each other. For example, the 7075 alloy, which is relatively difficult to forge, has nearly ten times the strength of 1100 aluminum at 700° F (370° C).

The elongation of all the alloys rises sharply at temperatures corresponding approximately with the onset of recovery or recrystallization. With increasing temperatures, the elongation values reach a maximum, then begin to drop—first gradually as grain growth occurs, then sharply when melting begins. For these reasons, elevated-temperature elongation values serve

Forging Light Alloys

Aluminum and Aluminum Alloys

Within the family of structural metals and alloys, the aluminum alloys are most readily forged to precise, intricate shapes. The most significant reasons for this are: (1) the alloys are ductile, (2) they can be forged in dies heated to essentially the same temperatures as the workpiece, (3) they do not develop scale during heating, and (4) they require low forging pressures.

The major factors influencing the forgeability of aluminum alloys are the solidus temperature and deformation rate. Most of the alloys are forged at about 100° F (55° C) below the solidus temperature. The risk of incipient melting exists when conditions of forging promote significant temperature increases. Such increases are caused by either too rapid or too large forging reductions. To minimize this risk, forging temperatures are often adjusted downward for increasing amounts and/or increasing rates of reduction.

Aluminum alloys can be forged in any type of forging equipment used for other metals. Single-stroke hydraulic presses are often preferred for forging shapes requiring large reductions, because lubrication can be applied easily and the pressing speeds are low enough to avoid harmful temperature increases. Because aluminum does not form a scale when heated, direct metal-to-die contact can cause seizing and galling. Hence, it is important to have good, continuous lubrication.

Two conditions influence the heat-treated properties of aluminum alloys: grain flow and final grain size. Wrought aluminum alloy bars generally exhibit directional ductility, the maximum in the longitudinal

direction of the bar and the minimum in the transverse direction. The transverse ductility can be improved by forging if the bar is deformed to produce flow in the transverse or lateral direction.

Heat-treatable aluminum alloys are generally strengthened slightly by forging in the hot-working range. This is due, in part, to the accompanying reduction of grain size and, to a certain extent, to precipitation hardening that occurs at the working temperatures. Maximum strength then is obtained by solution heat treating, quenching, and aging. Aluminum forgings are rarely used in the as-forged condition.

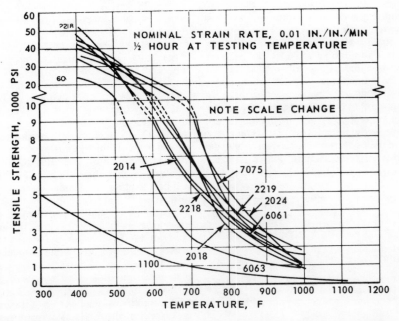

FIGURE 3-66. Elevated-temperature tensile strengths for several aluminum alloys.

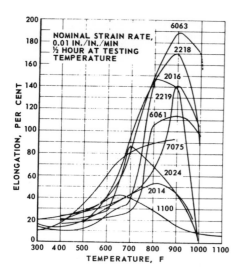

FIGURE 3-67. Elongation at elevated temperatures for aluminum and several aluminum alloys.

as a guide to selecting forging temperatures by indirectly identifying the temperatures for recovery, recrystallization, grain growth, and melting.

Magnesium Alloys

Forgeability of magnesium alloys is influenced by three important factors: solidus temperature, deformation rate, and grain size. Like aluminum alloys, magnesium alloys are forged at temperatures that are often within 100° F (55° C) of the solidus temperature. The high-zinc magnesium alloy (ZK60) will sometimes contain small amounts of the low-melting eutectic that forms during ingot solidification. This eutectic melts upon heating to temperatures just above 600° F (315° C) and can cause severe rupturing when the alloy is forged above this temperature. To minimize this problem, mill suppliers generally homogenize cast ingots at elevated temperatures for extended periods to redissolve the eutectic and to restore the higher solidus temperature. The commercial alloys containing aluminum, zirconium, and thorium do not form low-melting eutectics and therefore are not subject to hot shortness at such low temperatures. Wrought magnesium alloys exhibit excellent forgeability in slow hydraulic presses, but sometimes crack when forged in hammers. Even in presses, the faster moving flash metal is sensitive to cracking. Billets containing coarse grains are particularly sensitive to cracking during rapid deformation. For this reason, coarse-grained ingots are generally extruded before forging.

Effect of Forging Variables on Mechanical Properties The room-temperature mechanical properties of magnesium alloy forgings, especially ductility, are strongly dependent on forging procedures. In general, both longitudinal and transverse ductility improve with decreasing grain size and with increasing amounts of work. The following data show how longitudinal and transverse elongations are affected by grain size in typical AZ80 alloy forgings:

Grain size	Elongation, % Longitudinal	Transverse
Coarse ASTM 0–1	3–4	1–2
Fine ASTM 5–7	9–12	5–7

Although the basic strength properties of magnesium alloys are determined by alloy composition, forging plays an important role in establishing property uniformity and maximum ductility. It is important to provide as much flow in the transverse direction as possible during forging. This is necessary because wrought magnesium alloy bars exhibit highly directional ductility. By providing transverse metal flow, the transverse ductility is improved and the longitudinal ductility is kept high. The following transverse elongation values are typical of the AZ80 alloy:

Condition	Elongation, % Longitudinal	Transverse
Alloy bar	10	3
Forging receiving small transverse metal flow (about 15% reduction) ...	10–11	4–5
Forging receiving severe transverse metal flow (about 60% reduction)	····	6–8

Forging procedures for magnesium alloys are similar in many respects to those used for aluminum. Presses are favored for forging, and die temperature control is very important. If die temperatures are much below forging temperatures, surface cracking becomes a potential problem. When die temperatures exceed forging temperatures, the metal flows past impressions and promotes underfilling. Alloys requiring forging temperatures above 900° F (480° C) are heated in inert or reducing atmospheres to prevent burning. Sulfur dioxide gas is frequently used for this purpose.

Because magnesium alloys are comparatively brittle during cold trimming, flash is generally removed from forgings by band sawing. Magnesium forgings may be warm trimmed at temperatures corresponding to

TABLE 3-8. Influence of Strain Rate on the Elevated-Temperature Yield Strengths of Two Aluminum Alloys

Alloy	Strain rate, in./in./s	Yield strength(a), ksi, at: 500 °F (260 °C)	600 °F (315 °C)	700 °F (370 °C)
2014 ...	0.01	35	25	5
	1.0	42	30	···
7075 ...	0.01	30	20	5
	1.0	38	24	···

(a) Data for both alloys are for the solution-treated and aged condition (T6).

minimum forging temperatures. Warm trimming is generally confined to forgings with comparatively heavy sections in the flash-line regions to avoid bending and warping.

Titanium and Titanium Alloys

Forging Behavior. Titanium alloys can be forged into shapes similar to those forgeable from steel and other metals. However, various titanium compositions exhibit different degrees of forgeability based on their forging temperature range, forging pressure requirements, sensitivity to strain rate, and susceptibility to cracking. Generally, for the same amount of metal flow, more forging power is required than that required for alloy steels. Forging experience indicates that an alloy such as Ti-6Al-4V requires one and one half to two times the machine capacity needed to forge low-alloy steel into comparable shapes. For the above reasons, the particular forging characteristics of individual titanium alloys need to be considered with care in order to successfully produce the preferred shapes and properties.

Forging temperature and pressure considerations must be given a high priority in either the open die or impression die forging of titanium alloys. The pressure required to forge titanium alloys rises at a faster rate as the working temperature of the metal decreases than does the pressure required for forging alloy steel. This marked effect of working temperature on the forging pressure is characteristic of titanium alloys; there is some variation in this characteristic with alloy content, as shown in Figure 3-68. Also, forging pressure increases in an approximately linear relationship with the logarithm of the strain rate. This effect is illustrated in Figure 3-69. Moreover, because titanium alloys exhibit rapidly increasing strengths with increasing strain rates, more energy is required, for example, in hammer forging than in press forging at comparable temperatures. On the other hand, better temperature control of the workpiece is pos-

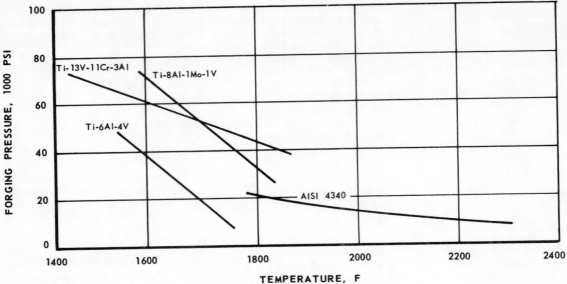

FIGURE 3-68. Comparative influence of temperature on forging pressure of titanium alloys and a low-alloy steel. Source: Ref. 15.

sible—resulting in less heat loss—when fast-acting, high-strain-rate machines are used.

Several techniques have been developed to reduce the severity of the previously mentioned characteristics. For example, die preheating and glass lubricants can be used to reduce the effects of die chilling. A glass lubricant can also act as a thermal insulator for the workpiece. Furthermore, hot die forging (die temperature near that of the workpiece) and isothermal forging (dies and workpiece at the same temperature) have been developed to afford the following potential advantages:

● Elimination of die chilling, which allows a reduction in the number of preforming and blocking dies required for forging a given part and allows forging to closer dimensional tolerances than in conventional forging.
● Use of a slow ram speed—e.g., a hydraulic press—since die chilling is not

a problem. This lowers the strain rate and the flow stress of the forged material and therefore reduces forging pressure, allowing for the production of larger parts in a given machine or reduced forging temperature requirements for production of a given part.

Metallurgical Considerations. All titanium forging alloys are produced commercially as cylindrical ingots by double vacuum-arc melting; some titanium ingots

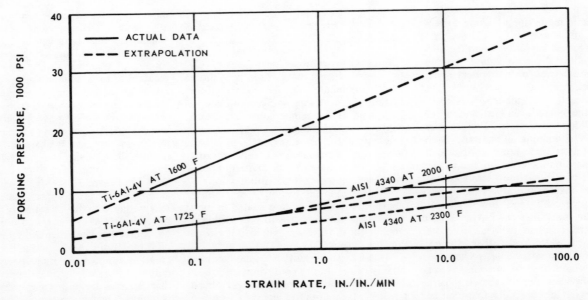

FIGURE 3-69. Comparison of forging pressures for two different materials by extrapolation of forging pressure curves. Source: Ref. 15.

are triple melted to produce larger size ingots or for the production of very high quality material. These products rarely contain segregations of other materials that might cause wide variations in forgeability. Ingots are first cogged to large round, square, rectangular, or octagonal shaped billets—a step commonly performed by the ingot producer—to prepare shapes suitable for further forging and to develop an initial preferred macrostructure in materials supplied to forgers. Smaller billet shapes produced from cogged ingots by hot rotary-forging machines or by hot rolling are also provided as forging stock. Such products supplied by producers for closed die forging are available in a variety of compositions and grades that have somewhat different forging characteristics; information on their metallurgy will assist in understanding the preferred forging parameters.

Types of Titanium Alloys

There are three basic types of titanium alloys—alpha, alpha-beta, and beta alloys—that exhibit characteristics related to alloy content and processing variables. Figure 3-70 illustrates one of the important characteristics imparted by alloy content—the stabilization of the dominant phases at different temperatures. The various alloying constituents also impart different degrees of strengthening to the alpha and beta phases.

Aluminum, for example, is a strong alpha-phase strengthener and, as shown in Figure 3-70, raises the transformation temperatures. The beta-phase stabilizers strengthen both phases and lower the transformation temperatures. The strength of alloys decreases with increasing temperature; strength is quite low in materials that are stressed above their beta transformation temperatures (beta transus). Initial breakdown forging of titanium alloys commonly is done at temperatures above the beta transus because strength is low and the beta phase (body-centered cubic structure) is quite ductile; forging pressure requirements are not as high as at lower temperatures.

Except for the beta titanium alloys, which can retain a metastable beta structure upon cooling from above the beta transus, beta-phase finish forging is in general less desirable than finish forging at lower temperatures, because a coarse-grained, transformed beta structure with low ductility can be obtained. However, if the proper amounts of deformation can be imparted, as well as good working temperature control and preferred cooling rates,

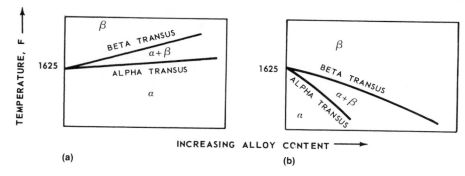

FIGURE 3-70. Effect of alloying constituents on the phase relationship in titanium alloys. Alpha phase (α) has an HCP structure, while beta phase (β) has a BCC structure. (a) Alpha-stabilizing behavior (aluminum, tin, oxygen, nitrogen, carbon). (b) Beta-stabilizing behavior (chromium, iron, manganese, molybdenum, vanadium). Source: Ref. 16.

TABLE 3-9. **Forging Parameters and Characteristics for Selected Titanium Alloys**

Alloy composition, wt %	Alloy type(a)	Temperature, °F (°C) Beta transus(b)	Temperature, °F (°C) Die forging range	Required forging pressure, ksi	Resistance to cracking
Unalloyed(c)	Alpha alloys	1760 (960)	1550–1700 (840–925)	65–75	Excellent
Ti-5Al-2.5Sn(d)		1900 (1040)	1725–1850 (940–1010)	75–85	Good
Ti-3Al-2.5V	Near-alpha, alpha-beta alloys	1715 (935)	1550–1700 (840–925)	65–75	Good
Ti-8Al-1Mo-1V		1860 (1015)	1750–1850 (950–1010)	75–85	Fair–good
Ti-6Al-2Sn-4Zr-2Mo		1825 (995)	1700–1800 (840–980)	75–85	Fair–good
Ti-6Al-4V(d)	Alpha-beta alloys	1825 (995)	1550–1800 (840–980)	75–85	Good–excellent
Ti-6Al-6V-2Sn		1735 (945)	1550–1675 (840–910)	65–75	Excellent
Ti-6Al-2Sn-4Zr-6Mo		1750 (950)	1625–1700 (885–925)	· · ·	Good–excellent
Ti-5Al-2Sn-4Cr-4Mo-2Zr	Near-beta, alpha-beta alloys	1625 (885)	1450–1700 (790–925)(e)	· · ·	Good–excellent
Ti-10V-2Fe-3Al		1475 (800)	1400–1600 (760–870)(e)	· · ·	Excellent
Ti-3Al-8V-6Cr-4Mo-4Zr	Beta alloys	1475 (800)	1500–1600 (815–870)	· · ·	Excellent
Ti-13V-11Cr-3Al		1325 (720)	1600–1800 (870–980)	85–100	Excellent

(a) Note that alpha-beta alloys with a lean beta-stabilizer content are sometimes classified as near-alpha alloys, while alpha-beta alloys with a rich beta-stabilizer content are classified as near-beta alloys. (b) ±25 °F (±15 °C); oxygen and other alloying elements can affect transus temperature. (c) Also known as commercially pure, or CP; actually an alloy with oxygen. (d) High-purity grades with low oxygen contents are known as ELI grades. (e) These alloys can be processed either with alpha-beta or beta processing (forging temperature) schedules.

beta forging is acceptable for some alloys. It is generally preferable to finish forge alpha and alpha-beta alloys below the beta transus temperatures. The lower finish temperature forging technique is practiced to develop preferred microstructures, which in turn enables the development of optimum combinations of mechanical properties by subsequent annealing or strengthening heat treatments.

Table 3-9 lists several important titanium forging alloys, their alloy type, typ-

ical beta transus temperature, typical die forging temperature range, forging pressure requirements, and qualitative remarks regarding their resistance to cracking. Commonly practiced conventional die forging parameters are described for principal alloy types in the following sections.

Alpha Alloys. The principal forging alloys in this group, unalloyed titanium and Ti-5Al-2.5Sn, represent extremes of forgeability; unalloyed titanium is easily forged (although, depending on the grade

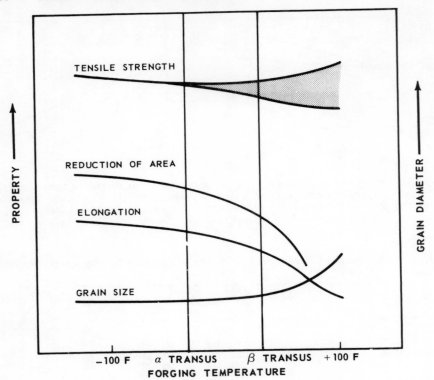

FIGURE 3-71. General influence of forging temperature on grain size and room-temperature mechanical properties of forged alpha titanium alloys. Source: Ref. 16.

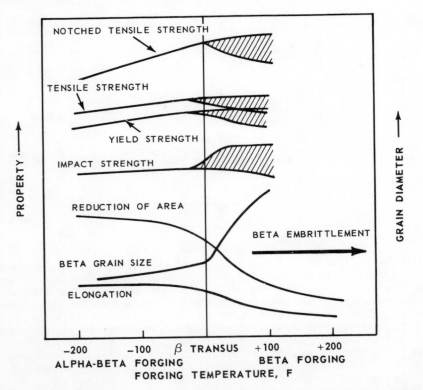

FIGURE 3-72. General influence of forging temperature on beta grain size and room-temperature properties of forged alpha-beta titanium alloys. Source: Ref. 16.

of material and the shape to be forged, not so easily as some alpha-beta and beta alloys), whereas Ti-5Al-2.5Sn is forged with more difficulty. For alloy Ti-5Al-2.5Sn, the recommended maximum forging temperature is 1900° F (1040° C); beta-grain coarsening begins at about 1950° F (1065° C). A lower maximum forging temperature is recommended for unalloyed titanium (Table 3-9). The ultimate strength level is not influenced greatly by forging temperature, but as-forged ductility is decreased significantly at the higher temperatures (above the beta transus), as shown in Figure 3-71. The best combinations of forgeability and mechanical properties are obtained by working within the alpha-beta field (i.e., between the alpha and the beta transus temperatures). Since this represents a narrower forging temperature range than that for the alpha-beta alloys, the alpha alloys are usually reheated often during forging. Proper die sequencing is essential to provide enough reduction in all parts of the forging shape, because the alpha alloys develop improved mechanical properties primarily through the working and refinement of the microstructure. Alpha alloys are not heat treatable to high strength levels.

Alpha-Beta Alloys. The Ti-6Al-4V alpha-beta alloy is by far the most commonly used titanium material; forgings make up a large quantity of the total amounts used. As shown in Table 3-9, there are several other alpha-beta-type alloys, ranging in alloy content from the near-alpha compositions (e.g., Ti-8Al-1Mo-1V) to the near-beta compositions (e.g., Ti-5Al-2Sn-4Cr-4Mo-2Zr). The near-alpha alloys are noted for their excellent elevated-temperature strength and are the most difficult to forge. The near-beta compositions are noted for their capacity for deep hardenability and are readily forgeable. Forging temperature ranges vary accordingly from high (for alloys with a high aluminum content and little beta) to low (for alloys with high beta-stabilizer content and moderate aluminum). For example, the near-alpha alpha-beta alloy with high aluminum content, Ti-8Al-1Mo-1V, has a forging temperature range of 1750 to 1850° F (950 to 1000° C), whereas near-beta alpha-beta alloys such as Ti-5Al-2Sn-4Cr-4Mo-2Zr are forged at a much lower temperature range, generally over the 1400 to 1700 °F (750 to 925 °C) range.

As shown in Figure 3-72, alpha-beta alloys, like alpha alloys, can be forged to develop improper microstructures and, consequently, undesirable properties if forging is accomplished much above the beta transus. The initial forging of alpha-beta materials may be done above the beta transus, but finishing with a considerable amount of working within the alpha-plus-beta two-phase region leads to the development of preferred combinations of properties.* The relationships between the forging reductions accomplished within the alpha-beta temperature range, which develop various quantities of equiaxed alpha within the microstructure and refine the structure, and the properties obtainable are shown for the Ti-6Al-4V and Ti-7Al-4Mo alpha-beta alloys in Figure 3-73.

There are several alternative schedules available in the selection of forging temperatures and reduction sequences for achieving a wide range of mechanical properties in alpha-beta alloys. These depend on the alloy composition, the prior structure of the billet, the type of forging to be produced, and the amount of reduction the various parts of the forging will receive. Moderate strengths with high ductility are available after properly forging and annealing. Combinations of very high strengths with moderate ductility levels are possible with solution and aging treatments following correct forging schedules.

Beta Alloys. The beta titanium alloys are not noted for their elevated-temperature strength except under conditions of high strain rate. This characteristic of beta alloys suggests to the forger that deformation is more easily accomplished in press forging than, for example, in hammer forging. Nevertheless, even hammer forging can be accomplished if the material is maintained at the proper working temperatures. For beta alloys, this is above the beta transus (Table 3-9). Generally, grain coarsening is not a severe problem in working these materials above the transus temperature, nor is there a problem with embrittlement, so long as sufficient work is imparted. Without sufficient work, grain growth may proceed rapidly at the higher temperatures.

The highly alloyed Ti-13V-11Cr-3Al beta alloy, the first commercial beta alloy, is quite difficult to forge and is currently little used. Alloys with a lower beta-stabilizer content, such as the Ti-3Al-8V-6Cr-4Mo-4Zr, have a lower working temperature range and are more easily forged. The near-beta alpha-beta alloy, Ti-10V-2Fe-3Al, is often considered within the beta alloy class; it is readily forged at beta temperatures (i.e., entirely above the beta transus). The Ti-10V-2Fe-3Al material also may be forged (for example, in finishing

*Schedules for the beta forging of alpha-beta alloys are available, but it is beyond the scope of this presentation for detailed descriptions.

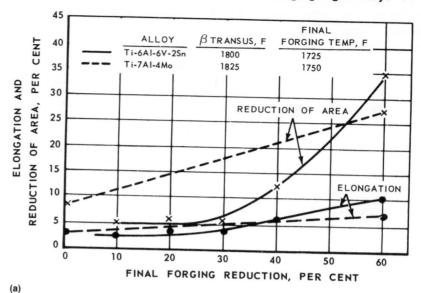

(a)

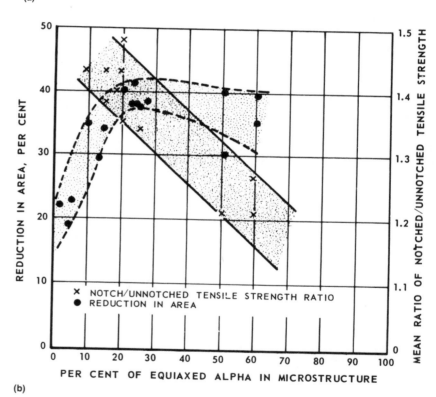

(b)

FIGURE 3-73. (a) Effect of final forging reduction below the beta transus on alpha-beta alloys on ductility. Forty to sixty percent final reduction in the alpha-beta field yields a preferred microstructure. Source: Ref. 17, 18. (b) Dependence of ductility and notched/unnotched tensile strength ratio on quantity of equiaxed alpha in the microstructure. Source: Ref. 19.

operations) at alpha-beta temperatures. This material is particularly amenable to hot die and isothermal forging practices. Generally, the maintenance of the working dies at elevated temperatures to prevent die chilling of the workpiece and, if practical, the working of the beta alloys at moderately low strain rates result in the suc-

TABLE 3-10. Room-Temperature Tensile Properties of Alpha, Alpha-Beta, and Beta Titanium Alloys From a Common Forging Configuration

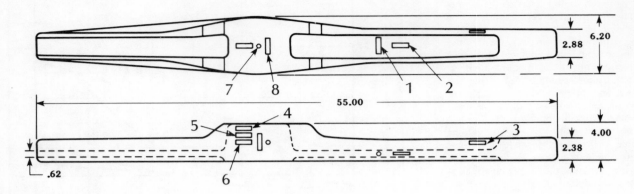

Location	0.2% Yield strength, ksi	Ultimate tensile strength, ksi	Elongation, %	Reduction in area, %
	Ti-5Al-2.5Sn			
	1350 °F (730 °C), 2h, air cooled			
	(alpha)(a)			
1	122.8	131.7	11.0	25.0
2	131.6	144.5	14.0	28.0
3	124.8	133.3	14.0	35.0
4	125.8	135.6	18.0	29.0
5	122.4	133.0	14.0	28.0
6	120.3	130.5	14.0	27.0
7	117.3	128.5	16.0	27.0
8	117.0	130.1	15.0	22.0
	Ti-8Al-1V-1Mo			
	1650 °F (900 °C), 1 h, air cooled + 1100 °F (595 °C), 8 h, air cooled			
	(near-alpha, alpha-beta)(b)			
1	137.5	143.4	18.0	29.4
2	144.1	152.7	17.0	35.4
3	139.9	151.7	21.0	35.4
4	120.9	132.1	12.0	17.1
5	116.2	126.8	8.0	24.9
6	113.6	126.2	14.0	18.5
7	117.9	130.5	12.0	15.2
8	116.8	128.1	8.0	14.2
	Ti-6Al-4V (0.13% O$_2$)			
	1750 °F (955 °C), 1 h, water quench + 1000 °F (540 °C), 4 h, air cooled			
	(alpha-beta)(c)			
1	147.0	159.0	13.0	37.0
2	146.0	158.2	13.8	35.7
3	147.8	160.0	13.8	48.5
4	136.0	151.0	13.0	37.0
5	130.0	144.8	13.5	40.5
6	127.8	144.0	10.0	25.1
7	129.6	143.8	10.0	24.5
8	124.0	142.0	6.5	18.8
	Ti-6Al-6V-2Sn			
	1675 °F (913 °C), 1 h, water quench + 1050 °F (565 °C), 4 h, air cooled			
	(alpha-beta)(c)			
1	188.0	199.5	7.5	29.0
2	189.1	200.9	8.0	24.2
3	185.8	198.0	12.0	31.5
4	159.5	172.3	11.0	27.3
5	153.0	166.2	11.0	22.5
6	146.6	161.2	13.0	28.7
7	148.9	161.3	10.0	38.5
8	153.0	166.8	8.0	23.3

TABLE 3-10 (continued).

Location	0.2% Yield strength, ksi	Ultimate tensile strength, ksi	Elongation, %	Reduction in area, %
	Ti-13V-11Cr-3Al **1335 °F (725 °C), 1 h, air cooled + 1450 °F (790 °C), 1/2 h, air cooled + 900 °F (480 °C), 15 h, air cooled** **(beta)(c)**			
1	171.8	182.0	3.4	7.8
2	166.8	178.0	6.0	8.6
3	170.0	181.0	5.0	7.8
4	171.8	182.0	3.5	7.8
5	165.0	176.0	5.0	6.2
6	166.0	176.0	4.0	8.6
7	163.2	174.0	4.0	4.7
8	163.0	174.0	6.0	12.2

(a) Annealed. (b) Duplex annealed. (c) Solution treated plus aged.

cessful forging of beta alloys. Beta alloys may be solution treated and aged to high strength levels after forging.

Forged Titanium Properties. An important feature of forged titanium materials is that there is only a moderate change in tensile properties with grain direction, provided that die design and die-filling practices are optimized. This is illustrated in Table 3-10 by as-forged tensile data for representative alpha, alpha-beta, and beta alloys. This table presents typical tensile properties obtained in a representative structural configuration after preferred forging schedules and heat treatments (appropriate for each alloy type) were given to the full-size parts. Comparable strengths and ductilities are obtainable from both thick and thin sections from this closed die configuration.

General Forging Considerations. There are a number of factors that make titanium alloys more difficult to forge than steels. The metallurgical behavior of the alloys not only imposes certain controls and limitations on the forging operation, but also influences all of the steps in manufacturing forged parts. Particular care should be exercised throughout the processing cycle to avoid contamination by oxygen, nitrogen, carbon, and/or hydrogen, which can severely impair ductility and toughness properties and the overall quality of a forged part.

Design principles for alpha and alpha-beta titanium alloys fall between those for alloy steels and hot/cold-work alloys. The designs should provide an adequate amount of deformation during forging, but the reductions need not be confined to narrow limits. The size of a titanium forging is limited only by the equipment available.

Producibility of close-tolerance, precision titanium alloy forgings has been proven. However, owing to such factors as excessive die wear, the need for expensive tooling, and problems of microstructure control and contamination, the cost of close-tolerance forging is usually prohibitive. Successful precision forging, therefore, is confined to small forgings, such as blades and fittings, that do not have complex flow patterns. Very-close-tolerance forgings in some moderately large parts are currently being developed using hot die and isothermal forging techniques.

Forging High-Temperature Superalloys

"Creep-resisting" nickel- and iron-base alloys have inevitably been linked with the development of gas turbines. In the 1940s, the early years of production, nickel-base alloys containing chromium and hardened by additions of titanium and aluminum were used predominantly for the manufacture of gas turbine blades. In the late 1940s and early 1950s, the trend developed toward greater use of creep-resisting iron-base alloys for disks. Initially, these alloys were austenitic, and some were warm worked—for example, 16Cr-25Ni-6Mo and G 18B. In the United States, these alloys tended to be replaced by alloys containing iron, nickel, chromium, titanium, and possibly molybdenum, such as Discaloy, Tinadur, and A-286. Throughout the early 1960s, the austenite alloys were replaced by a series of creep-resisting modified 12% chromium martensitic stainless steels, which were the most popular materials in the United Kingdom.

Although the martensitic alloys were used to a certain extent by the U.S. engine building industry, there was a tendency to use the austenitic iron-base systems containing increasingly high levels of alloys. Alloys such as M-308, V-57, and 901 were developed for wheels and hubs and, at this stage, 901 contained a higher percentage nickel content than iron, with about 43% nickel, 35% iron, 15% chromium, 5% molybdenum, and 3% titanium as the major alloying elements. Later, Waspaloy was used as a disk material; in the mid-1960s, 718, Astroloy, and René 95 were developed for this purpose. Presently, a modified type of IN-100 is used.

Table 3-11 lists the typical chemical compositions of several alloys. During the last decade, all of the major aircraft engine builders in the Western world have tended to use superalloys for turbine wheels and often for the disks in the later stages of the compressor. The term superalloy is used in reference to a complex solid-solution-hardened alloy, usually FCC, further strengthened by a precipitation-hardening mechanism and generally used in elevated-temperature applications where strengths and oxidation resistance are important. This discussion presents the developments in those systems that have been used extensively in the production of disks and is not intended to serve as a comprehensive discussion of all the superalloys developed.

For the past 30 years, the forging industry—in particular, that sector supplying the aerospace market—has been required to continuously improve its understanding of the technology of the superalloy systems in order to produce rotating components that can meet the demands of the engine builders; these demands have become increasingly exact-

TABLE 3-11. Nominal Compositions of Forged Superalloys Used for Rotor Applications

Alloys	C	Mn	Si	Cr	Ni	Co	Mo	W	Fe	Ti	Al	B	Zr	Other(s)
G18B	0.40	0.80	0.80	13.0	13.0	10.0	2.0	2.50	Bal.	...	...	...	...	3.0 Nb
19-9DL	0.30	1.15	0.50	18.5	9.0	...	1.40	1.40	Bal.	0.25	...	...	...	0.40 Nb
19-9DX	0.30	1.15	0.50	18.5	9.0	...	1.60	1.40	Bal.	0.60	...	...	...	...
16-25-6	0.06	1.30	0.70	16.0	25.0	...	6.00	...	Bal.	...	...	...	...	0.15 N_2
Discaloy	0.04	1.00	0.75	13.5	26.0	...	2.70	...	Bal.	1.75	0.15	0.005	...	...
Tinadur	0.04	1.00	0.75	14.5	26.0	...	...	...	Bal.	2.25	0.15	...	...	...
A-286	0.03	1.00(a)	0.75(a)	15.0	26.0	...	1.30	...	Bal.	2.15	0.20	0.005	...	0.30 V
M-308	0.08	1.00(a)	0.75(a)	14.0	33.0	...	4.00	6.50	Bal.	2.00	0.25	0.005	0.10	...
V-57	0.04	0.25(a)	0.25(a)	1.45	27.0	...	1.25	...	Bal.	3.00	0.25	0.010	...	0.50 V
901	0.03	0.25(a)	0.25(a)	12.5	42.0	...	5.60	...	Bal.	2.90	0.20	0.014	...	...
718	0.04	0.20(a)	0.20(a)	18.5	52.0	...	3.00	...	Bal.	1.00	0.60	0.004	...	5.20 Nb + Ta
René 41	0.07	0.30(a)	0.50(a)	19.0	Bal.	11.0	10.0	...	5.00(a)	3.10	1.50	0.005	...	...
Waspaloy	0.03	0.15(a)	0.15(a)	19.5	Bal.	13.5	4.50	...	2.00(a)	3.00	1.30	0.006	0.06	...
Astroloy	0.04	0.15(a)	0.15(a)	15.0	Bal.	17.0	5.25	...	0.50(a)	3.50	4.00	0.025	0.06(a)	...
René 95	0.07	0.15(a)	0.20(a)	13.0	Bal.	8.0	3.50	3.50	0.50(a)	2.50	3.50	0.010	0.05	3.50 Nb + Ta
In-100	0.18	0.15(a)	0.20(a)	10.0	Bal.	15.0	3.00	...	...	4.70	5.50	0.014	0.06	1.00 V

(a)Maximum. Source: Ref. 20.

ing. During the last three decades, forging has been transformed from an industrial skilled art into a technology that involves consideration of the influence of deformation conditions from a metal physics aspect.

There has been an increase in the total weight of forged superalloys used in engines, and these materials are currently used for disks, shafts, and hubs, both in the turbine and compressor rotating sections of gas turbines. Superalloys are also used extensively in stator rings, spacers, burner cans, blades, and compressor and turbine cases. Today, superalloys are no longer used primarily because of their creep resistance at or around the temperatures found at turbine inlets. Increased understanding of the alloy systems (and, in particular, of structure-property relationships) has permitted the upgrading of forgings by mechanical and thermal treatments to satisfy requirements for increased strength in areas other than high-temperature creep.

After the mid-1960s, the development of new superalloys for rotating components slowed down. An increasing number of difficulties were encountered in the design of significantly improved, new forgeable alloys. Costs were escalating rapidly, and the potential return from alloy development was diminishing. Therefore, the existing alloy systems were reexamined in the light of the technology available at the time and the changing property requirements of the users. What were the structural and compositional factors responsible for particular properties? Traditionally, emphasis has been placed on chemical composition, but could further improvements be made by suitable mechanical and thermal processing of an already optimized chemical composition? Could some of the detrimental character-

istics which, at that time, limited performance be improved? These questions have been and are still being asked, and results have proved useful.

Forging Behavior: General Trends

The two basic material characteristics that greatly influence forging behavior of superalloys are flow stress and ductility. Because these alloys were designed to resist deformation at high temperatures, it is not surprising that they are very difficult to hot work; ductility is limited and the flow stress is high. Furthermore, any addition in alloying that improves service qualities usually decreases workability. Increased precipitates require an increased deformation temperature because the alloys are usually worked with the precipitates dissolved. The higher concentration of dissolved alloying elements (40 to 50% total) gives rise to higher flow stress, higher recrystallization temperature, and lower solidus temperature, thus narrowing the useful temperature range for hot forming. Where ductility is defined as the amount of strain to fracture, the ductility of these alloys is influenced by the deformation temperature, strain rate, prior history of the material, composition, degree of seg-

TABLE 3-12. Critical Melting, Precipitation, and Forging Temperatures for Several Nickel-Base Superalloys

Alloy	First melting temperature °F	°C	Precipitation temperature °F	°C	Maximum forging temperature °F	°C	Minimum forging temperature °F	°C
Hastelloy X	2300	1260	1400	760	2225	1215	1550	840
Inconel 718	2300	1260	1550	840	2100	1150	1650	900
Waspaloy	2250	1230	1800	980	2150	1180	1800	980
Incoloy 901	2200	1200	1800	980	2150	1180	1800	980
Inconel X-750	2350	1290	1750	950	2200	1200	1850	1010
M-252	2200	1200	1850	1010	2150	1180	1775	968
Hastelloy R-235	2300	1260	1900	1040	2200	1200	1850	1010
René 41	2250	1230	1950	1065	2175	1190	1875	1025
Udimet 500	2250	1230	2000	1100	1925	1050	2200	1200
Udimet 700	2250	1230	2050	1120	2125	1163	1900	1040
Astroloy	2250	1230	2050	1120	2175	1190	2000	1100
Hastelloy W	...	...	...	...	2200	1200	1900	1040
Nickel 200	...	...	...	...	2200	1200	1600	870
Inconel 600	...	...	...	...	2100	1150	1900	1040
Inconel 751	...	...	...	...	2200	1200	1900	1040
Hastelloy C	...	...	...	...	2250	1230	1850	1010
Nimonic 90	...	...	...	...	2100	1150	1850	1010
Nimonic 115	...	...	...	...	2150	1180	2000	1100
Unitemp 1753	...	...	...	...	2150	1180	1850	1010
Unitemp AF 2-1DA	...	...	...	...	2150	1180	1950	1065
Udimet 710	...	...	...	...	2150	1180	1950	1065
René 95	...	...	...	...	2050	1120	1950	1065
MAR-M-421	...	...	...	...	2100	1150	1900	1040

Source: Ref. 21.

regation, cleanliness, and the stress state imposed by the deformation process.

Temperature limits for forging nickel-base superalloys are determined largely by melting and precipitation reactions. Critical melting and precipitation temperatures for superalloys and recommended forging ranges are given in Tables 3-12 through 3-14. More specific data on superalloy ductility versus temperature are given in Figures 3-74 through 3-77.

An intermediate temperature region of low ductility is likely to be encountered in attempts to forge metals near a temperature between regimes of low- and high-temperature deformation. The region of low ductility often occurs at temperatures around 0.5 of the melting point as measured on the Kelvin scale. The dividing temperature has a physical basis. At hot-working temperatures, self-diffusion rates are high enough for recovery and recrystallization to counteract the effects of strain hardening.

TABLE 3-13. Forging Temperature Ranges of Several Iron-Base Superalloys

Alloy	Maximum forging temperature		Minimum forging temperature	
	°F	°C	°F	°C
A-286	2200	1200	1750	950
V-57	2150	1180	1650	900
M-308	2150	1180	· · ·	· · ·
19-9DL	2150	1180	1200	650
W-545	2000	1100	1700	930
Discaloy	2200	1200	· · ·	· · ·
16-25-6	2100	1150	· · ·	· · ·
AFC-260	2075	1135	1750	950
Pyromet 860	2050	1120	1900	1040

Source: Ref. 21.

For Inconel 600, slight changes in the relative amounts of sulfur and those elements which tend to tie up sulfur cause large ductility differences. Furnace atmospheres and lubricants used in forging the superalloys should be free of sulfur.

TABLE 3-14. Forging Temperature Ranges of Several Cobalt-Base Superalloys

Alloy	Maximum forging temperature		Minimum forging temperature	
	°F	°C	°F	°C
V-36	2250	1230	1600	870
S-816	2250	1230	1600	870
L-605	2275	1245	1700	930
J-1570	2175	1190	1825	995
J-1650	2150	1180	1850	1010

Source: Ref. 21.

Metals containing large amounts of alloying elements develop dendritic segregation during solidification. In nickel-base alloys the age-hardening elements, aluminum and titanium, freeze late and form larger and more stable γ' hardening precipitates. Carbides and other intermetallic particles also concentrate in the interstices of the dendrites. The presence of stable intermetallics, such as carbides, can initiate

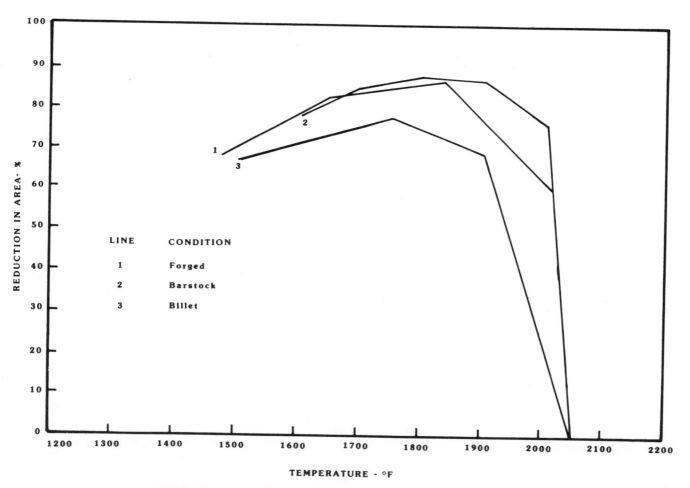

LINE	CONDITION
1	Forged
2	Barstock
3	Billet

FIGURE 3-74. Hot ductility of A-286 iron-base superalloy. Source: Ref. 22, 23.

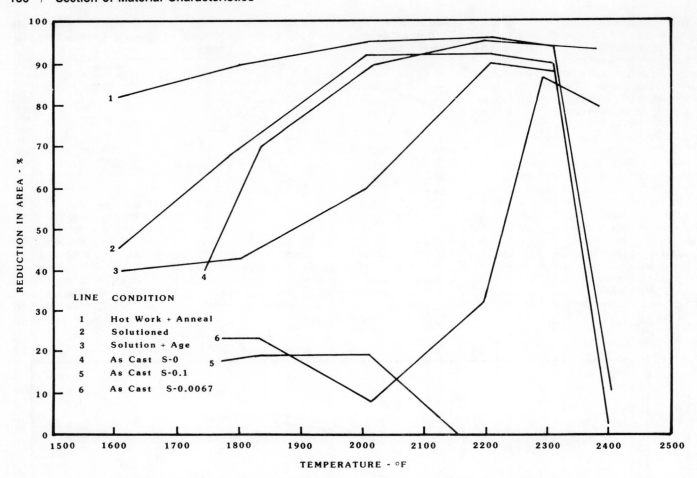

FIGURE 3-75. Hot ductility of Inconel 600 as measured in the Gleeble test. Source: Ref. 24, 25.

the formation of minute ruptures in forging. The segregates cause poor forgeability by reducing ductility at high temperatures. For that reason, metals produced by vacuum-arc remelting or electroslag remelting usually have better forging properties. Diffusion heat treatments minimize the severity of segregation. Homogenization treatments are generally applied to cast heat-resistant ingots prior to their conversion to wrought products.

Cleanliness of the superalloys probably has the greatest influence on hot forgeability. Those containing reactive elements, such as aluminum and titanium, are commonly produced by consumable-electrode melting in vacuum or under a protective slag cover. These practices minimize oxygen and nitrogen contamination. Oxide and nitride stringers markedly impair the forgeability of air-melted alloys.

The stress state induced by the process modifies ductility. If the process maintains compressive stresses in all parts of the deforming workpiece, cavity formation cannot begin and ductile fracture does not occur. If, however, the process allows secondary tensile stresses to develop, cavity formation can begin and may lead to fracture. Forging practices should be designed to minimize the development of tensile stresses during deformation.

Design Considerations for Forged Superalloys

Given a better understanding and subsequent control of critical processing parameters, it has been possible to modify many alloys in order to improve the particular limiting mechanical property in the component design. Typically, this involves optimization of one or several of the following properties:

- Creep strength
- Tensile strength
- Low-cycle fatigue (LCF) response
- High-cycle fatigue (HCF) response
- Creep rupture behavior
- Cyclic rupture (creep-fatigue interaction) behavior

Sometimes, changes in the thermomechanical processing (TMP) sequence for an *existing* alloy may also lead to the desired mechanical property levels. The idea of "tailoring" existing alloys in order to upgrade specific property characteristics is advantageous because:

- The development of new alloys is expensive and full characterization is time consuming.
- Existing alloys should present no significant change in thermal expansion coefficient, thus minimizing replacement design changes.
- Such alloys have been tested in the production and operation of engines in some related conditions.

In optimizing specific mechanical properties in superalloys via TMP, it is nec-

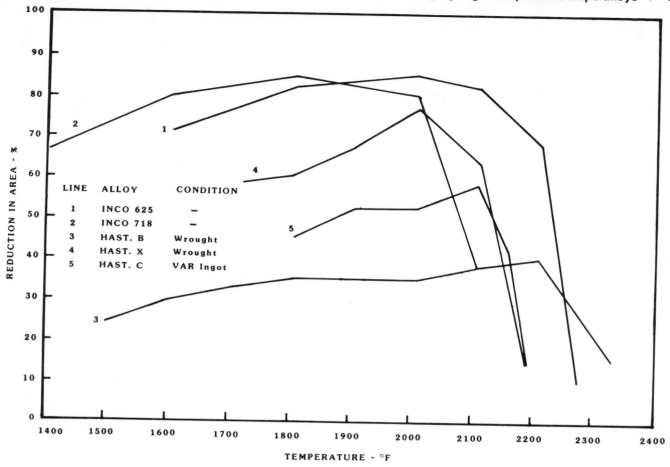

FIGURE 3-76. Hot ductility of several nickel-base alloys. Source: Ref. 22, 26, 27.

essary to consider the processing interactions of (1) the degree of deformation, (2) the temperature at which such deformation is carried out, (3) the subsequent solution treatment, and (4) the precipitation times and temperatures used, if the correct microstructure and properties are to be obtained. Because many of the important structural features cannot be observed optically, electron microscopy must be utilized, and either the replica or thin-foil transmission technique employed to resolve features of significance. Both matrix and grain-boundary conditions may be important. Consideration must be given to the carbide morphology; the particle size and distribution of the hardener precipitate; grain-boundary denudation; presence and morphology of any intermetallics, e.g., η, δ, μ, σ, and Laves phases (A_2B); and matrix condition, i.e., fully recrystallized, mixed recovery and recrystallization, or residual strain energy in the form of warm work, and the resulting dislocation substructure.

Important Processing and Material Variables for Superalloy Forging

Temperature and severity of deformation are the most important processing variables to establish for superalloys. However, the measurement of "hot workability" should also be considered, especially in the context of thermomechanical processing (TMP). The usual objective is to measure some interrelation of strain to fracture, temperature, and strain rate, and then select an appropriate peak ductility temperature (or other variable). With TMP it is still essential to obtain this information, but the independent variable, i.e., temperature, is selected on the basis of its effect on properties, not on the basis of maximum achievable deformation.

Selection of strain rate influences radiation and conduction heat losses and the percentage of deformation heating retained in the workpiece during conventional hot forging. In the present state of

the art, the as-deformed microstructure is the measure from which forging temperature and strain rate are typically selected. For superalloys, the following are the most important microstructural effects:

- Dynamic recrystallization is the most important softening mechanism during hot working of superalloys.
- Grain boundaries are preferred nucleation sites for recrystallization.
- The rate of recrystallization decreases as the temperature and/or the extent of deformation decreases.
- Precipitation that may occur during the recrystallization can inhibit the softening process. Recrystallization cannot be completed until the precipitate coarsens to a relatively ineffective morphology.

These trends are illustrated in Figures 3-78 and 3-79 for Inconel 718. Comparison of Figures 3-78(a) and (b) shows a refinement of grain size due to recrystallization. The restoration process is not recovery at these heavier strains. It cannot be proven

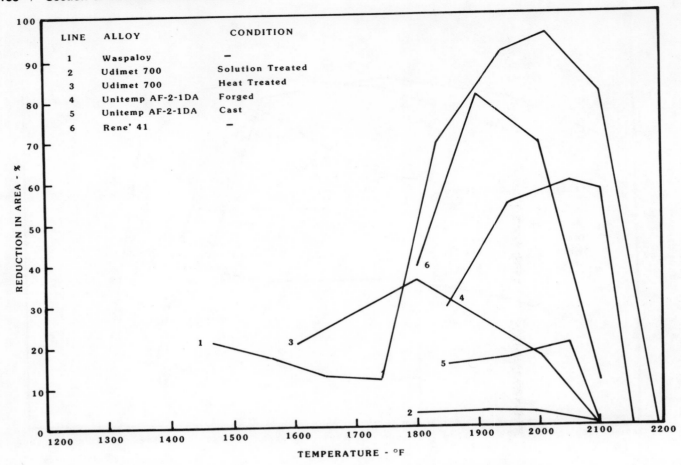

LINE	ALLOY	CONDITION
1	Waspaloy	—
2	Udimet 700	Solution Treated
3	Udimet 700	Heat Treated
4	Unitemp AF-2-1DA	Forged
5	Unitemp AF-2-1DA	Cast
6	Rene' 41	—

FIGURE 3-77. Hot ductility of several nickel-base alloys. Source: Ref. 22, 28, 29.

that recrystallization is dynamic, but ob- servations agree with the theory that it is dynamic. Figure 3-78(b) illustrates the preferred nucleation along grain bound- aries (and twin boundaries as well).

The rate of recrystallization decreases as the temperature decreases. The Inconel 718 structure in Figure 3-78(b) was forged at an average upsetting rate of approxi- mately 0.5 s⁻¹. The metal remained above any precipitation temperature, and recrys- tallization is essentially complete. In con- trast, Figure 3-78(d) shows the structure after forging approximately 0.05 s⁻¹. The total deformation heat was generated over a longer time span, and chilling was a greater factor. Consequently, recrystalli- zation is incomplete. Figure 3-79 shows the electron microstructure in a fine-grained recrystallized region of Inconel 718. Due to the presence of carbide and/or Ni₃Cb particles at triple points, it is inferred that the temperature dropped during deforma- tion and that the precipitate halted the progress of recrystallization.

Subsequent "solution" treating at 1750 °F (950 °C) precipitates Ni₃Cb on all existing as-forged boundaries. Figure 3-78(c) shows no growth in the completely recrystallized fine-grained structure. Figure 3-78(e) shows relatively extensive growth restricted to the precipitate-free remnants of as-forged warm-worked grains. Other observations indicate the following for Inconel 718 and similar alloys:

- Coarse particles accelerate recrystalli- zation by supplying surfaces for nu- cleation.

- Medium particles (200 Å) do not affect recrystallization, but may inhibit sub- sequent grain growth.

- Fine particles (50 Å) retard recrystalli- zation when present in approximately 0.1 vol% by stabilizing the subgrains and that the reduced nucleation rate may result in a relatively coarse recrystal- lized grain size.

Similar studies of other superalloy sys- tems are needed to verify these trends, however.

It is recognized that differences in strain rate (and temperature history) and the re- sultant differences in microstructure are important. In Inconel 718, the stress-rup- ture ductility is particularly sensitive (Ta- ble 3-15). In other alloys, it may be a dif- ferent property of a different temperature range. In the absence of total understand- ing, empirically designed experiments must be relied on.

Phase equilibria are also an important material consideration when selecting pro- cessing conditions. In the initial stages of TMP development, control of the bound- ary carbides was the prime objective. Phase studies had shown that the γ' was dis- solved above 1925 °F (1050 °C) and the M₆C above 2050 to 2100 °F (1120 to 1150 °C). Figure 3-80 shows a ring rolled with much of the carbon in solution. Dur- ing subsequent 1950 °F (1065 °C) solution treatment, a thick layer of M₆C precipi-

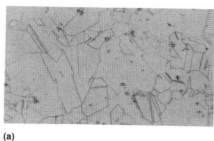

(a)

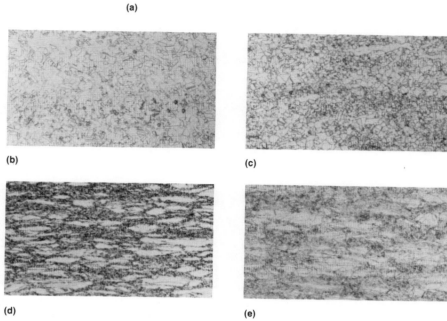

(b)

(c)

(d)

(e)

FIGURE 3-78. Inconel 718 microstructures demonstrating the effect of adiabatic heating upon recrystallization. (a) Heated 2 h at 1825 °F (995 °C); $T/T_m \sim 0.85$ to 0.90. (b) Forged at a high strain rate. (c) Plus 1750 °F (955 °C) solution; 15% rupture elongation. (d) Forged at a slow strain rate. (e) Plus 1750 °F (955 °C) solution; 25% rupture elongation. All magnifications 70×. Source: Ref. 30.

tated at grain and twin boundaries. Due to reduced carbon in solution, the heat-treated structure has clean twins and much decreased boundary precipitation. The ductility was vastly improved (Table 3-16).

Control of the carbide equilibria has progressed furthest in Waspaloy. Figure 3-81 shows the different $M_{23}C_6$ morphology in two forgings from one bar. The intercellular structure was obtained using a 1900 °F (1040 °C) forging temperature, whereas the spherical structure resulted from an 1875 °F (1025 °C) temperature. Both structures could be obtained in an 1875 °F (1025 °C) forging by a 25 °F (15 °C) difference in solution-treat temperature. It is assumed that this variation is due to the amount of carbon in solution at the beginning of the 1550 °F (840 °C) stabilize cycle. This variation is too minute to verify by analysis. Actual practice is built on a rather shaky hypothesis; new knowledge of phase equilibria resulting

from new equipment and better experiment design is needed.

Introduction of In-Process Anneals. High-temperature cycles have been widely employed in TMP cycles for a variety of reasons, e.g., more uniform γ' solvus, dissolving of brittle intermetallics, and control of in-process grain size. The industry has generally been able to ignore critical grain growth because sufficient carbides were present to limit the maximum size.

Powder products will force a return to the critical growth concept. High-temperature anneals are tempting because they minimize the original powder boundary influence. Figure 3-82 shows two IN-100 compacts forged at 2150 °F (1175 °C), one with a 2250 °F (1230 °C) intermediate anneal and one without. The critical growth was disastrous to the tensile ductility. In the manufacture of contour shapes, some nonuniformity of strains will be inevitable

and TMP development must build in some flexibility to accommodate this need.

Reproducibility of Manufacturing Practices. After specifying a TMP cycle, can the manufacturing process be adequately controlled? Table 3-17 lists data collected on one run of ten Waspaloy forgings in an 18,000-ton hydraulic press. The heating time was much more closely controlled in the finish than in the upset. Transfer and waiting times had similar averages in finish and upset, but the sigma (standard deviation) is relatively large in both. The finish deformation cycle is much slower.

The factor of interest is the small sigma of the deformation cycle as compared to the handling cycle. Machines are much more production-oriented than are people. Semiautomated manufacturing processes can more fully achieve the total potential offered by TMP.

Heating facilities are another limiting factor in TMP control. René 95 has an in-process anneal designed to control one of the two segments of the aim duplex structure (Figure 3-83). The aim is represented by the 2075 °F (1135 °C) structure, shown in Figures 3-83(c) and (d). Note that a ±15 °F (±8 °C) variation results in significantly different structures. If one is working from a furnace with a ±25 °F (±15 °C) certification, the TMP cycle obviously will not be reproduced.

Another factor important in the design of TMP cycles is the economic penalty associated with mistakes. Table 3-18 shows the loss in ultimate strength that occurs in the stronger segment (the hub) of a René 95 forging due to a restrike. The damage is not visible by optical microscopy, but is revealed only by destructive testing. TMP places a heavier burden on the mutual trust and between consumer and producer.

Effects of Processing on the Microstructure and Properties of Superalloys

There are two approaches to superalloy processing. The first is to form the required component by the most economical method and to rely on subsequent heat treatment to produce the anticipated mechanical properties. (The only metallurgical objective in forming would be to refine and control the grain size and to break up any as-cast structure.) The acceptability of this approach depends on the nature of the subsequent solution treatment processes. If these are such as to remove prior dislocation substructures and take into solution any strengthening precipitates—i.e., solution treatments above the γ' or γ″ sol-

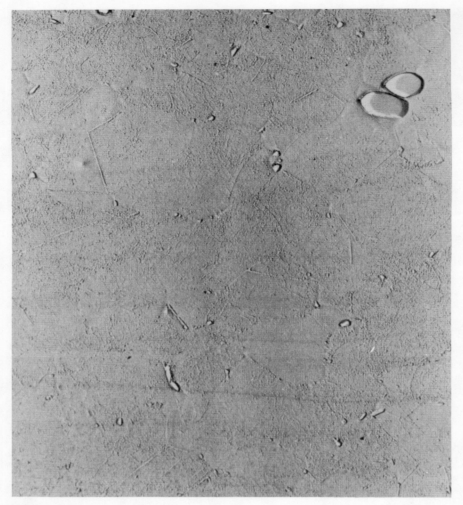

FIGURE 3-79. Electron microstructure of Inconel 718 as-forged at a slow rate from an 1825 °F (995 °C) initial temperature. 17,000×. Source: Ref. 30.

TABLE 3-15. Effect of Strain Rate on Stress Rupture of Inconel 718(a)

Forging temperature °F	°C	Average forging strain rate, s⁻¹	Elongation in 1200 °F (650 °C) stress rupture, %
2025	1105	. . . . . .0.5	6
2025	1105	. . . . . .0.05	25
1825	995	. . . . . .0.5	15
1825	995	. . . . . .0.05	25

(a) These were pancake forgings, some of which are shown in Figure 3-81. Source: Ref. 30.

vus (e.g., 1975 °F, or 1080 °C, for Waspaloy; or 2135 °F, or 1170 °C, for Udimet 700)—then this is the logical approach, particularly when the as-formed components are obtained from a number of manufacturers and the substructures may not be identical. However, such heat treatments may not develop the maximum mechanical properties of the alloy.

The second approach is to aim the processing at a preselected microstructural goal. This is particularly the case where solution treatments below the γ' or γ" solvus temperatures are required (e.g., 1850 °F, or 1010 °C, for Waspaloy; or 1740 °F, or 950 °C, for Inconel 718) and the degree of recrystallization and retention of prior dislocation substructures during solution treatment is sensitive to the nature of the prior thermomechanical processing. Improved strength, fatigue, and creep properties can be achieved by such an approach. However, the properties are now process-history-sensitive and the success of this approach depends on the ability of the designer to specify, and the manufacturer to provide, components in which the thermomechanical processing has

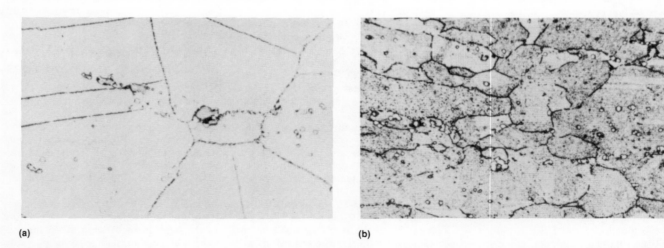

(a)

(b)

FIGURE 3-80. Micrographs at 385× of typical structures in René 41 rolled rings. Heat treated at 1950 °F (1065 °C), 4 h, air cooled; 1400 °F (760 °C), 16 h, air cooled. (a) Ring rolled from 2050 °F (1120 °C) furnace. (b) Ring rolled from 1975 °F (1080 °C) furnace. Source: Ref. 30.

TABLE 3-16. Effect of Rolling Temperature on Tensile Properties for René 41

Rolling temperature °F	°C	0.2% yield strength, ksi	Ultimate tensile strength, ksi	Elongation, %	Reduction in area, %
Room-temperature tensile					
2050	1120	. . . . 136.0	150.1	3.0	5.4
1975	1080	. . . . 138.5	203.1	21.0	21.1
1400 °F (760 °C) tensile					
2050	1120	. . . . 117.3	160.1	8.0	10.9
1975	1080	. . . . 118.3	148.1	20.0	31.5

Source: Ref. 30.

TABLE 3-17. Manufacturing Results for One Run of Ten Waspaloy Forgings

	Average	Sigma
Upset operation		
Heating time (min)	116.0	21.8
Transfer time, furnace to die (s)	28.0	4.9
Time lapse awaiting top die contact (s)	15.0	9.7
Duration of deformation cycle (s)	8.6	1.9
Finish operation		
Heating time (min)	95.0	3.2
Transfer time, furnace to die (s)	29.0	3.0
Time lapse awaiting top die contact (s)	16.0	5.7
Duration of deformation cycle (s)	19.0	3.1

Source: Ref. 30.

been closely controlled. Otherwise, the variability in prior microstructural condition will result in a largely unpredictable scatter in tensile strength and fatigue life.

The specific metallurgical objectives of thermomechanical processing are to provide either a fine recrystallized structure or a warm-worked structure. The extreme example of a recrystallized fine-grained structure is, of course, one that has superplastic properties, the development of which allows greater flexibility in the forming operation and the forging of alloys that, under conventional forging conditions, are considered to be unworkable. The superplastic properties of fine-grained

recrystallized structure have provided the basis of creep forming methods and also partially powder metallurgical processes such as the Gatorizing process for the fabrication of turbine disks. The requisite fine-grained structure may also be brought about by forging or extruding just above the recrystallization temperature and within the $(\gamma + \gamma')$ phase field, where a large volume fraction of coarse γ' particles stabilizes the grain size by suppressing grain-boundary migration and increasing the number of sites for nucleation.

This thermomechanical processing technique essentially forms the basis of the "Minigrain" process, in which a uniform

dispersion of η (in Nimonic 901, or Waspaloy) or δ (in Inconel 718) is used. Subsequent superplastic deformation may then be carried out, if required, below these recrystallization temperatures, followed by grain coarsening and subsequent aging heat treatments in components in which a high-temperature creep strength is required.

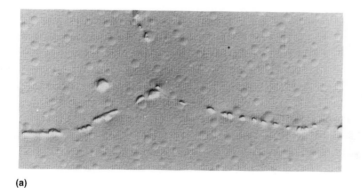
(a)

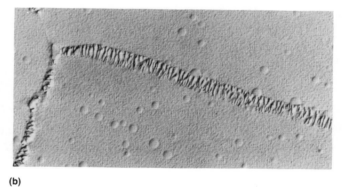
(b)

FIGURE 3-81. Varied morphologies of $M_{23}C_6$ in Waspaloy. Heat treated at 1865 °F (1018 °C), 4 hr, oil quenched; 1550 °F (845 °C), 4 h, air cooled; 1400 °F (760 °C), 16 h, air cooled. (a) Spheroidized grain boundary. (b) Cellular grain boundary. Source: Ref. 30.

(a)

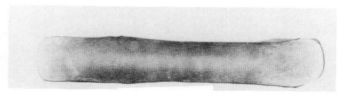
(b)

FIGURE 3-82. Macrosections at 1× of forged IN-100 powder compacts. (a) With intermediate 2250 °F (1230 °C) anneal. (b) Without intermediate anneal. Source: Ref. 30.

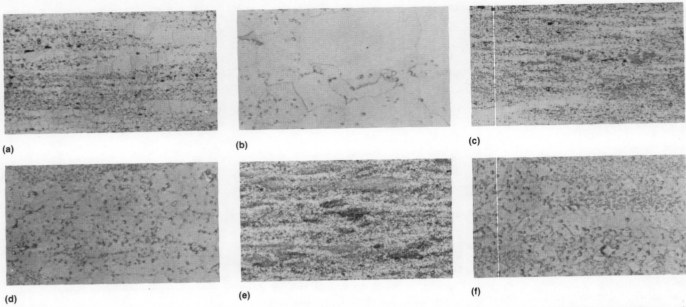

FIGURE 3-83. Sensitivity of René 95 structure to in-process annealing. (a) 2090 °F (1145 °C) anneal. 75×. (b) 2090 °F (1145 °C) anneal. 350×. (c) 2075 °F (1135 °C) anneal. 75×. (d) 2075 °F (1135 °C) anneal. 350×. (e) 2060 °F (1125 °C) anneal. 75× (f) 2060 °F (1125 °C) anneal. 350×. Source: Ref. 30.

TABLE 3-18. Effects of Restriking on Tensile Properties for René 95

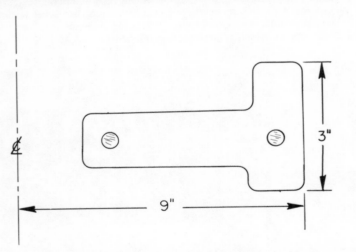

Location	Restrike	0.2% yield strength, ksi	Ultimate tensile strength, ksi	Elongation, %	Reduction in area, %
Room-temperature tensile					
Hub tangential	No	190.0	234.0	13.0	15.2
Hub tangential	Yes	193.6	224.8	11.5	13.7
Rim tangential	No	178.4	228.8	11.0	12.2
Rim tangential	Yes	177.8	224.6	13.5	13.7
1200 °F (760 °C) tensile					
Hub tangential	No	174.0	223.0	15.0	18.1
Hub tangential	Yes	175.4	213.2	10.0	10.8
Rim tangential	No	161.6	216.6	10.0	15.2
Rim tangential	Yes	166.0	213.0	12.0	14.4

Source: Ref. 30.

Warm working in the $(\gamma + \gamma')$ or $(\gamma + \gamma'')$ phase fields as a thermomechanical processing operation, followed by direct aging, can markedly improve properties as a result of (1) a more uniform carbide precipitation on dislocations, (2) the presence of dislocation networks on the scale of the γ' interparticle spacing, and (3) enhanced nucleation of fine γ' arising from an enhanced vacancy concentration. Some warm-worked structures also show a recrystallized band of small grains at the deformed grain boundaries (Figure 3-84) whose growth is restricted by the larger γ' particles—the so-called "necklace" structure. In René 95, for example, such a structure gives better high-temperature properties than those provided by fine-grained material.

The oxide dispersions in mechanically strengthened alloys are also very effective in pinning grain boundaries during hot- or warm-working operations and prevent the rapid grain growth that may otherwise occur. The resultant grains also tend to be elongated along the direction of deformation, which provides an increased creep strength in that direction.

A small amount of work has been carried out on thermomechanical processing of superalloys by shock deformation; as with warm working, improvements in properties occur as the result of the fine dislocation substructures that are induced. In Inconel 718, for example, the interaction between preexisting γ'' and disloca-

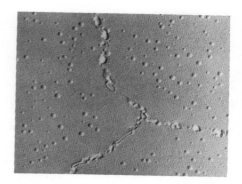

FIGURE 3-84. Microstructure of forged and heat-treated Waspaloy. Alloy was finish forged at 1760 °F (960 °C). TEM replica shows coarse and fine γ′ throughout the structure. Note the fine, recrystallized ("necklace") grain structure at the grain boundaries. Source: Ref. 31.

tion substructure gives rise to a more homogeneous γ″ particle distribution during post-deformation aging.

A final potential advantage of these thermomechanical processing treatments is that they may allow the precipitation of TCP and other phases in forms that are not deleterious to the mechanical properties. This is already being realized to a certain extent in nickel/iron-base superalloys where warm working below the η or δ solvus temperatures leads to the precipitation of these phases in a globular form that is effective in pinning grain boundaries and thus leads to grain refinement. Similar considerations apply to nickel-base super alloys where the effects of TCP compounds must be correlated with their morphology (globular or platelike), location (inter- or intragranular), and volume fraction. Again, these variables may be controlled by mechanical working such that the precipitation of TCP compounds has a beneficial, rather than a deterimental, effect on superalloy properties.

- Forging pressure increases with increasing reduction.
- Forgeability is not influenced greatly by deformation rate.
- Die-chilling effects place practical limits on section sizes obtainable in impression die forgings.

The only significant difference between the two systems is their oxidation behavior, which influences heating, lubricating, and handling procedures.

Forging Design Considerations. Because the metallurgical quality of molybdenum and molybdenum alloy forgings depends greatly on the control of forging reductions, the designs are more critical than those for aluminum or steel forgings. The designs are more generously contoured to provide both smooth metal flow and more readily controlled reductions.

Most die forging experience has been with circular shapes (cones, rings, disks) varying in size up to about 16-in. diameters, but structural shapes, blades, and shafts have also been successfully forged. A typical molybdenum impression die forging is shown in Figure 3-85.

Forging Refractory Metals

Molybdenum and Molybdenum Alloys

Nominal compositions and forging characteristics of molybdenum and several molybdenum alloys are given in Table 3-19. As a rule, molybdenum alloys can be forged over a wide range of temperatures. Because the alloys have higher recrystallization temperatures than unalloyed molybdenum, the forging temperatures are usually higher.

Molybdenum alloys are prepared commercially by two basic methods: vacuum consumable-electrode arc melting, and pressing and sintering. Pressed and sintered billets are forged directly.

Forging Behavior. Because of their coarse, radially oriented grains, arc-melted ingots are generally brittle when subjected to tensile deformation. For this reason, ingots are normally extruded at warm-working temperatures with at least a 2:1 ratio. The most common practice for achieving good forgeability consists of extrusion at hot/cold-working temperatures with at least a 4:1 extrusion ratio, and recrystallization of the extruded bar to refine the grain size. Wrought molybdenum and molybdenum alloy billets wtih a fine grain structure are forged successfully with large reductions over a wide range of temperatures and deformation rates.

Effect of Forging Variables on Mechanical Properties. The strength of molybdenum alloy forgings is developed primarily by work hardening. Thus, control of mechanical properties requires carefully planned forging sequences. These can be readily established because molybdenum exhibits classical strain-hardening behavior; i.e., hardness increases with increasing reduction and decreases with increasing forging temperature.

There are limits, however, to which these alloys can be cold worked to achieve a desired level of elevated-temperature properties. The amount of cold work influences the recrystallization behavior of the alloys, which in turn establishes the maximum service temperatures. At a given forging temperature, the recrystallization temperature generally decreases with increasing reduction. Conversely, at a given level of reduction, the recrystallization temperature generally increases with increasing forging temperature.

Forging Practices. The metallurgical principles involved in the forging of molybdenum are similar to those for the work-hardenable austenitic stainless steels and superalloys (e.g., 16-25-6, 19-9-DL). Principles common to both systems are:

- Cold work is necessary to increase strength.
- Recrystallization temperatures decrease with increasing cold work.
- Forgeability increases with decreasing grain size.

Columbium and Columbium Alloys

Nominal compositions and forging characteristics of columbium and several columbium alloys are given in Table 3-20. Relatively little forging has been done on columbium alloys. Most of the impression die forging experience has been with unalloyed columbium and the Cb-1Zr alloy.

Limited experience indicates that columbium and several of its alloys (such as Cb-1Zr and Cb-33Ta-Zr) are readily cold workable and may be forged directly from as-cast ingot. The more highly alloyed compositions containing tungsten or molybdenum (such as Cb-15W-5Mo-1Zr) generally require hot breakdown fabrication by extrusion to overcome undesirable features of the cast structure and to ensure adequate forgeability. After initial breakdown, most of these alloys are ductile below room temperature. This has been evidenced by the ability to roll many of the alloys with up to 90% reduction at or near room temperature. Columbium and its alloys are generally prepared commercially in the form of cylindrical cast ingots by consumable-electrode, vacuum-arc, or electron-beam melting.

Forging Behavior. Primary working (i.e., cast ingot breakdown) of the high-strength columbium alloys requires temperatures from about 2000 to 3000 °F (1100 to 1650 °C). Secondary working opera-

TABLE 3-19. Nominal Compositions and Forging Characteristics of Molybdenum and Several Molybdenum Alloys

Nominal composition, wt %	Unalloyed molybdenum	Mo-0.5Ti	Mo-0.5Ti-0.08Zr	Mo-30W
Consolidation methods(a)	AC, PM	AC	AC	AC, PM
Approximate solidus, °F (°C)	4730 (2610)	4700 (2600)	4700 (2600)	4800 (2650)
Approximate minimum recrystallization temperature, °F (°C)	2100 (1150)	2400 (1315)	2600 (1425)	2300 (1260)
Minimum hot working temperature(b), °F (°C)	2400 (1315)	2700 (1480)	3000 (1650)	2500 (1370)
Normal forging temperature range, °F (°C)	1900–2400 (1040–1315)	2100–2600 (1150–1425)	2200–2700 (1200–1480)	2100–2400 (1150–1315)
Lowest reported forging temperature, °F (°C)	1100 (590)	1800 (980)	1200 (650)	2200 (1200)
Forgeability:				
Presses	Good	Fair	Good	...
Hammers	Good	Good	Good	Fair
High-energy-rate machines	Good	...	Good	...

(a) AC, consumable electrode arc-cast; PM, powder metallurgy methods. (b) Minimum hot-working temperature is defined as the lowest temperature at which the alloy will begin to recrystallize during forging. Below these temperatures, the alloys are cold worked.

FIGURE 3-85. Typical impression die forging of molybdenum showing generously contoured circular design. The forging has a 15-in. O. D. and weighs 37 lb.

tions such as impression die forging and rolling can then be carried out at temperatures from about 2500 °F (1370 °C) to as low as 1000 °F (535 °C). Because most of the alloys can withstand large amounts of hot/cold work without intermediate annealing, once in the wrought condition, it should be possible to select forging temperatures that represent the best compromise between forgeability and forging pressure requirements.

Tantalum and Tantalum Alloys

Table 3-21 gives the nominal compositions and forging characteristics of tantalum and a tantalum alloy that have been successfully forged. Applications for these compositions have been primarily in the form of sheet; thus, most of the forging experience has been concerned with the conversion of ingots to sheet or bar. Die and ring forgings have been successfully produced on a limited scale from unalloyed tantalum and from the Ta-10W alloy.

Unalloyed tantalum and its alloys are prepared commercially by powder metallurgy, consumable-electrode arc melting, and electron-beam melting. Billets prepared by powder metallurgy techniques are not amenable to direct forging, but the materials listed in Table 3-21 can be directly forged from cast ingots.

Forging Behavior. To reduce the resistance to deformation and to overcome undesirable effects of the coarse-grained cast structure, initial breakdown forging operations are almost always performed at fairly high temperatures in the hot/cold-working range—generally between 2000 and 2400 °F (1100 and 1300 °C). After about 50% reduction, the forging temperatures for several of the alloys can be reduced below 2000 °F (1100 °C) with no difficulty.

The forging behavior of the Ta-10W alloy, for which there has been the most production experience, is representative of most of the single-phase tantalum alloys.

Interstitial impurities such as carbon, oxygen, and nitrogen have a deleterious effect on forgeability. The forgeability of the Ta-10W alloy is drastically reduced as the carbon level increases.

Tungsten and Tungsten Alloys

Table 3-22 gives the compositions and general characteristics of tungsten and two tungsten alloys. Tungsten-base materials, like the other refractory alloy systems, can be classified into two broad groups: unalloyed metal and solid-solution alloys (typified by additions of molybdenum or rhenium) and dispersion-strengthened alloys (typified by additions of thoria).

These classifications are convenient because they categorize the alloys not only according to their metallurgical behavior, but also by the applicable consolidation methods. Thus, while the solid-solution alloys and unalloyed tungsten can be produced by either powder metallurgy or

TABLE 3-20. Nominal Composition and Forging Characteristics of Columbium and Several Columbium Alloys

Nominal composition, wt%	Designation	Approximate solidus °F	Approximate solidus °C	Approximate minimum recrystallization temperature °F	Approximate minimum recrystallization temperature °C	Approximate minimum hot-working temperature(a) °F	Approximate minimum hot-working temperature(a) °C	Forging temperature, °F (°C) Normal range	Forging temperature, °F (°C) Lowest reported	Forgeability(b)
99.2Cb	Unalloyed	4475	2470	1900	1040	1500	815	RT-2000 (1100)	RT	Excellent
Cb-1Zr	FS-80	4350	2400	1900	1040	2100	1150	RT-2300 (1260)	RT	Excellent
Cb-10W-5Zr	CB-752	4400	2430	2200	1200	2500	1370	2200–2600 (1200–1430)	2200 (1200)	Moderate
Cb-10Ta-10W	SCb-291	4710	2600	2100	1150	2400	1315	1700–2200 (925–1200)	1600 (870)	Good

(a) Minimum hot-working temperature is defined as the lowest temperature at which the alloy will begin to recrystallize during forging. Below these temperatures, the alloys are cold worked. (b) Based on ingot breakdown and breakdown rolling experience.

TABLE 3-21. Nominal Compositions and Forging Characteristics of Tantalum and a Tantalum Alloy

Nominal composition, wt%	Approximate solidus °F	°C	Approximate minimum recrystallization temperature °F	°C	Approximate minimum hot-working temperature(a) °F	°C	Normal forging temperature range °F	°C	Forgeability(b)
99.8Ta	5425	3000	2000	1100	2400	1315	RT-2000	RT-1100	Excellent
Ta-10W	5495	3030	2400	1315	3000	1650	1800–2300	980–1260	Good

(a) Minimum hot-working temperature is defined as the lowest forging temperature at which the alloys begin to recrystallize during forging. Below these temperatures, the alloys are cold worked. (b) Based on ingot breakdown and breakdown rolling experience.

TABLE 3-22. General Characteristics of Tungsten and Tungsten Alloys

Nominal composition, %	Consolidation methods(a)	Approximate solidus °F	°C	Typical forging temperature range °F	°C	Typical recrystallization temperature °F	°C	Typical stress relief temperature °F	°C
Unalloyed	PS, AC, EB, PAS	6170	3410	2200–3000	1200–1650	2500–2900	1315–1600	2200–2400	1200–1315
W-2ThO$_2$	PS	6170	3410	2400–2500	1315–1370	3000–3200	1650–1760	2200–2400	1200–1315
W-15Mo	AC, PS	5970	3300	2000–2500	1100–1370	2700–2900	1480–1600	2300–2500	1260–1370

(a) PS, pressed and sintered; AC, arc cast; EB, electron beam melted; PAS, plasma arc sprayed and sintered.

melting techniques, forging billets of ThO$_2$ dispersion-strengthened alloys can be produced only by powder metallurgy methods.

Forging Behavior. Like molybdenum, the forgeability of tungsten and tungsten alloys is directly related to billet density, grain size, and interstitial content, and therefore to the method of billet production. The general influences of billet processing history on the forgeability of tungsten materials are summarized in Table 3-23.

TABLE 3-23. Influence of Billet History on Forgeability of Tungsten and Tungsten Alloys

Billet condition	Comments on forgeability
Pressed and sintered	Billets up to about 6 in. in diameter can be forged in any direction; as billet centers become weaker with increasing diameters, larger billets usually must be extruded or side forged before they can be upset by forging
Slip cast and sintered	Not forgeable
Arc cast	Usually not forgeable
Arc cast and extruded	Forgeability increases with increasing extrusion ratio; a ratio of 4:1 is approximately minimum for reasonable forgeability; can be both side and upset forged
Electron beam melted	Limited data indicate that as-cast tungsten can be directly forged if grain size is fine; otherwise, ingots are usually not forgeable

Tungsten and several W-Mo alloys have been successfully forged over a wide range of temperatures, from as low as 1800 °F (980 °C) to 3500 °F (1925 °C). Temperatures in the vicinity of 3000 °F (1650 °C) are favored for initial breakdown of pressed and sintered billets, but finish forging is usually done at temperatures ranging from 2600 °F (1425 °C) downward to 2000 °F (1100 °C). In rocket nozzle forging, a billet height reduction of approximately 80% at forging temperatures below the critical recrystallization temperature is considered essential. This imparts the fine-grained, fully wrought, fibrous structure that is characteristic of high strength and good ductility at low transition temperatures.

Forging Practices. Metallurgical principles in the forging of tungsten are much the same as those for molybdenum. Tungsten is generally forged in the hot/cold-working temperature range, where hardness and strength increase with increasing reductions. Both systems exhibit increasing forgeability with decreasing grain size.

Tungsten requires considerably higher forging pressures, which tax the stress limits of most forging die materials. For this reason, it is often necessary to use in-process recrystallization annealing treatments to reduce load requirements for subsequent forging steps. The need for lateral support during forging is greater for tungsten than for molybdenum, and the design of preliminary forging tools is more critical. This is particularly true for pressed and sintered billet stock, which has some porosity and less than theoretical density.

Forging Design Considerations. The greatest use for tungsten forgings has been in rocket nozzle applications. Production forging of tungsten nozzle inserts has been largely limited to billets prepared by powder metallurgy techniques because of their earlier availability than cast tungsten.

The impression die forging of tungsten was originally confined to generously contoured, solid shapes. Even slightly hollowed cones required back extrusion, which often led to cracking. Thus, size capability was restricted to short, conical shapes in the vicinity of 10 in. in diameter and ring shapes up to about 15 in. in diameter. This problem has been largely overcome by improved billet uniformity and the development of improved forging practices. By careful die staging and combined methods of fabrication such as upsetting, back extrusion, and ring rolling, size capabililty has been increased and thinner sections with improved tolerances and surface finish are possible.

Beryllium Grades

Two common grades of beryllium, their basic compositions, and the forms in which they are available for forging are listed in Table 3-24. These beryllium grades, which contain BeO in nominal amounts ranging from about 0.8 to about 4.3 wt%, may be considered as alloys of beryllium, in that they harden and possess other characteristics of alloys. Otherwise, the nearest approach to a commercial beryllium-base alloy system would be beryllium-aluminum alloys containing from about 24 to 43% aluminum.

Beryllium exhibits many desirable en-

TABLE 3-24. Compositions of Several Grades of Beryllium Available for Forging

Grade	Nominal composition, wt%									Forms available(a)
	Be, min	BeO, max	Al, max	C, max	Fe, max	Si, max	Mg, max	Others, max		
Nuclear and structural	98.0	2.0	0.16	0.15	0.18	0.08	0.08	0.04		P, B
Instrument	92.0	4.25	0.20	0.50	0.50	0.15	0.10	0.10		B

(a) P, power, B, hot-pressed block.

gineering characteristics: high strength, high modulus of elasticity, low density, and oxidation resistance up to about 1400 °F (760 °C). However, the metal has seen only limited applications because it is expensive, brittle, and toxic.

Beryllium billet stock can be produced either by consumable-electrode, vacuum-arc melting, or vacuum hot pressing of powder. Because hot-pressed block is more readily workable than cast ingot, it has been the principal form of billet stock used for forging. Beryllium bock in sizes up to 6 ft in diameter and weighing up to about 5 tons has been produced. Billet manufacture can be bypassed by canning loose beryllium powder in evacuated steel containers and hot forging to consolidate the powders during deformation. This technique has been used to produce a variety of forged parts weighing from about one pound to over a ton.

Forging Behavior. In many respects, beryllium behaves like magnesium during plastic deformation. Both metals are characterized by a close-packed hexagonal crystalline structure, which imparts low ductility at room temperature. The plasticity of beryllium is dependent on grain size (brittle when coarse, more ductile when fine). Because it is strain-rate-sensitive, it exhibits better forgeability in presses than in hammers. Also, like magnesium, beryllium exhibits strong anisotropic flow characteristics. Essentially, this means that once flow has begun in one direction, extensive plastic flow in that direction can occur. This increasing plasticity in one direction is usually accompanied by reduced plasticity in other directions.

Because beryllium recrystallizes in the vicinity of 1400 °F (760 °C), it exhibits hot-working behavior above and cold-working behavior below this temperature. The similarities between magnesium and beryllium end when their ductililties in the hot-working range (above their respective recrystallization temperatures) are compared. Beryllium does not exhibit the sharp rise in ductility that magnesium does.

Forging Practices. Figure 3-86 shows an example of the canned-powder billet shape, forged conventionally at 1900 °F (1040 °C), which is the relative size for producing a forging (a flat-bottom cup). Vacuum hot-pressed beryllium billets can be readily forged in impression dies, provided that a positive method for keeping the billet in compression is employed to avoid cracking of the material from the high tensile stresses normally induced during forging.

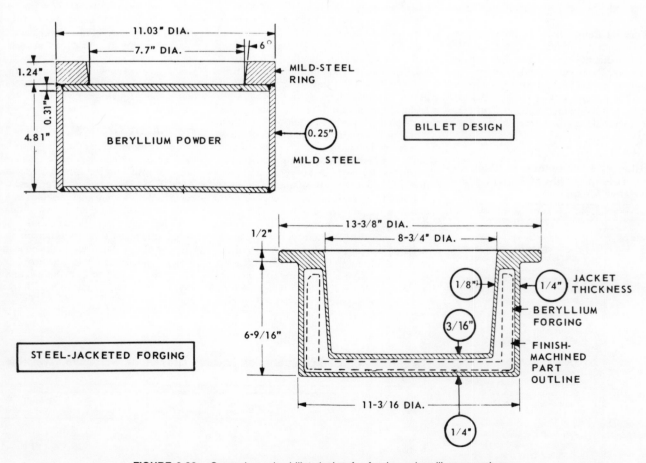

FIGURE 3-86. Canned-powder billet design for forging a beryllium cup shape.

References and Bibliography

12. Semiatin, S.L., Workability in Forging, in *Workability Test Techniques*, G.E. Dieter, Ed., American Society for Metals, 1983.

13. Sabroff, A.M., Boulger, F.W., and Henning, H.J., *Forging Materials and Practices*, Rheinhold Book Corp., 1968.

14. *Aluminum Forging Design Manual*, 1st ed., Forging Division, Aluminum Association, New York, Nov. 1967.

15. Henning, H.J., Sabroff, A.M., and Boulger, F.W., "A Study of Forging Variables," Technical Documentary Report No. ML-TDR-64-95, Contract No. AF 33(600)-42963, Battelle Memorial Institute, Columbus, OH, March 1964.

16. Sabroff, A.M., Boulger, F.W., Henning, H.J., and Spretnak, J.W., *Fundamentals of Forging Practice*, Supplement to Technical Documentary Report No. ML-TDR-64-95, Contract No. AF 33(600)-42963. Battelle Memorial Institute, Columbus, OH, March, 1965.

17. Colton, R.M., "The Effects of Fabrication Procedure on the Properties of Ti-6Al-6V-2Sn-0.5Fe-0.25Cu," Watertown Arsenal Laboratories Monograph Series, Sept. 14–15, 1959.

18. Hamer, J.E., "Investigation to Develop Optimum Properties in Forged Ti-7Al-4Mo," WADD TR 60-489, Crucible Steel Company, Final Report on Contract AF 33 (616)-6122, Oct. 1960.

19. Henning, H.J. and Frost, P.D., "Titanium-Alloy Forgings," DMIC Report No. 141, Defense Metals Information Center, Battelle Memorial Institute, Columbus, OH, Dec. 1960.

20. Wilkinson, N.A., Technological Considerations in the Forging of Superalloy Rotor Parts, in *Forging and Properties of Aerospace Materials*, Book 188, The Metals Society, London, 1978.

21. Altan, T., Boulger, F.W., Becker, J.R., Akgerman, N., and Henning, H.J., *Forging Equipment, Materials, and Practices*, MCIC-HB-03, Metals and Ceramics Information Center, Battelle's Columbus Laboratories, Columbus, OH, 1973.

22. Cremisio, R.S. and McQueen, N.J., Some Observations of Hot Working Behavior of Superalloys According to Various Types of Hot Workability Tests, in *Superalloys—Processing*, Proceedings of the 2nd International Conference, MCIC-72-10, Metals and Ceramics Information Center, Battelle's Columbus Laboratories, Columbus, OH, 1972.

23. "Manufacture of Large A-286 Turbine Disk," Kobe Steel Internal Report.

24. Yamaguchi, S., *et al.*, Effect of Minor Elements on Hot Workability of Nickel Base Superalloys, *Metals Technology*, Vol. 6, May 1979, p. 170.

25. Weiss, B., Grotke, G.E., and Stickler, R., Physical Metallurgy of Hot Ductility Testing, *Welding Research Supplement*, Vol. 49, Oct. 1970, p. 471-s.

26. Beiber, A.L., Lake, B.L., and Smith, D.F., A Hot Working Coefficient for Nickel Base Alloys, *Metals Engineering Quarterly*, Vol. 16 (No. 2), May 1976, pp. 30–39.

27. Savage, W.F., Apparatus for Studying the Effects of Rapid Thermal Cycles and High Strain Rates on the Elevated Temperature Behavior of Materials, *Journal of Applied Polymer Science*, Vol. VI, Issue 21, 1962, p. 303.

28. Owczarski, W.A., *et al.*, A Model for Heat Affected Zone Cracking in Nickel Base Superalloys, *Welding Journal* (Suppl.), Vol. 45, April 1966, p. 145-s.

29. "Manufacture of Large Waspaloy Turbine Disk," Kobe Steel Internal Report.

30. Couts, W.H. and Coyne, J.E., TMP: Its Effect on Turbine Hardware, in *Superalloys—Processing*, Proceedings of the 2nd International Conference, MCIC-72-10, Metals and Ceramics Information Center, Battelle's Columbus Laboratories, Columbus, OH, 1972.

31. Bailey, R.E., Some Effects of Hot Working Practice on Waspaloy's Structure and Tensile Properties, in *Superalloys—Processing*, Section J, Report MCIC-72-10, Metals and Ceramics Information Center, Battelle's Columbus Laboratories, Columbus, OH, 1972.

- Sellars, C.M. and Tegart, W.J.M., Hot Workability, *International Metallurgical Reviews*, Vol. 17, 1972.
- Capeletti, T.L., Jackman, L.A., and Childs, W.J., Recrystallization Following Hot-Working of a High-Strength Low-Alloy (HSLA) Steel and a 304 Stainless Steel at the Temperature of Deformation, *Metallurgical Transactions*, Vol. 3, 1972.
- Luton, M.J. and Sellars, C.M., Dynamic Recrystallization in Nickel and Nickel-Iron Alloys During High Temperature Deformation, *Acta Metallurgica*, Vol. 17, Aug. 1969.
- Donachie, M.J., Pinkowish, A.A., Danesi, W.P., Radavich, J.F., and Couts, W.H., Jr., Effect of Hot Work on the Properties of Waspaloy, *Metallurgical Transactions*, Vol. 1, Sept. 1970.
- Gladman, T., McIvor, I.D., and Pickering, F.B., Effect of Carbide and Nitride Particles on the Recrystallization of Ferrite, *Journal of the Iron Steel Institute*, Vol. 209, May 1971.
- Speight, M.V. and Healey, T., Effects of Temperature and Alloy Composition on the Coarsening of Metal Carbide Particle, *Metal Science Journal*, Vol. 6, 1972.
- Oblak, J.M. and Owczarski, W.A., Cellular Recrystallization in a Nickel-Base Superalloy, *Transactions of the Metallurgical Society of AIME*, Vol. 242, Aug. 1968.
- VanDerMolen, E.H., Oblak, J.M., and Kriege, O.H., "Control of γ′ Particle Size and Volume Fraction in the High Temperature Superalloy Udimet 700, *Metallurgical Transactions*, Vol. 2, June 1971.
- Bridge, J.E., Jr. and Maniar, G.N., Effects of Molybdenum on the Diffusion of Aluminum and Titanium from Ni_3 (Ti, Al) Intermetallic Compounds into a Ni-Mo Matrix, *Metallurgical Transactions*, Vol. 3, April 1972.
- Hammond, C. and Nutting, J., The Physical Metallurgy of Superalloys and Titanium Alloys, *Forging & Properties of Aerospace Materials*, Book 188, The Metals Society, London, 1978.
- Couts, W.H., *The Superalloys*, John Wiley & Sons, New York, 1972.
- Edington, J.W., *Metals Technology*, Vol. 3, 1976.
- Moore, J.B. and Athey, R.L., *Design News*, Vol. 5, 1970.
- Knott, A.R. and Symonds, C.H., *Metals Technology*, Vol. 3, 1976.
- Oblak, J.M. and Owczarski, W.A., *Metallurgical Transactions*, Vol. 3, 1972.

- Shamblen, C.E., Allen, R.F., and Walker, F.E., *Metallurgical Transactions*, Vol. 6A, 1975.

- Benjamin, J.S., *Metallurgical Transactions*, Vol. 1, 1970.

- Meyers, M.A. and Orava, R.N., *Metallurgical Transactions*, Vol. 6A, 1976.

- Moll, H.J., Maniar, G.N., and Muzyka, D.R., *Metallurgical Transactions*, Vol. 2, 1971.

- *Forging Industry Handbook*, Forging Industry Association, Cleveland, 1966.

- Daykin, R.P. and DeRidder, A.J., Primary Working of Superalloys, *Superalloys—Processing*, Proceedings of the 2nd International Conference, MCIC-72-10, Metals and Ceramics Information Center, Battelle's Columbus Laboratories, Columbus, OH, 1972.

- Sakakibara, M. and Sekino, S., Hot Workability of Inconel 600 and Hastelloy X, *Superalloys—Processing*, Proceedings of the 2nd International Conference, MCIC-72-10, Metals and Ceramics Information Center, Battelle's Columbus Laboratories, Columbus, OH, 1972.

- Schey, J.A., *Introduction to Manufacturing Processes*, McGraw-Hill, New York.

- McQueen, H.J. and Fulop, S., Hot Torsion Testing of Waspaloy, in *Forging and Properties of Aerospace Materials*, Book 188, The Metals Society, London, 1978.

- Fulop, S. and McQueen, H. J., Mechanisms of Deformation in Hot Working of Nickel-Base Superalloys, in *Superalloys—Processing*, Proceedings of the 2nd International Conference, MCIC-72-10, Metals and Ceramics Information Center, Battelle's Columbus Laboratories, Columbus, OH, 1972.

4 MANUFACTURE OF FORGINGS

Editor: Thomas G. Byrer, Manager,
Metalworking Section,
Battelle's Columbus Laboratories,
Columbus, Ohio

Associate Editors: Dr. S.L. Semiatin,
Principal Research Scientist,
Battelle's Columbus Laboratories, Columbus, Ohio;
and Donald C. Vollmer, Researcher,
Battelle's Columbus Laboratories, Columbus, Ohio

Introduction

The manufacture of forged products is fundamentally a process of forming metal, under impact or pressure, to economically produce a desired shape with improved mechanical properties. This controlled plastic deformation, whether performed hot, warm, or cold, is accomplished by several forging methods that are all basically related to hammering and pressing.

The particular forging method and equipment used in a given instance is dependent on factors such as the quantity of parts to be produced, the characteristics of the material, and the configuration to be forged. Forging, relative to other metalworking processes, results in metallurgically sound, uniform, and stable products that will have optimum properties as operating components after processing and assembly.

In North America, these important engineered products are made in about 300 plants that are located largely in the principal industrial centers of the United States and Canada. Plants vary in capability from those producing a modest quantity of forgings by a single forging method that are designed for the efficient handling of small orders, to those producing large quantities of forgings that are equipped to perform a wide variety of high-volume forging operations.

Large or small, the forge plant and its organization reflect the various operations into which the fundamental manufacturing steps are divided, as shown in the flowchart in Figure 4-1. This section discusses the operations necessary to produce forgings, including preliminary engineering, die sinking, stock preparation, heating for forging, typical forging sequence, flash removal (trimming), heat treating, cleaning, finishing, and inspection. Because of its widespread use in the industry, the forging hammer and the operations related to it have been chosen to illustrate the steps in producing forgings.

Preliminary Engineering

Forging is the fine blend of art and science, requiring many critical decisions by the forging engineer far in advance of production. Forging production starts at the forging engineer's drawing board, where tool design and processing methods are determined. It is almost mandatory that the forging engineer and the component designer consult regularly to arrive at the optimum forging design that will meet end-product requirements. This is particularly desirable in light of the current rapid development of new forging alloys with distinct forging characteristics, new forging techniques, development of new die materials, new methods of die sinking, new research in die lubricants, and varied types of equipment, thus allowing the forger to make larger and/or more complex parts.

The forging engineer must select the most efficient forging methods for this job, as well as the proper equipment size, and he must determine the ultimate design of the dies. Some of the more important factors influencing these decisions (and consequently forging costs) include:

- Part configuration and tolerances of the forged product
- Material and forging stock selection
- Applicable specifications
- Weight of the forged part
- Quantity to be produced
- Mechanical properties
- Availability of equipment
- Design of the forging dies

Part Configuration and Tolerances. The general shape of the proposed forging is dictated primarily by its function as an operating component. Ordinarily, the general configuration is established by the component designer, although the forging engineer frequently can suggest cost-sav-

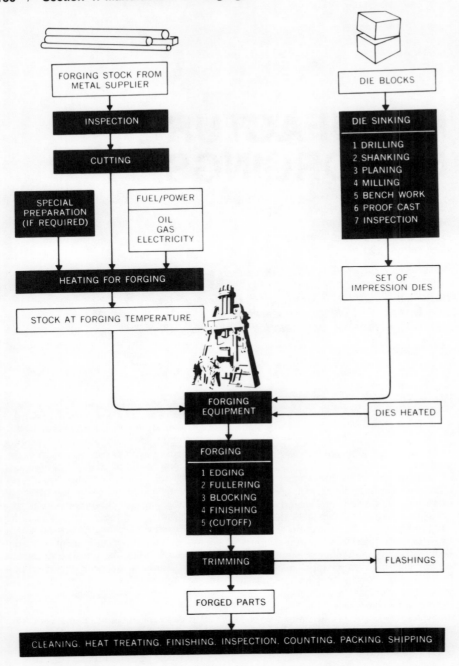

FIGURE 4-1. Flow chart of typical operations in the production of hot forgings.

ishing, and repair. Considerations such as these are taken into account by the forging engineer and are reflected in the cost estimate.

In addition, designs with small radii and restrictive tolerances can create higher die costs. Additional die machining time is almost always necessary, and in extreme cases, extra steps in the forging process must be contemplated.

Material and Stock Selection. Materials for forging include an extremely broad range of ferrous and nonferrous metals, superalloys, and refractory materials, many of which are available to the forgings producer in various forms such as rolled blooms, billets, bars or rods, cast or sintered ingots, extruded shapes, and metal powder. Selection of the exact grade and form of material for a particular forging application is primarily dependent on the physical and mechanical properties desired and usually is determined after consultation between component designer and forging engineer.

Strength and fatigue resistance of the material, response to heat treatment, machinability, and corrosion resistance are among the important attributes of the finished forging that must be taken into consideration at the time of material selection. The forging engineer must also be concerned with the suitability of the proposed material for the required forging operations and supplementary processing.

Although most metals are capable of being forged, they vary widely in degree of forgeability. A forged part that might be made with little difficulty in a single die from easily forged material may require different equipment or successive forging operations, or both, if produced from a less forgeable material.

Weight of the Forged Part. The volume of material to be contained in the proposed forging design is of critical importance to the forging engineer. The amount or weight of stock required to make a forging of a given forged weight can vary considerably, depending on variations in cross section of the part and the design of the tooling. The amount of raw material required in relation to the finished weight of the part will generally be greater as the cross section of the part varies and as the area between projections increases. This is due to greater losses in flash. Where conditions warrant, preshaping of the stock by upsetting, rolling, drawing, or fullering and the use of successive blocker dies can minimize stock loss due to excess flash and can increase die life. With the new developments in precision forging, start-

ing modifications, particularly where tolerances, draft angles, and corner and fillet radii are involved.

Unnecessarily tight tolerances, thin webs and ribs, small draft angles, and sharp corner and fillet radii result in needless manufacturing problems and high costs. Close straightness tolerances may cause an additional hand straightening or cold

restrike operation (requiring additional dies and possibly checking fixtures for large parts), and special flash-grinding operations may be necessary to meet restrictive flash extension tolerances. Tight tolerances can also shorten the life of forging dies, because dies commonly are designed to take advantage of the dimensional allowance to compensate for die wear, poling material weight is of critical significance.

Quantity. The forging method, size and type of equipment, and the number and sophistication of forging dies necessary for the job are governed to a considerable extent by the number of forgings to be produced. As quantities increase, it is usually practical to build increasingly more elaborate tooling. The increased efficiency of more elaborate dies and processing can result in the elimination of hand operations and requirements for subsequent machining. This can favorably affect the piece price to the extent that the higher tooling cost is more than offset, resulting in a net savings. It is necessary, of course, to evaluate each forging on the basis of specific cost and anticipated production volume. Early knowledge of the customer's anticipated annual quantity requirements greatly assists the forging engineer in planning tooling for optimum overall economy.

The mechanical properties in a finished forging and the suitability of these properties to the functional requirements of the part can be materially affected by factors such as the type of forging method, the amount of hammering and pressing used in producing the part, placement of the parting line, and orientation of the stock in the dies. Because the forging operation determines the direction of grain flow, dies are designed and operations are planned to provide forgings with specific directional characteristics. Special operations commonly are included to produce the desired grain flow characteristics when this cannot be accomplished in the course of normal production. In forgings that undergo severe deformation, requiring progressive operations in successive sets of dies, the grain structure of the original billet is usually sufficiently refined to ensure a uniform wrought structure throughout. However, in simple shapes in which metal movement is insufficient, it may be necessary to prework the stock between flat dies or blocker dies for the sole purpose of achieving the desired metallurgical structure.

Forging Die Design. The forging engineer draws on knowledge and skills gained through experience. He is familiar with the characteristics of metal flow, capabilities and capacities of available equipment, characteristics of die materials, die sinking equipment and methods, stock preparation, forging techniques, die lubrication requirements, and heat treating.

He must design the forging die to take advantage of the most efficient die sinking procedures for producing the desired part

configuration. At the same time, his planning determines the grain flow of the part and the refinement of the grain, and it is his responsibility to see that it will be properly positioned for maximum strength at points of greatest stress. Planning of the die impressions for the preliminary steps of edging, fullering, and blocking is particularly important, because proper preworking ensures that metal flow, under the pressure of the forging process, will proceed without disrupting the continuity of the fiber structure.

For forgings of modest size and complexity, die impressions for edging, fullering, and blocking, as well as the finishing impression, commonly are included in a single set of dies. However, for larger forgings and forgings of intricate design, two or more sets of dies may be required to perform the necessary operations on available equipment. In this event, two or more different pieces of equipment may be necessary, and additional heating of the forging stock between operations may be required. The sinking of forging dies can begin only when all factors affecting the quality of the forging and the economics of production have been considered and have been reduced to detailed drawings.

Stock Preparation

Rolled products, particularly bars and billets, which are the form of material used in the overwhelming majority of forging applications, will be used as representative examples to highlight the typical steps in stock preparation and forging. After forging stock has been received at the forge plant from the materials supplier and has been counted, weighed, and stacked, test samples are often sent to the laboratory for verification that the material meets specified metallurgical requirements. This test procedure may consist of chemical analyses, determination of physical and mechanical properties, and checks to ensure proper grain structure, fiber formation, and cleanliness. In addition, the surface of the stock is inspected to ensure that the material is free from laps, seams, blisters, and other imperfections that could interfere with quality forging.

Bars of stock commonly are received at the forge plant in lengths ranging from 10 to 20 ft and stock diameters that do not exceed $1\frac{1}{2}$ in. Forgings frequently are made directly from the end of the bar and are severed from the bar by a cutoff blade in the dies after the forging is completed. In these instances, the original bar of stock need only be cut to convenient lengths of 6 to 8 ft, so that they can be accommo-

dated in the working space around the forging equipment.

In many situations, however, and whenever larger diameter stock is involved, bars are cut into precise segments called "multiples," which contain the volume of material necessary to produce the forging. Individual multiples are necessary when stock must be placed on end between the dies and compressed to produce radial grain flow—which is desired for many forgings, including certain types of gear blanks. In some forge plants, tong holds are then added by welding a short length of smaller bar to the multiple.

Forge shops may use one of several alternative methods for cutting stock, depending on size and hardness of the material and required surface condition of the cut ends. Carbon and low-alloy stock up to 6 in. square having a hardness no higher than 250 HB usually may be sheared quickly and efficiently at room temperature, with the exception that some materials are inherently unsuited to cold shearing regardless of their hardness. When stock hardness is too high and for materials that are not suited for cold shearing, the bar may be heated to increase its plasticity and may be sheared hot. Figure 4-2 shows an assortment of sheared products.

If shearing is not suitable, bars, billets, and blooms may be cut into multiples with power hacksaws, automatic circular sawing machines, band saws, or abrasive wheel cutoff machines. In certain critical applications, burrs left from the sawing operation are removed from either or both ends of the cut stock to ensure a forging with less likelihood of defects. A special radius machine may be used to form a constant radius on the end of the multiple, but small quantities usually are deburred on a grinding wheel or disc.

Heating for Forging

Forging operations such as heading, extrusion, and swaging are being performed at room temperature with increasing regularity as advances in forging technology permit. Cold forging methods are also being adapted to other more complex forging operations. In these cases, reduced metal plasticity is accepted and overcome to obtain other advantages. More conventionally, however, materials to be forged are heated within the limitations of metallurgical requirements to temperature ranges in which optimum plasticity occurs.

Wide variations exist between forging temperature ranges for various materials. For example, copper and its alloys can be forged cold, but commonly are forged at

FIGURE 4-2. An assortment of sheared products. Courtesy of Advanced Machine Design Co.

FIGURE 4-3. Heating stock for forging in a slot-type forge furnace. Courtesy of Harris-Thomas Drop Forge Co.

in stock, availability of various fuels, and the several types of forging equipment in use, produce a substantial variety of equipment and procedures that can be used to heat stock for forging in impression dies. In every case, properly controlled heating is necessary to ensure that forgings subsequently produced will develop structural integrity and mechanical properties to their full potential. Temperature and atmospheric conditions within the forge furnace can affect material structure and surface condition, respectively.

Heating for forging can be accomplished by means of electric or fuel-fired furnaces, by electrical induction or resistance processes, or by special gas burner techniques (Figure 4-3). Regardless of the method of heating, precautions must be taken to prevent excessive scaling, decarburization, burning, overheating, or rupturing of the forging stock.

Typical Forging Sequence

Usually, a new set of dies is released to the forging equipment for tryout only after die proofs of the finishing impression have been approved by the customer and the forging engineer and after a final check by both the tool and die making department and the inspection department has been made. As the dies are released, an operations sheet is issued by the production department. This form usually covers in detail the sequence of forging operations to be used, recommended stock size, number of pieces on the initial order, target dates for production, and perhaps requests for sample forgings. With the exact forging procedure spelled out on the operations sheet, every effort is made to simulate actual production conditions during tryout.

Die Tryout. Dies are installed in the forging equipment and adjusted until the match surfaces reveal perfect alignment. Before the tryout can begin, dies must be heated to prevent cracking and to facilitate proper metal flow during forging. In some forge plants, this is done in a die heating furnace prior to setting the dies in the hammer or press. In most cases, however, dies are warmed while in position in the forging equipment. This is usually accomplished by placing pieces of hot metal between the die faces or by surrounding the dies with gas burners. Dies are generally heated to a range of 250 to 400 °F (120 to 200 °C) and sometimes as high as 700 °F (370 °C), depending on the application.

With dies in position and heated, a multiple is delivered from the furnace, and actual tryout of the tooling begins. The first

700 to 850 °F (370 to 450 °C). Low-carbon and low-alloy steels, comprising the largest percentage of all forgings produced, usually are heated to a range of 2000 to 2300 °F (1100 to 1260 °C). Aluminum ordinarily is forged at 700 to 850 °F (370 to 450 °C); magnesium at approximately 600 to 700 °F (315 to 370 °C); stainless steels from 2000 to 2250 °F (1100 to

1230 °C); titanium and its alloys from 1350 to 1950 °F (730 to 1065 °C); most nickel-base superalloys at 1900 to 2100 °F (1040 to 1150 °C); columbium, molybdenum, and tantalum between 2100 and 2500 °F (1150 and 1370 °C); and tungsten above 2500 °F (1370 °C).

Variations in forging temperature among different materials, as well as differences

FIGURE 4-4. Inspector carrying a sample "platter" of two connecting rods for careful inspection. Courtesy of Alcoa.

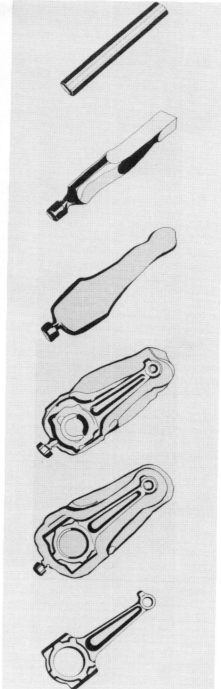

FIGURE 4-5. Sequence of steps in working metal to develop automobile connecting rods.

piece is forged in strict accordance with instructions on the operations sheet and is immediately checked for overall thickness, match, and general conformance to all applicable dimensional tolerances.

At this point, a member of the inspection department known as the "hot inspector" is on hand to check the part dimensions and to inspect it for possible defects or improperly filled contours. The slightest indication that the forging does not conform to print specifications, or that metal flow in any of the die cavities is not exactly as desired, means that corrective changes must be made before production can begin. Only by sampling during die tryout can the critical points of metal flow be studied and any visible defects evaluated and corrected, as shown in Figure 4-4.

If any corrections are necessary, additional samples are produced. Only when the inspector, forging engineer, and equipment operator are satisfied that all forging specifications are properly met is approval given to proceed with production.

Forging Sequence. The forging of automotive connecting rods in a forging hammer provides a graphic example of various steps (Figure 4-5) used in working metal through progressive stages in the development of a forging. The sequence begins with round bar stock. When the stock has reached the proper forging temperature, it is delivered to the hammer from the furnace.

Preliminary hot working properly proportions the metal for forming the connecting rod. The fullering and edging operations improve grain structure, reduce the cross-sectional area of the stock between the crank pin end and the piston end of the connecting rod, and gather the metal for other sections.

The blocking operation forms the connecting rod into its first definite shape. This involves hot working of the metal in several successive blows of the hammer, thus compelling the workpiece to flow into and fill the blocking impression in the dies. Flash is produced and appears as flat, unformed metal around the edge of the connecting rod. The exact shape of each connecting rod is obtained by the impact of several additional blows in the hammer that force the stock to completely fill every part of the finishing impression. The flash remains to be trimmed.

Flash Removal. Upon completion of the forging operation, flash may be removed from a forging in several ways—with trim dies in a mechanical press or, in special circumstances, by sawing or grinding. In most cases, particularly where forgings are produced in volume and for forgings that have a complex outline, special dies called "trim dies" are used to remove the flash mechanically. A set of trim dies consists of a sharp shearing edge, produced to the exact contour of the forging at the flash line, and a punch, also contoured to fit the forging. Such tools may be set up in a trim press near the forging equipment to handle forgings that are trimmed hot immediately after forging. Small forgings in particular frequently may be moved to a separate department in the plant and may be allowed to cool before trimming.

The trimmed connecting rod forgings are now ready for heat treating and machining. Subsequent machining and finishing are held to a minimum by forging in impression dies to close weight and dimensional tolerances.

Heat Treating

Many impression die forgings produced by hot forging methods are given some type of heat treatment after completion of final forging operations and before machining or end use. Proper heat treatment can produce a forging with optimum grain size, microstructure, and mechanical properties.

Depending on the customer's requirements, forgings may be shipped as forged, receive final heat treatment, or may only be annealed or normalized at the forge plant preparatory to rough machining and final heat treatment elsewhere. When heat treating facilities are available in forge plants, they normally include equipment for normalizing, annealing, hardening with either water or oil quench, and tempering. Batch-type and rotary and pusher-type continuous heat-treat furnaces are in common use. Heat treatment and its importance in meeting customer specifications is detailed elsewhere in this section.

Cleaning

Because of the high temperatures required for forging and heat-treating operations, forgings produced from most materials acquire a thin coating of hard, abrasive oxide commonly called scale. Scale is formed as a result of the chemical action between free oxygen in the furnace and the heated metal.

When subsequent machining, plating, painting, or coating operations are contemplated, it is generally necessary to remove scale before the forgings are used or further processed. This is usually accomplished by blast cleaning, tumbling, pickling, or, as with aluminum, caustic-nitric etch. Blast cleaning (often called shot blasting) is the most commonly used method.

Blast Cleaning. During blast cleaning, forgings are bombarded with high-velocity streams of steel or iron shot, hardened wire clips, aluminum oxide grit, or sand that is propelled by centrifugal force or compressed air. The particles abrade the surface of the forging, mechanically chipping, breaking, and dislodging the scale. Most blast cleaning machines are batch-type equipment and differ from one another principally in the method in which the forgings are positioned for blasting.

Forged materials with particularly hard, difficult-to-remove scale are cleaned as effectively as materials with loose, non-adherent scale by merely extending the period of time they are exposed to shot blasting. After blasting, forgings display a pleasing, dull finish that is well suited for subsequent inspection procedures and machining operations. Surfaces are clean and free from smudges or other products of chemical reaction and are therefore receptive to oiling, greasing, or application of other protective coatings. Because shot blasting removes the scale mechanically and without moisture, it avoids the danger of pitting by chemical action. Also, because forgings need not be tumbled violently in the process, their edges and corners are not damaged.

Tumbling is one of the oldest and simplest cleaning methods. Forgings are placed in a barrel apparatus together with abrasive and small bits of metal, and the barrel is rotated. The forgings, along with the abrasive, roll and tumble upon themselves, and the resulting abrasive action removes the scale.

With the tumbling process, a slight peening effect, which may be particularly desirable for some applications, often takes place on the surface of the forgings. Generally, tumbling is used on smaller forgings without extremely long or thin sections and where tolerances on corners and edges of the forging are not critical.

Pickling. Forgings cleaned by the pickling process are dipped into a heated acid bath in which scale is removed by chemical action. Organic chemicals known as "inhibitors" generally are used during pickling to retard the action of the acid on the metal beneath the scale. Forgings generally are bathed in an alkaline cleaning compound prior to pickling to remove oil or grease, which interferes with the action of the acid. After pickling, forgings are rinsed in clean, hot water to remove all traces of acid.

A hot caustic solution (sodium hydroxide) usually is used to clean aluminum forgings. After the forgings have been immersed in the caustic for the required period of time, the etching action of the solution is neutralized by bathing the forgings in nitric acid. A hot water rinse follows. The process is commonly called "caustic-nitric etch."

FIGURE 4-6. Dimensional accuracy of precision forged parts is checked in final inspection.

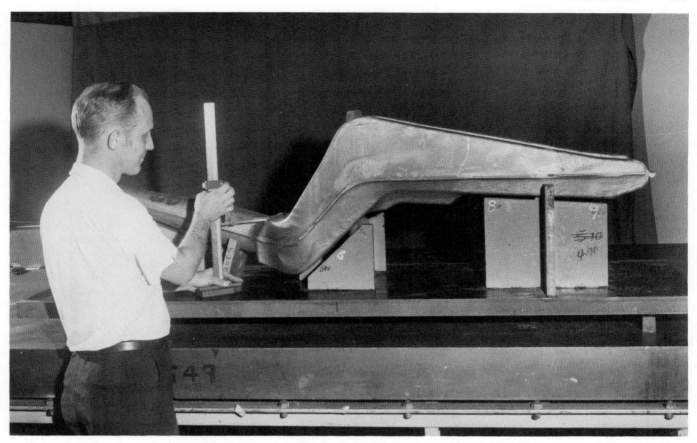

FIGURE 4-7. Final dimensional check on 882-lb flap carriage forging that actuates the high-lift inboard trailing-edge wing flaps of the Boeing 747. It is forged of consumable electrode vacuum remelted 300M alloy steel. Courtesy of Ladish Co.

Finishing Operations

When forgings have been heat treated and cleaned, they are near the end of the sequence of forging and related operations. The few steps that remain are performed cold and consist primarily of minor dimensional corrections. Collectively, these steps usually are referred to as finishing operations.

Coining. When forgings require closer tolerances than can be economically produced in the forging die, these restrictive tolerances can often be met by coining or sizing. Coining may be performed either hot or cold, depending on the nature of the requirements, and can be accomplished either before, during, or after heat treating. However, cold coining is often preferred, because material shrinkage is not then a factor and excellent surface finishes can be obtained. The coining operation is performed in a trim press or special coining press of the appropriate size. Although special dies can be designed for coining irregular shapes and contours, flat surfaces such as bosses are most suitable for coining and can be sized to tolerances within a few thousandths of an inch.

Straightening. During trimming, heat treating, cleaning, or handling, forgings occasionally become slightly warped, twisted, or bent. Parts frequently can be straightened manually and checked with a template, gauge, or fixture. On occasion, the straightening operation is mechanized, using a fixture designed for the purpose and combining the straightening operation with the sizing operation in a coin press. Cylindrical parts such as tubes, axles, and various types of shafts ordinarily are straightened in machine rolls when sufficient quantities are involved. Smaller quantities are often straightened by manual manipulation in a press.

Inspection

Inspection of forgings begins with the first forging produced and is carried out during all phases of the forging operation. "Hot inspection," which begins during die tryout, is continuous, and the inspector has authority to stop the work for necessary corrections if product standards are not being met.

At frequent intervals during the workday, forgings are taken at random from production, quickly cooled, cleaned, and subjected to detailed inspection and gauging of overall quality and critical dimensions. The general pattern of die wear is noted. This enables the inspection department to anticipate and prevent production of off-quality forgings.

Forgings receive a visual inspection for dimensional accuracy and surface condition after finishing operations and miscellaneous touch-up work, such as grinding, have been performed (Figures 4-6 and 4-7). A wide variety of additional inspection procedures may be used, as required, for checking surface condition and structural integrity, including dye penetrant testing, magnetic particle inspection, and ultrasonic tests. Depending on requirements, forgings are also tested for compliance with physical and mechanical property specifications. Inspection and other quality control procedures are discussed in a later section.

Forging Processes and Methods

Contributing Authors: Steve Wasco, Chief Manufacturing Development Engineer, Edgewater Steel Company, Oakmont, Pennsylvania; Al Orris, Engineering Manager, Aluminum Precision Products, Inc., Santa Ana, California; Dr. Pal Raghupathi, Associate Section Manager, Battelle's Columbus Laboratories, Columbus, Ohio; and Dr. S. L. Semiatin, Principal Research Scientist, Battelle's Columbus Laboratories, Columbus, Ohio

The means by which metal may be shaped by plastic deformation span a wide range of equipment and techniques. In practice, specific production methods commonly are classified according to the type of production equipment employed. Accordingly, basic forging techniques are discussed initially here, followed by a survey of the important forging methods.

Compression Between Flat Dies

Compression between flat dies is also called "upsetting" when an oblong workpiece is placed on end on a lower die and its height is reduced by the downward movement of the top die. Friction between the end faces of the workpiece and the dies is unavoidable. It prevents the free, lateral spread of the ends of the workpiece and results in a barrel shape, as shown in Figure 4-8. This operation is the basic type of deformation employed in flat die forging.

Compression Between Narrow Dies

Although upsetting between parallel flat dies is one of the most important basic forging operations, it is limited to deformation that is symmetrical around a vertical axis. If preferential elongation is desired, the dies must be made relatively narrow in a manner similar to Figure 4-9. When compressing a rectangular (or round) bar, frictional forces in the axial direction of the bar are smaller than in a direction perpendicular to it. Therefore, most of the material flow is axial, as shown in Figures 4-9(b) and (c). Because the width of the tool still presents frictional restraint, spread is unavoidable. The proportion of elongation-to-spread depends on the width of the die. A narrow die will elongate better, but die width cannot be reduced indefinitely, because cutting instead of elongation will result (Figure 4-9c).

The direction of material flow can be more positively influenced by using dies with specially shaped surfaces. As shown in Figure 4-10, material can be made to flow away from the center of the dies or to gather there. In the instance of "fullering," the vertical force of deformation always possesses (except for the exact center) a component that tends to displace material away from the center (Figure 4-10a). An "edging" or "roller" die has concave surfaces. Thus, it forces material to flow toward the center and the thickness there can be increased above that of the original bar dimensions (Figure 4-10b). "Drawing" is an operation similar to fullering, but is performed on the ends rather than between the ends of the bar. Drawing has the same purpose of reducing the cross section of the bar and increasing length.

The material flows in the direction of minimum resistance, and, even with shaped tools, it is inevitable that some of the material will flow sideways and increase the original width. To eliminate spread, the workpiece must be turned 90° and shaped again. A succession of such steps is a very effective means of reducing the cross section and is widely practiced for preforming the stock (drawing out).

Rolling

Compression between narrow dies (drawing out) is a discontinuous process, and if a reduction of thickness is desired over a long workpiece, a number of strokes must be executed while the workpiece is gradually moved in an axial direction. This task can be made continuous by rolling (Figure 4-11). The resemblance between Figure 4-9 and Figure 4-11 is apparent. The width of the die is now represented by the length of the arc of contact (L in Figure 4-11b), and the elongation achieved in a given part will depend on the length of this contract arc. Larger rolls cause greater lateral spread and less elongation because of the greater frictional difference presented in the arc of contact, whereas smaller rolls elongate more.

Impression Die Forging

In all cases discussed thus far, the material was free to move in at least one direction; therefore, all dimensions of the workpiece could not be closely controlled. Three-dimensional control requires impressions in the dies, and thus is derived the term "impression die" forging.

In the simplest example of impression die forging (Figure 4-12), a cylindrical (or rectangular) workpiece is placed in the bottom die (Figure 4-12a). The dies contain no provision for controlling the flow of excess material. As the two dies are brought together, the workpiece undergoes plastic deformation until its enlarged sides touch the side walls of the die impressions (Figure 4-12b). At this point, a small amount of material begins to flow outside the die impressions, forming what is known as "flash." In the further course of die approach, this flash is thinned gradually. As a consequence, it cools rapidly and presents increased resistance to deformation. In this sense, the flash becomes a part of the tool and helps to build up high pressure inside the bulk of the workpiece. This pressure can aid material

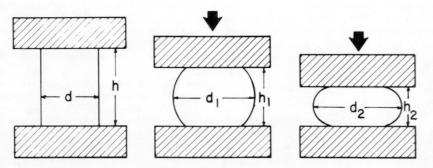

FIGURE 4-8. Schematic illustration of compression of cylindrical workpiece between flat dies showing barreling effect caused, in part, by friction on the interface. Source: Ref. 32.

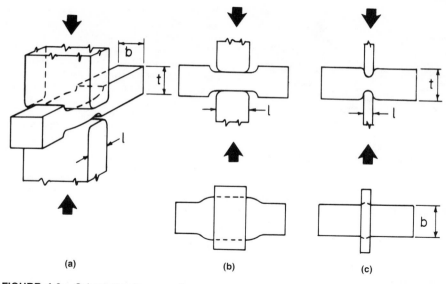

FIGURE 4-9. Schematic diagram of compression between flat, narrow dies. See text for discussion of (a) to (c). Source: Ref. 32.

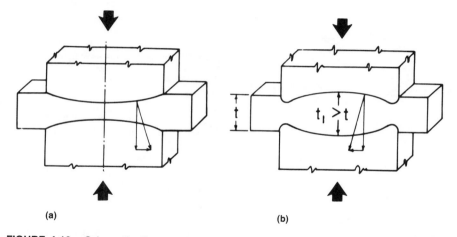

FIGURE 4-10. Schematic diagram showing compression between (a) convex and (b) concave die surfaces (fullering, and edging or rolling, respectively). Source: Ref. 32.

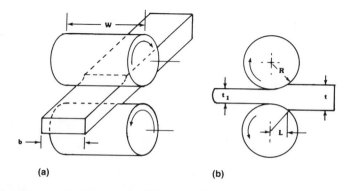

FIGURE 4-11. Rolling between cylindrical rolls. See text for discussion of (a) and (b).

flow into parts of the impressions previously unfilled so that, at the end of the stroke, the die impressions are nearly filled with the workpiece material (Figure 4-12c).

The process of flash formation is also illustrated in Figure 4-13. Flow conditions are very complex even in the case of the very simple shape shown. During compression, the material is first forced into the impression that will later form the hub, then the material begins to spread, and the excess begins to form the flash (Figure 4-13a). At this stage, the cavity is not yet completely filled, and success depends on whether a high enough stress can be built up in the flash to force the material (by now gradually cooling) into the more intricate details of the impressions. Dimensional control in the vertical direction is generally achieved by bringing opposing die faces together, as shown in Figure 4-13(b).

The flash presents a formidable barrier to flow on several counts. First, the gradual thinning of the flash causes an increase in the pressure required for its further deformation. Second, the relatively thin flash cools rapidly in intimate contact with the die faces, and its yield stress rises correspondingly. Third, the further deformation of the flash entails a widening of the already formed ring-like flash, which in itself requires great forces.

Thus, the flash can ensure complete filling even under unfavorable conditions, but only at the expense of extremely high die pressures in the flash area. Such dependence on excessive die pressures to achieve filling is usually undesirable, because it shortens die life and creates high power requirements. Forging design aims at easing metal flow by various means, making extremely high pressures in the flash superfluous. To reduce the pressure exerted on the die faces, a flash gutter usually is formed (Figure 4-13b) to accommodate the flash and to permit the dies to come together to control the vertical (or thickness) dimension.

The flash will be removed later in the trimming operation and thus represents a loss of material. Although flash is usually a by-product necessary for proper filling of the die cavity, too large a flash is not only wasteful, but also imposes very high stresses on the dies.

Impression die forging, as described above, accounts for the vast majority of all commercial forging production. Although the terms are often used interchangeably in this handbook and elsewhere, the method known as closed die forging is a special form of impression die forging that does not depend on the for-

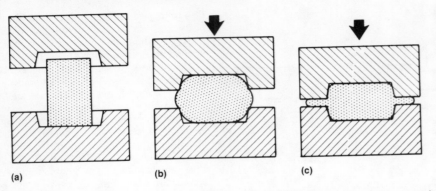

FIGURE 4-12. Schematic diagram of compression in simple impression dies without special provision for flash formation. See text for discussion of (a) to (c).

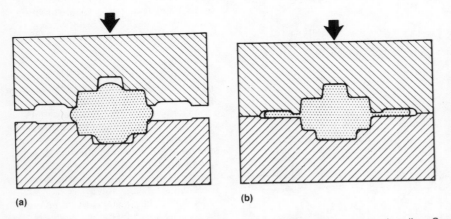

FIGURE 4-13. The formation of flash in a conventional flash gutter in impression dies. See text for discussion of (a) and (b).

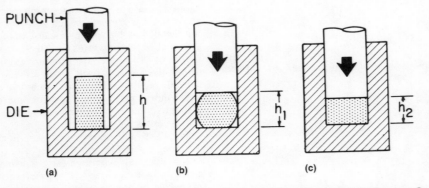

FIGURE 4-14. Compression in a totally enclosed impression: true closed die forging. See text for discussion of (a) to (c).

(Figure 4-15a). If provision is made to split the container into two die halves, it is possible to force undercut shapes (Figure 4-15b), or to upset a head on the end of a bar (Figure 4-15c), as is done on the horizontal forging machine (upsetter).

At first thought, true closed die forging would seem to present no difficulties as to metal flow. The metal has no choice; it must fill the die cavity irrespective of its shape or dimension. However, closed die forging is considerably more demanding with respect to die design than is impression die forging during which flash is formed. For example, in impression die forging on the hammer, some problems in material distribution may be rectified by taking more blows to finish the forging. There is seldom such a possibility in closed die forging, because the operation is ordinarily performed on forging machines (upsetters) and presses and is often planned for completion without failure in one stroke.

Closed die forging poses additional requirements. In the absence of flash formation, careful control of the volume of the workpiece is necessary to achieve complete filling of the forging without generating extreme pressures in the dies due to overfilling of the cavity. Another potential problem is the trapping of gas and lubricant. Vents in the dies are sometimes provided to prevent excessive pressure buildup and to permit complete filling of the impression. Other features of the process are discussed below in the section "Flashless Forging."

Extrusion

In extrusion, the workpiece is placed in a container and compressed by the movement of the ram until pressure inside the workpiece reaches the flow stress. At this point, the workpiece is upset and completely fills the container. As the pressure is further increased, material begins to leave through an orifice and forms the extruded product. The orifice can be, in principle, anywhere in the container or the ram (Figure 4-16). Depending on the relative direction or motion between the ram and extruded product, the terms forward or direct extrusion (Figure 4-16a) and reverse or backward extrusion (Figure 4-16c) are used.

The extruded product may be solid or hollow. In the latter, the outer diameter of the workpiece can be reduced as shown in Figure 4-17(a), or allowed to remain at its original dimension (Figure 4-17b). Tube extrusion is typical of forward extrusion of hollow shapes (Figure 4-17a), whereas

mation of flash to achieve complete filling of the die impression. In true closed die forging (also known as "flashless forging"), the material is deformed in a cavity that allows little or no escape of excess material. In the simplest case, deformation is first of the compression-type (Figure 4-14b). Barreling of the specimen is limited by contact with the container walls so that pressure inside the workpiece rises, and all parts of the cavity are filled (Figure 4-14c). As far as material flow is concerned, it is immaterial whether the cavity penetrates into the container or the punch

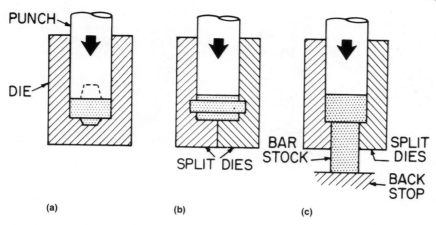

FIGURE 4-15. Compression in a totally enclosed impression. See text for discussion of (a) to (c).

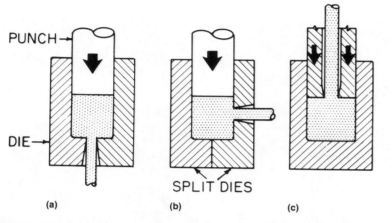

FIGURE 4-16. Schematic diagram of extrusion of solid shapes. (a) Forward (direct). (b) Radial. (c) Backward (reverse).

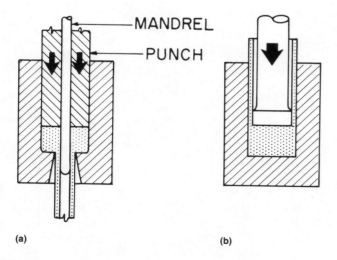

FIGURE 4-17. Schematic diagram of extrusion of hollow shapes. (a) Forward (direct). (b) Backward (reverse).

reverse or backward extrusion is used for the mass production of containers (Figure 4-17b).

Piercing

Piercing is a method used to produce hollow bodies. It is closely related to reverse extrusion, but is distinguished from extrusion by greater movement of the punch relative to movement of the workpiece material. The dimensions of the container, workpiece, and punch diameter determine whether metal flows radially (Figure 4-18a), or whether it is displaced axially between the container wall and the punch surface (Figure 4-18b). The various basic processes by which metal may be deformed (covered in the previous section) translate into specific forging methods—each distinguished by the type of forging equipment and tooling employed.

Preforming in Forging Practice

Proper design of the forging is essential to successful and economical production, but is by no means sufficient in itself. The shape of a forging is very seldom simple enough to permit forging in a single die impression. Attempts to bring the workpiece to final configuration in a single die impression through repeated blows usually result in several potential problems. A large excess of material has to be forged into flash; also, resistance to deformation is very high because of excessive cooling. Productivity is low, and overheating of the die can result in premature die failure. In addition, the quality of the forging may suffer due to less than optimum grain orientation and poor surface quality caused by entrapped scale.

Quality and economy of production can be improved if the finishing impression of the die is relieved of the bulk of the work of deformation and is called upon only to bring the forging to its final contour. This is why preforming of the forging stock by bending or rolling, or by working in preliminary dies, is usually desirable.

Although preforming constitutes an additional operation and contributes to the total cost of production, this is offset by the gains of higher productivity, increased die life, and improved quality, all of which are important in quantity production, but may be of less significance if production is limited. Therefore, forging in only a final die impression may occasionally be a practicable solution for extremely small runs.

Blocking. The duties of the finishing impression can be relieved if the majority

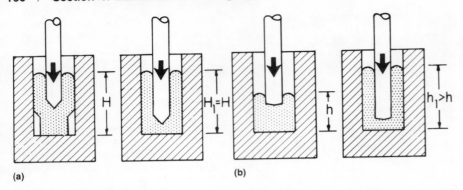

(a)   **(b)**

FIGURE 4-18. Schematic diagram of piercing sequence. (a) With radial metal flow. (b) With rising metal flow.

of deformation is done in a die impression of similar, but simplified, shape (the blocking impression). The guiding prin-ciple in designing the shape of the block-ing impression is ease of metal flow in the blocking and finish-forging operations.

Radii usually are generous, particularly where large quantities of material are to be displaced. However, blocker design depends on the configuration of the forg-ing, and few general rules can be stated. Blocking can be applied to both hammer and press forging, although differences may exist in the design and function of the blocking impression.

Distributing the Material. Many forg-ings are characterized by unequal material distribution. In these instances, economy of forging can be improved by displacing the material to conform roughly to final shape requirements. A good example is the connecting rod with its slender stem and heavy ends. Methods of material distri-bution are numerous and encompass the whole field of forging technology.

Preforming operations are commonly

FIGURE 4-19. Preforming by upsetting.

FIGURE 4-20. "Drawing out" by ingot manipulation in a hydraulic press. Courtesy of Scot Forge.

done in the hammer, but preforming is also done on presses, forging machines (upsetters), forge rolls, or special machines. In the simplest preforming operation, the cut stock is placed on end in the hammer and upset with the dual purpose of increasing the diameter and removing scale (Figure 4-19). When only one end of the stock is reduced, the operation is called "drawing out" and is normally done with a drawing tool impression or forging rolls (Figure 4-20).

Fullering finds application when the cross-sectional area needs reducing in the center of the forging stock. Edging or rollering is used to increase the cross section of the forging stock. Spread of the material during edging and fullering is unavoidable, and the stock is normally rotated 90° between successive blows to keep its width within desired limits. Fullering and edging may be followed by bending where required.

Many forgings require the accumulation of stock at one end. This may be achieved by forging on a forging machine (upsetter). Alternately, of course, a stock may be chosen with sufficient cross section to provide enough material for the largest section of the part. Smaller sections are then extruded or drawn out.

Roll forging (Figure 4-21) is capable of producing most shapes required in the preforged form, corresponding to the products made by fullering, edging, and even bending. In its various forms, the process

is economical and offers high productivity.

Hammer and Press Forging. Forging on the hammer is carried out in a succession of die impressions using repeated blows. A typical example is the forging of connecting rods two-at-a-time. The bar is first fullered to reduce the sections that will become the I-beams, followed by a roller operation, blocking, and forging to final shape. Trimming of the flash is done cold in separate tools on a press.

The quality of the forging and the economy and productivity of the hammer forging process depend on the tooling and the skill of the operator. Other things being equal, the skill of the operator is less important on the press and forging machine than on the hammer.

In a press, the stock is usually hit only once in each die impression, and therefore design of each impression becomes more important. The forging sequence consists of preforming, blocking, and finishing, followed by a combination hot trim and punching operation in a trimming press.

Hammers and presses may be used in combination. For crankshafts, a tong hold is first forged on a small press, then preforming is often carried out in three impressions on a hammer. The billet is rounded in the first roller impression, the mass distributed in the second roller, and finally the crank throws are pushed out in the "bender." The preformed stock is transferred to a mechanical press, where

it is blocked and finished. After trimming, the cranks are twisted into position on a special press.

Ring Rolling

The seamless rolled ring process offers a unique product: homogeneous circumferential grain flow, ease of fabrication and machining, and versatility in material, size, mass, and geometry. Seamless rolled rings have circumferential texture that is characterized by generally high tangential strength and ductility, which imparts required resistance to tangential stresses. In addition, properties in the length direction generally comply with any specification requirements.

Seamless rolled rings are produced in thousands of different cross-sectional shapes (see Figure 4-22), measuring from several inches to over 20 ft in diameter. Some of these rings weigh as little as a pound, while others can exceed tens of thousands of pounds.

Seamless rolled rings are often less expensive than similar closed die forgings, which may require large draft angles and more costly tooling. In many cases, several small circular parts can be combined and produced as a single rolled ring. Frequently, it is economically advantageous to cut small curved parts (segments of a circular part) into individual segments from a rolled ring. Machining of a ring before cutting into segments often imparts considerable machining savings. Rolled rings have found numerous applications for torque-reaction loads such as gears, couplings, and rotor spacers and for internal and external pressure applications such as components for pressure vessels and valves.

Ring rolling is an old hot forming process. As far back as 1842, records show that in Manchester, England, a rolling mill was constructed for R.B. Jackson and Company by Bodmer to produce seamless rings. Over the years, great advancements have been made in the relatively simple forming process of producing rings of infinitely variable weights, sizes, and cross-sectional shapes from a hot doughnut-shaped blank.

Today, ring rolling has developed from an art into a strictly controlled engineering process, with extremely high technology being used to construct the current breed of ring rolling lines. This strict control extends throughout the blanking and handling processes to the ring rolling mill.

Of late, ring rolling has passed through a fairly rapid period of growth. This can be attributed to expansions in machine tool design, along with a growth in electronic

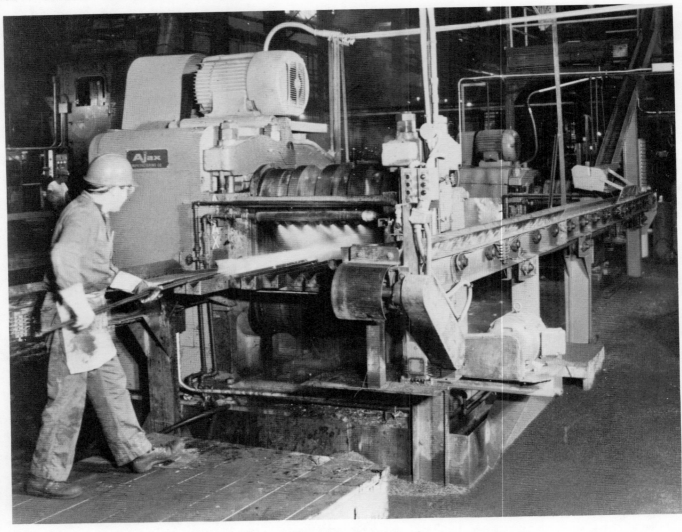

FIGURE 4-21. Roll forging operation. Courtesy of Ajax Manufacturing Co.

controls and the recent arrival of the microprocessor. With modern ring rolling mills, ring production can be fully automated from billet heating through to final rolled rings stacked on pallets. This growth of ring mill sophistication has led to the production of rolling mills that are far removed from those capable of radial rolling only, with ring heights held in a closed pass-type main roll, or taken off the mill for leveling to height under presses or hammers. This type of mill produced many of the early tyres and rings in the early 1900s.

By the mid-1900s, a combination of radial/axial rolling mills with digital displays of ring sizes emerged. During this period, work was taking place on drives for radial/axial units utilizing solid-state devices for varible-speed capabilities for both radial and axial units. During this period, at the small end of the ring market, rings of only a few pounds in weight were being rolled on mechanical table mills. These automatically produced fairly high volumes of gear blanks or small automotive-type bearing rings by radial rolling only.

With current ring rolling mills, the following features are available: (1) controlled feed rate of radial and axial units together, thus allowing for predetermining ring growth diametrically with axial reduction at a desired rate; (2) digital measurement of all ring sizes, including wall thickness; (3) with facilities to measure ring dimensions early in the rolling phase, one has the capability to compute volume, compare this with final required volume, and alert the operator to any excess, allowing the choice to absorb this excess on one or more of the ring axes.

Manufacture of a rolled ring requires the production of a blank or doughnut-shaped forging from a cut-to-weight billet, made either by press forging or, in some cases, by hammer forging on a mandrel. From this doughnut-shaped forging, ring rolling can progress (Figure 4-23).

Seamless rolled rings are produced on a variety of equipment, much of it modified by individual producers to serve the primary markets in which they compete. All methods give the user essentially the same product: a seamless section with substantially circumferential grain orientation. A preform is heated to forging temperature

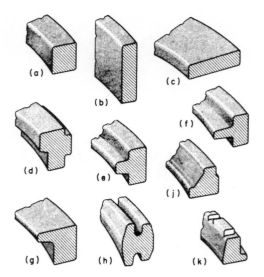

FIGURE 4-22. Typical cross sections that can be produced by ring rolling; shapes (h) and (k) must be completed in closed dies.

and placed over the I.D. roll of the rolling mill. By applying pressure to the wall as the ring rotates, the cross-sectional area is reduced, casuing the O.D. and I.D. to gradually expand.

Ring rolling machines are of two general types, horizontal and vertical, which differ primarily in the direction in which the workpiece rotates. The horizontal machines can accommodate a larger range of ring sizes. They are better adapted to the rolling of large rings more than 30 in. in diameter. They are also suitable for rolling small rings (Figure 4-24). Ring rolling equipment may be categorized into three basic types.

Radial mills (Figure 4-25) can be either vertical or horizontal, with an O.D. roll on either side of which is a pivoted guide roll. Opposing these three rolls is an I.D. roll mounted on slides and usually moved into the O.D. roll. The preform placed over the I.D. roll is brought into contact with the O.D. roll, causing the preform to rotate and grow in diameter as the wall is reduced. Length is not always controlled in the rolling cycle, and it may be necessary to control length on supplemental equipment.

Radial axial mills (Figure 4-26) are similar to radial mills, with the exception that length is controlled by the addition of axial rolls. Directly behind the I.D. roll, in the center line of the mill, is a movable carriage containing these rolls. These rolls, one or both driven, move in a vertical plane. During the rolling cycle, as diameter is increased, the axial carriage moves back and the top axial roll closes to a predetermined length.

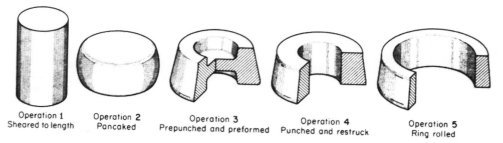

| Operation 1 | Operation 2 | Operation 3 | Operation 4 | Operation 5 |
| Sheared to length | Pancaked | Prepunched and preformed | Punched and restruck | Ring rolled |

FIGURE 4-23. Sequence of operations in producing a small ring in an automatic horizontal ring rolling machine.

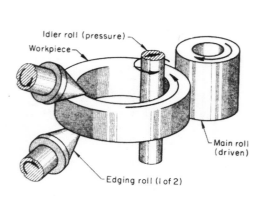

FIGURE 4-24. Horizontal ring rolling setup.

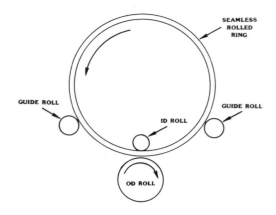

FIGURE 4-25. Radial mill configuration.

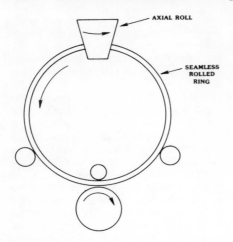

FIGURE 4-26. Radial axial mill configuration.

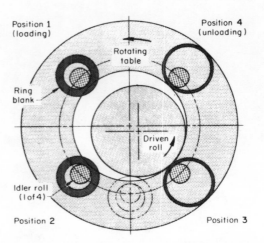

FIGURE 4-27. Principle of semiautomatic ring rolling, showing rotating table, drive roll, idler rolls, preformed ring, and processing positions.

Table or mechanical mills (Figure 4-27) are horizontal, having the driven O.D. roll in the center surrounded by four I.D. rolls mounted on a circular table moving on a different center and at a different speed than the O.D. roll. Cam-controlled guide rolls exert force on the O.D. of the ring during the short rolling cycle and square up the O.D. Axial length is determined by the O.D. roll pass, and diameter control is achieved by the adjustment of the I.D. rolls relative to the center or O.D. roll. With good weight control, these mills are capable of excellent dimensional repeatability, are able to contour O.D. or I.D., and are easily automated. Table mills are designed for lightweight, small-diameter, high-volume production.

Alternative methods of producing rings include the use of hammer forging of a doughnut on a saddle/mandrel (Figure 4-28).

Confined or enclosed systems are also sometimes used on horizontal ring rolling machines. The driven roll has flanges along its edges that confine the work metal and maintain the required width of the ring. Flanged rolls are less versatile than straight rolls, and they require a larger diameter profiled portion on the idler roll. Rolling in deep grooves requires a large-diameter flanged main roll and a disk on the idler roll that will enter the channel of the driven roll as ring rolling proceeds. This also requires that the punched hole in the blank be enlarged to admit the disk on the idler roll.

Rectangular cross sections can be rolled between straight rolls, but hollowness, ridges, and burrs are likely to occur on the side faces, which must then be worked in a hammer or press. As a result, it is usually better to use a confined system or edging rolls along with the straight rolls. Rolling with flanged driven rolls or straight rolls can be done with a simpler setup than rolling with edging rolls.

Edging rolls are used for most horizontal ring rolling. They are tapered (cone-shaped) to adapt their surface speed to that of the ring edge. The driven lower edging roll is mounted in the base of the machine frame, and the upper roll is located in a slide moving vertically in the upper sections of the frame; thus, the edging rolls can accommodate rings of different widths. In rolling profiles, the contours of the inner and outer surfaces of the ring are machined into the idler and driven rolls, and the side faces are controlled by the edging rolls, as shown in Figure 4-29. Deep outer profiles can be rolled only from adequately preformed blanks. Inner profiles, even of complex contours, can be rolled with little difficulty.

Wheel Mill. The differences between the processes for tires and for rolled steel wheels begin when the heated block reaches

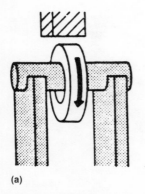
(a)

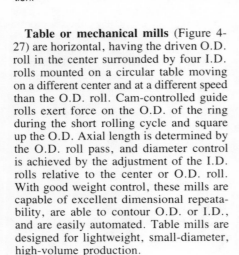
(b)

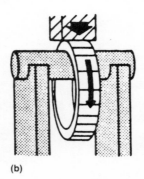
(c)

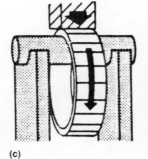
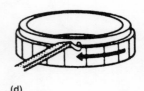
(d)

FIGURE 4-28. Sequence of operations for saddle forging of a ring. (a) Preform mounted on saddle/mandrel. (b) Metal displacement: reduce preform wall thickness to increase diameter. (c) Progressive reduction of wall thickness to produce ring dimensions. (d) Machining to near-net shape.

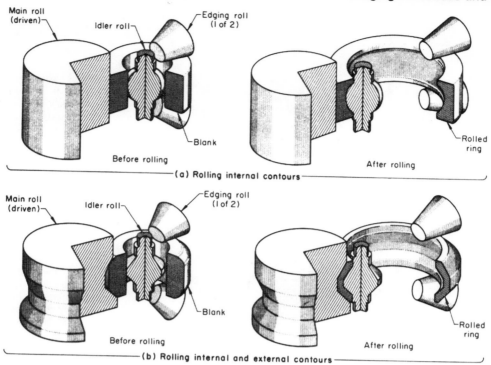

Main roll (driven) Idler roll Edging roll (1 of 2) Blank Before rolling After rolling Rolled ring

(a) Rolling internal contours

Main roll (driven) Idler roll Edging roll (1 of 2) Blank Before rolling After rolling Rolled ring

(b) Rolling internal and external contours

FIGURE 4-29. Rolling profiles on a horizontal ring rolling machine using driven, idler, and edging rolls.

the forging press. For the manufacture of wheels (Figure 4-30), the press is equipped with two sets of dies. In the first press operation, the bloom is formed and a hole for the bore of the wheel is punched. In the second press operation, the bloom is forged in dies to produce a blank. In this operation, the hub is extruded from the bloom width to the full hub length. This provides for thorough working of the metal in the hub. At the same time, the plate and rim are partially shaped. The forged blank, which is still flangeless, goes directly from the press to the mill without reheating.

In the roll setup for wheels (Figure 4-31), the main roll and the front guide rolls are similar to those used for tires. The pressure roll is not used, and the edging rolls are contoured rolls bearing on the plate and the rim. The back guide rolls are kept close together just back of the edging rolls and act as pressure rolls. The rolling action between these and the edging rolls reduces the section of the rim and rolls it out to form an additional plate and a new rim of larger diameter, but less cross section, complete with flange. As the wheel rotates, the plate and rim are reduced in section, and the diameter is increased. When rolled to size, the wheel is transferred to the dishing press, where the plate is coned and the wheel is stamped with its serial number. Figure 4-32 shows a pro-

duction sequence in rolling a large ring; Figure 4-33 shows other assorted large and small ring configurations.

Flashless Forging

Flashless forging is a type of impression die forging in which the dies are designed to include no flash areas. Because of this, volume control is critical and entails careful monitoring of the following:

- Preform or blank weight
- Preform and tooling temperatures, and thus the amount of thermal expansion
- Tool wear
- Die alignment

In addition to the above variables, several other features distinguish flashless forging from conventional impression die forging. These include the need for more careful die design, particularly for non-symmetrical parts. For this reason, flashless forging of symmetrical or other simple parts (such as that shown in Figure 4-34), in which metal flow is more easily controlled, is most common. In addition, for the process to be successful in a production environment, safety features (such as shear pins) must be added to forging equipment to allow perodic overloads (due to poor volume control, for example) and prevent jamming of the presses.

Flashless forging has developed into a commercial process over the last 20 years, with its greatest popularity in Europe, Japan, and the United States. Because it was developed primarily as a precision forging process, its main application is in cold and warm forging of steels primarily and other materials to a lesser degree. In addition to reduction of material waste (because there is no flash), this technique offers several other advantages. Better control of grain flow is achieved. The process eliminates grain flow that lies at 90° to the free surface, which occurs in the flash region of conventional impression die forgings. Lower material costs result. Because flash is eliminated, there is less material used to make parts compared to conventional forgings. There are also lower energy costs. Because less material is used to make the forging, less energy is needed to heat blanks for forging. Finally, the flash-trimming operation and equipment can be eliminated.

Precision Forging Processes

Precision die forgings (Figure 4-35), also known as no-draft forgings or pressings, are a relatively recent product evolution of conventional forging processes. They are distinguished from other types principally by their thinner, more detailed geometric

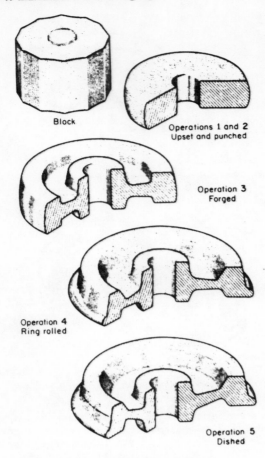

Block

Operations 1 and 2
Upset and punched

Operation 3
Forged

Operation 4
Ring rolled

Operation 5
Dished

FIGURE 4-30. Wheel process.

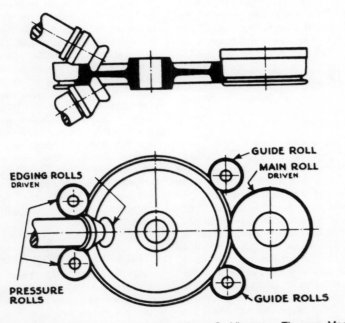

GUIDE ROLL

MAIN ROLL
DRIVEN

EDGING ROLLS
DRIVEN

PRESSURE
ROLLS

GUIDE ROLLS

FIGURE 4-31. Roll set-up for wheels. Adapted from G. Vieregge, Thyssen Maschinebau GmbH (Wagner Dortmund, West Germany).

features, virtual elimination of drafted surfaces and machining allowances, varying die parting line locations, and closer dimensional tolerances. These types of parts are most commonly manufactured from light metals such as aluminum and more recently from titanium for aerospace applications in which weight, strength, and intricate shaping are important considerations along with price and delivery.

Precision forgings are best suited for applications in which the basic characteristics of minimum weight forgings are desired or required and for applications in which a conventionally forged part would demand extensive machining. Most precision forgings are more costly than conventionally forged parts in the as-forged condition. However, they are usually less expensive than the same part machined from conventional or hand forgings, bar, or plate. Key consideration should be given to the comparative costs of the parts ready for use. Figure 4-36 compares cross sections of precision and conventional forgings.

Most precision forgings are designed to eliminate the need for machining, aside from drilling attachment holes. For extremely complex designs, it may be more economical to precision forge those sections of the part that are costly to machine and machine those areas that would be costly to precision forge. The objective is always to achieve the highest quality with the lowest cost.

The primary objective of precision forgings is to produce a finished or nearly finished part ready for use at a lower cost than a conventionally forged or completely machined part. Other advantages besides lower net cost include savings in time and reduction in weight and assembly work. In addition, reduced scrap loss helps conserve material and energy.

An example of the cost-saving characteristics of precision forging is illustrated in Figure 4-37 involving a 1-lb supersonic airframe part measuring 37 in. long with a U-shaped cross section that has a 5° twist from end to end. The chart compares the cost of producing the part three ways: as a hogout with extensive machining, as a conventional forging with additional machining, and as a precision forging. Although initial die costs were higher for precision forging, the per-piece cost, including the amortization of tooling, was considerably lower on a 200-piece production run. The cost figures compared include material, tooling, setup charges, and machining. Costs of producing parts similar to the wing nose ribs in Figure 4-38 as a machined hogout and a precision

FIGURE 4-32. Ring rolling sequence. Courtesy of Edgewater Steel Co.

forging are compared. Again, tooling for precision forging is higher than for the hogout, but the break-even point is reached quickly and the per-piece price drops rapidly.

Three production methods are compared with precision forging in Figure 4-39 with the same results. Net cost of the precision-forged piece becomes less than the machined hogout and conventional forging when production reaches approximately 130 pieces. From that point, savings are appreciable. Note that the cost comparisons cited in the Figures 4-37 through 4-39 reflect data compiled at the time the parts were made. Since then, costs may have changed. However, the relative differences in costs should still be valid.

Conventional die forgings (Figure 4-40) generally are formed between an upper and a lower die cavity. These are distinguished by a singular flash plane, either flat or curved, established by the contacting faces of the two dies. In addition, they have pronounced drafted surfaces normal to the parting plane, which allow the parts to be readily ejected from the cavities. Because the two halves of a conventional die must be indexed together laterally by means of guide pins or side and end locks while forming a part, an inherent mismatch of the upper and lower die cavities is always present, caused by the sliding clearances required by these guiding surfaces. This factor must be considered by means of a mismatch tolerance applied to all length and width dimensions of the forgings. Mismatch tolerances and draft angle allowances can, of course, be reduced by close control of die sinking and forging operations and by the use of mechanical ejectors within the die cavities. However, precision forging techniques generally eliminate the problems of mismatch and draft altogether.

A first step in precision forging begins quite naturally with those produced in conventional die halves, but refined to close tolerance and low draft by means of die inserts, by improved accuracy in die sinking procedures, and also by close control of process temperatures and pressures during the forging cycle. The latter is sometimes accomplished by use of an isothermal forging process in which the part and die are maintained at the same temperature under controlled pressure in a hydraulic press. This process is described in more detail elsewhere.

Modifications in die construction are key factors in precision manufacture. Two die designs called through die and wrap die techniques are discussed below to illustrate how precision forged parts are made.

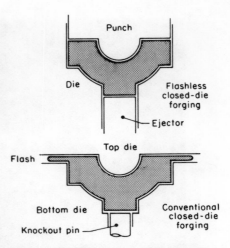

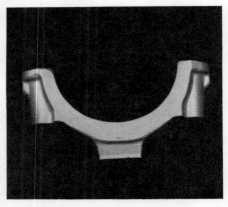

FIGURE 4-34. Connecting-rod cap made by flashless forging. Courtesy of Battelle's Columbus Laboratories, Metalworking Section.

FIGURE 4-33. Examples of various rolled ring sizes. Courtesy of Coulter Steel & Forge Co.

The through die construction depicted in Figure 4-41 derives its name from the fact that the outer periphery of the forging cavity is machined completely through the die. An upper and lower punch enter and forge the part entirely within this "ring." The top punch is then retracted out of the ring by the press stroke, and the completed forging is ejected by raising the lower punch attached to a knockout mechanism below.

Because the "ring die" forms the entire outer peripheral surface of the forging and also because any excess forging material is flashed vertically upward and downward off the ends of the part within the ring instead of horizontally outward between die halves, there is no longer any requirement for an external draft angle or a mismatch allowance. Also, the upper and lower punches slide snugly within the ring during the forging cycle, which reduces the mismatch of the internal features of the forging to a negligible amount and ensures

FIGURE 4-35. Typical precision die forgings. Courtesy of Aluminum Precision Products, Inc.

and strength of the die and operating features of the forging equipment. For the same reason, there is no need other than convenience for specified fillet sizes along the outer walls of the forgings where they join the horizontal web feature.

Large fillets frequently are required in this area on conventional forgings to aid in filling of the die cavities and also to reduce the possibility of flow-through defects. On through die forgings, however, small fillets are acceptable, simply because the problems of die fill and flow-through defects do not occur.

The final category and most versatile method of producing precision-shaped parts is known as wrap die forging. Two variations of this die assembly are shown in Figure 4-42(a) and (b). As with the through die arrangement, a through die or ring surrounds the part. However, instead of forming the forging itself, this ring serves to contain two or more removable insert dies called "wraps," whose inner surfaces shape the desired details around the periphery of the forging. The upper and lower punches fit snugly within these inserts during the forging operation, after which the ring and wrap assembly is separated by the press mechanisms. The wraps are

the dimensional integrity of all length and width dimensions.

The vertical flow path of metal in the outer walls of through die forgings naturally aids in filling the die cavity. In fact, there is theoretically no limit to the forgeable height or thinness of these outer walls. The limits are established only by the size

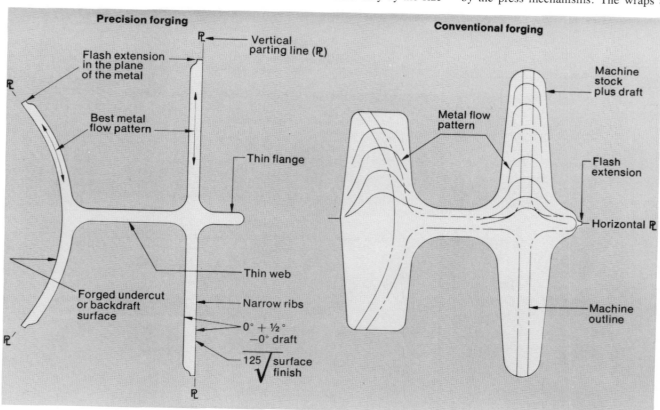

FIGURE 4-36. Cross sections of precision and conventional forgings. Courtesy of Alcoa.

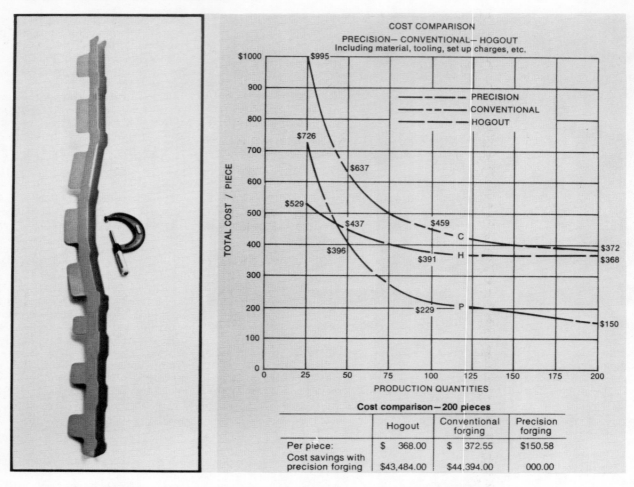

FIGURE 4-37. Cost comparison of channel-type precision forging. Courtesy of Alcoa.

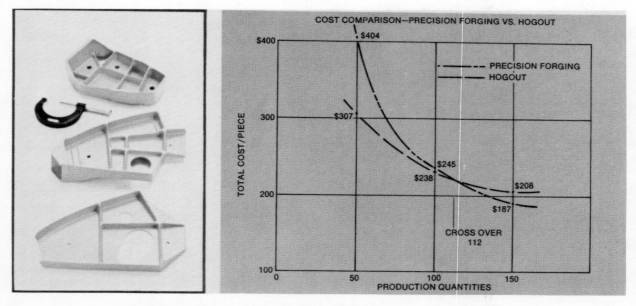

FIGURE 4-38. Cost comparison of wing nose rib forgings. Courtesy of Alcoa.

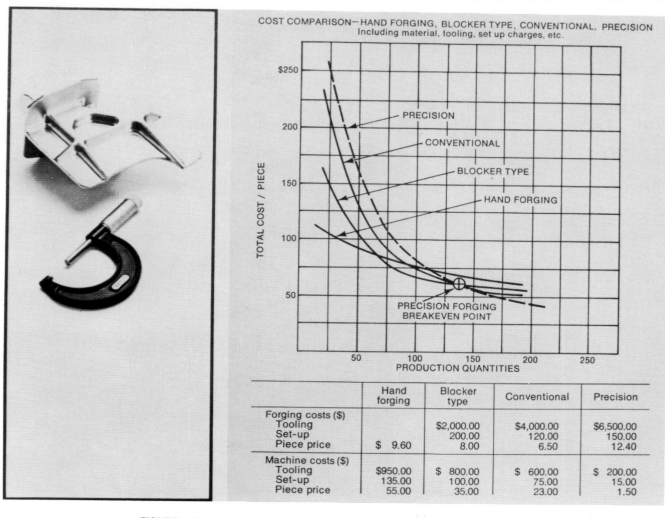

FIGURE 4-39. Cost comparison of wide-body jet forging. Courtesy of Alcoa.

	Hand forging	Blocker type	Conventional	Precision
Forging costs ($)				
Tooling		$2,000.00	$4,000.00	$6,500.00
Set-up		200.00	120.00	150.00
Piece price	$ 9.60	8.00	6.50	12.40
Machine costs ($)				
Tooling	$950.00	$ 800.00	$ 600.00	$ 200.00
Set-up	135.00	100.00	75.00	15.00
Piece price	55.00	35.00	23.00	1.50

then removed laterally from the forging, which in turn is lifted off the exposed bottom punch.

In the first die assembly (Figure 4-42a), the ring is held by clamps stationary to the bed of a hydraulic press. The bottom punch, which supports and carries the wrap segments, however, is free to move vertically within the wrap ring by means of an attachment to a controllable ejector mechanism below. The top punch attached to the top holder also moves vertically into and out of the ring by the action of the main press ram. After forging, the top punch is cleared from the ring, and the bottom punch and wrap combination is raised to the top surface of the ring where die separation and part removal occur.

An alternate method (Figure 4-42b) is accomplished with the bottom punch and wrap assembly always remaining on the bed of the press while the wrap ring is raised or lowered over the assembly by means of a separate controllable press mechanism located below or above. The upper punch, attached to the top holder as before, enters the ring and forges the part by the action of the main ram of the press.

Both ring dies are constructed with tapered inner wall surfaces, usually 3° from the vertical, which under forging pressure serve to bind the wrap inserts tightly around the part and prevent leakage of forged material between the segments. Each of the two methods described has its own distinct advantages and disadvantages in operation, although there is no difference in the types of parts that can be produced.

As the figures indicate, wrap dies form the outer peripheral surfaces of parts not only without draft or mismatch but also with external features such as negative angles, undercuts and pockets, protruding bosses, swargs, and three dimensionally contoured surfaces as well.

The wrap die design also offers many possible orientations and locations for flashing and guttering within the wrap segments. This in turn permits selective control and variability of forging pressures within the die cavity for establishing preferential flow directions and velocities and aids in forging defect-free parts to the required thicknesses. These factors coupled with the previously mentioned practices of die sinking accuracy and temperature control make possible repeatable production of extremely thin, highly detailed, close tolerance, net shape forgings at competitive costs.

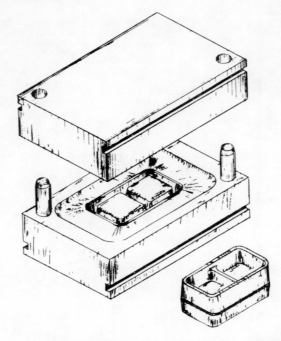

FIGURE 4-40. Conventional forging die. Courtesy of Aluminum Precision Products, Inc.

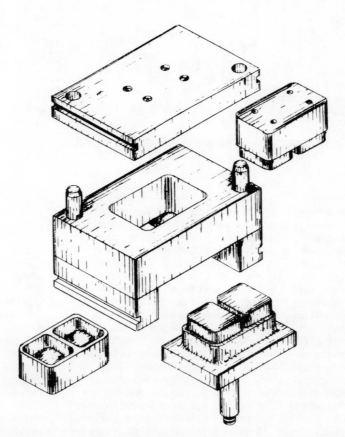

FIGURE 4-41. Through die design for precision forging. Courtesy of Aluminum Precision Products, Inc.

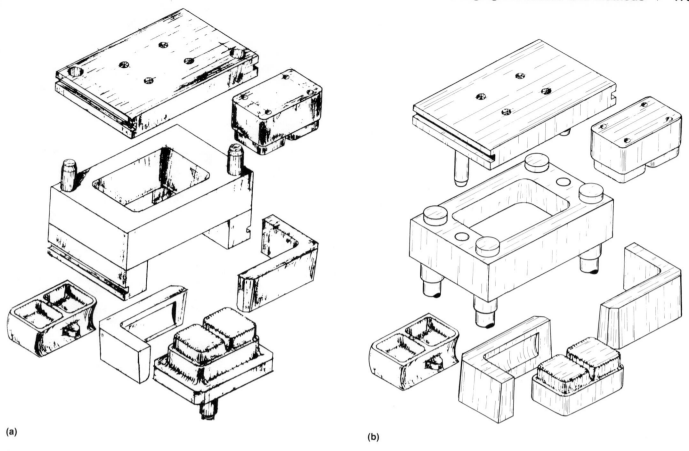

(a)

(b)

FIGURE 4-42. Wrap die variations for precision-forged part manufacture. See text for discussion of (a) and (b). Courtesy of Aluminum Precision Products, Inc.

Cold Forging

Cold forging is a very general term that covers processes such as extrusion, upsetting, ironing, and coining. The most widely accepted definition of a cold forging process is the forming or forging of a bulk material at room temperature. This definition is easy to understand and eliminates the normal confusion associated with the formal definition of forming above (hot) or below (cold) recrystallization temperatures. The emphasis on the current definition is that "no heating" of either the initial slug or the interstages is involved for the actual forming operation. The metallurgical phenomena taking place both during forming and intermediate processing (annealing, etc.), however, must be understood to make use of all the advantages of cold forging processes. A large number of cold forging processes are used in industrial practice. The most important will be considered here.

Forward Extrusion. Figure 4-43 shows three forward extrusion processes. The material flows in the same direction as the machine movement. In this process, the billet is pushed through a container or die by means of a punch to provide exit sections of either circular or other cross section and reduction of thickness of hollow slugs or the manufacture of cans of either cylindrical cavities or cavities with different sections.

Backward Extrusion. Figure 4-44 shows the three backward extrusion methods. The material flows in the opposite direction of the machine movement. The workpiece is formed either in the cavity formed between the punch and the die, or in the cavity of the punch.

Side Extrusion. In this technique, the material flows lateral to the direction of the machine movement. Some processes in this category are shown in Figure 4-45.

Upsetting/Heading. In these processes, either the workpiece diameter is increased or a flange is formed on a solid or hollow part, as shown in Figure 4-46.

Ironing. In this process, the wall thickness of hollow cans or tubes is reduced (Figure 4-47). The force is applied by a relatively long punch acting on the bottom of the preform. Unlike hollow forward extrusion, the workpiece wall is under tension.

Nosing. With this process, the end of a can or its radius can be reduced, as in Figure 4-48.

Radial Forging. In this method, radially moving tools forge the workpiece to the desired shape (Figure 4-49).

Combined Processes. Many of the above processes can be combined in a single operation to produce the desired workpiece. Some interesting process combinations are shown in Figure 4-50.

Materials Suitable for Cold Forging. The list of materials that can be cold forged is very large and is increasing. Some of the most common are listed in Table 4-1. The degree of formability varies widely among materials. With increasing content of alloying elements, the amount of deformation possible in a single stage is generally reduced.

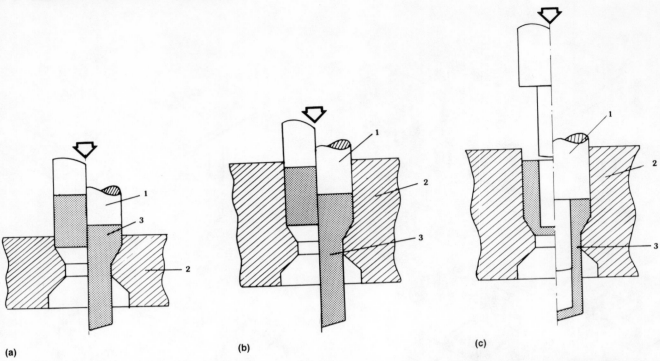

FIGURE 4-43. Forward extrusion processes. (a) Open die. (b) Trapped. (c) Hollow. 1, Punch; 2, die; 3, forging. Source: Ref. 33.

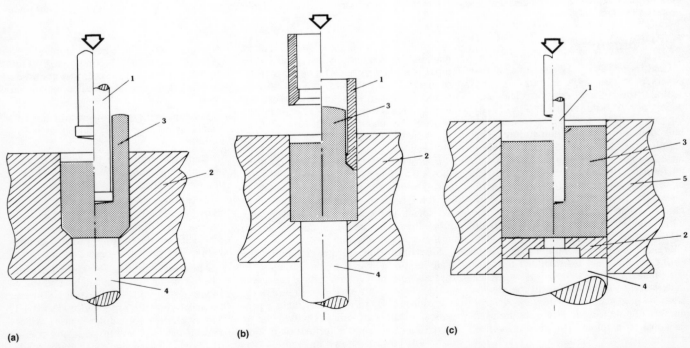

FIGURE 4-44. Backward extrusion processes. (a) Can. (b) Rod. (c) Indenting/piercing. 1, Punch [indentor, in (c)]; 2, die [support block, or piercing die, in (c)]; 3, forging; 4, ejector; 5, die. Source: Ref. 33.

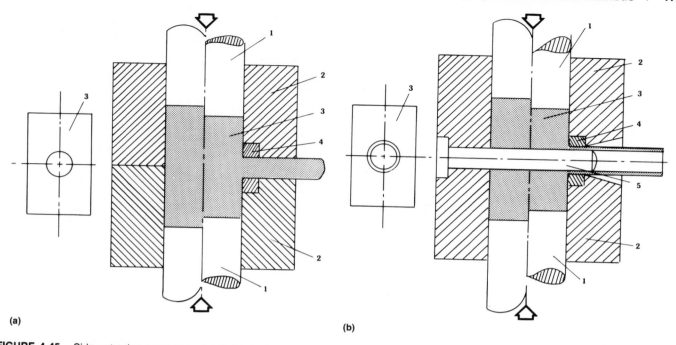

FIGURE 4-45. Side extrusion processes. (a) Solid (rod). (b) Hollow (tube). 1, Punch; 2, die; 3, forging; 4, die insert; 5, mandrel. Source: Ref. 33.

Shapes That Can Be Cold Forged. The simplest shapes that can be cold formed are shapes with rotation symmetry, that is, circular or round parts. Some parts in this category are shown in Figure 4-51. Shapes with a high degree of symmetry are the second best types suitable for cold forging; examples are squares, hexagons, etc. Some sample parts are shown in Figure 4-52. Nonsymmetrical shapes are the most difficult to cold forge. Due to nonsymmetrical flow, the die design becomes extremely critical for such parts. Some of the nonsymmetrical parts successfully forged are shown in Figure 4-53.

Process Sequence in Cold Forging. Except in rare instances, parts cannot be forged in one stroke. Several stages are required to produce parts due to limitations on the formability of materials, die load-ing, press loading and characteristics, and the possibility of combining processes. If the formability limit is reached, the work-piece must be annealed at an intermediate stage before proceeding with the next operation. Surface coatings, such as phosphating, oxalating, etc., and lubrication may be necessary for some materials between stages as well. Hence, the design of process sequence is based on many years

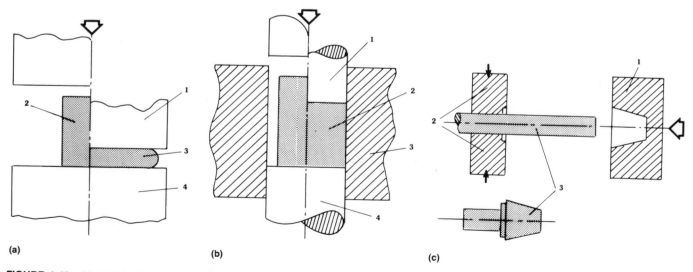

FIGURE 4-46. Upsetting processes. (a) Simple. 1, Top die; 2, billet before upsetting; 3, forging after upsetting; 4, bottom die. (b) Sizing. 1, Punch; 2, forging; 3, die; 4, counter punch. (c) Heading. 1, Punch; 2, die; 3, forging. Source: Ref. 33.

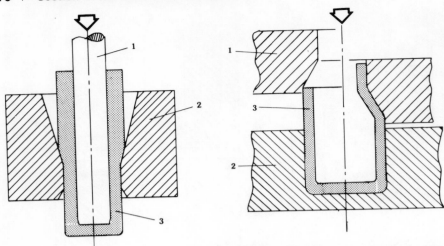

FIGURE 4-47. Schematic of the ironing process. 1, Punch; 2, die; 3, component. Source: Ref. 33.

FIGURE 4-48. Schematic of the nosing process. 1, Nosing punch (die); 2, holder die; 3, workpiece. Source: Ref. 33.

of experience of the design engineer. The process sequences for two parts are illustrated in Figures 4-54 and 4-55. The intermediate annealing and surface treatments necessary are not indicated on the figures. The process sequence for forming

the bevel gear (Figure 4-55) shows the progress of cold forging technology in recent years to form very intricate shapes. Figure 4-56 shows some of the recent parts that have been produced by cold forging.

Tolerances and Surface Quality of Cold Forged Workpieces. For the three

basic processes of producing rod (forward), can (backward), and hollow sleeves (hollow forward), the achievable dimensional tolerances are given in Table 4-2. The tolerances are in the SI tolerance bands. A lower number indicates a closer dimensional tolerance. The corresponding machining operations that ensure such quality levels are also given in Table 4-2.

Another important quality characteristic is the surface roughness values that can be obtained in cold forging. The normal value of surface roughness, R_a, is in the range of 1.6 to 25 μm (64 to 1000 μin.). The lower values are more difficult to obtain and require careful processing (tool surface finish, etc.).

Cold forging processes also enable the use of cheaper materials to obtain the required strength by work hardening the material instead of the conventional method of hot forging (or machining) and heat treatment. An example given in Table 4-3 illustrates this point. A wheel hub produced by two different methods results in nearly the same yield strength. The savings in material and in finish machine time and the elimination of the finish heat treatment operation for the cold forged material (also cheaper than the material used for hot forging) highlight the major advantages of the cold forging process.

Trends in Cold Forging Technology. From traditional fastener manufacturing to the production of shells for ammunition, cold forging processes are finding application in almost all fields. The automobile industry continues to be the major user of this technology. Gears and internal and external splines are also being formed increasingly by this method. The current limitations on nonsymmetrical geometries are being overcome by a combination of cold and warm forging technologies. Another area of growth is cold forging (or forming) from thick plates; a typical example is shown in Figure 4-57.

Isothermal and Hot Die Forging

In isothermal forging, the dies and the forged material are at about the same temperature. Aluminum alloys are usually forged isothermally at around 800 °F (425 °C), because conventional die materials can be heated to and maintained at this temperature without any significant loss of strength and hardness. However, high-temperature materials such as steels, titanium, and nickel alloys are forged in the range of 1700 to 2300 °F (925 to 1260 °C). Therefore, isothermal forging of these alloys requires special tooling materials that

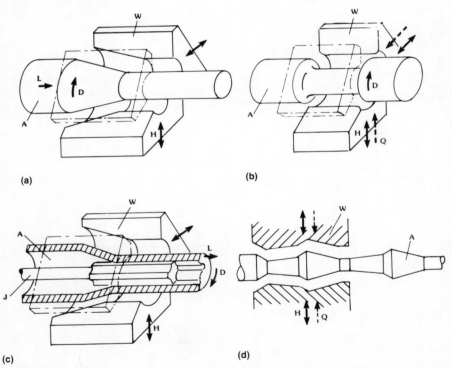

FIGURE 4-49. Schematic of the radial forging process. (a) Throughfeed process. (b) Plunge process. (c) Forging of inner profiles with mandrels. (d) External profile forging. A, workpiece; W, tools (die); J, mandrel; H, movement of die; D, rotary movement; L, longitudinal movement; Q, plunge feed. Source: Ref. 34.

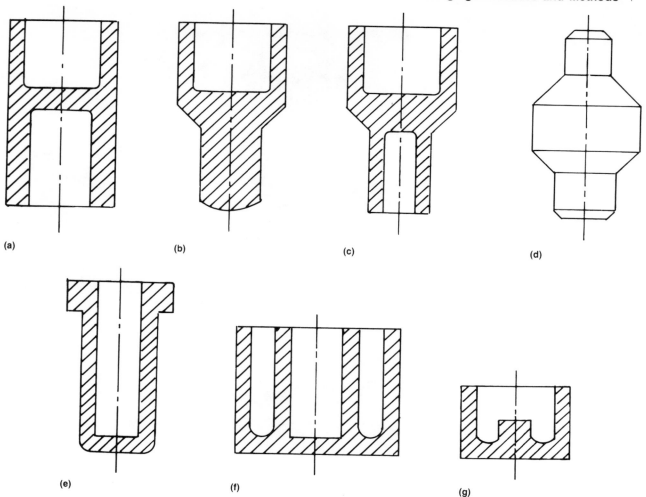

FIGURE 4-50. Schematic of process combinations. (a) Backward can/forward can. (b) Forward rod/backward can. (c) Hollow forward/backward can. (d) Forward rod/backward rod. (e) Backward can/flange upsetting. (f) Backward can/backward can. (g) Backward can/backward rod. Source: Ref. 35.

retain their strength at high temperatures, such as nickel-base superalloys and molybdenum alloys (Figure 4-58) and lubricants that can perform adequately at these temperatures.

In conventional forging practice, dies are heated to a temperature no more than approximately 600 to 800 °F (315 to 425 °C) to reduce die chilling, i.e., heat transfer from the hot material to the colder dies. Higher die temperatures cannot be used with conventional die steels, because they lose their strength and hardness above this temperature range. Die chilling is very significant because the flow stress of most high-temperature alloys (titanium and nickel, in particular) increases drastically with decreasing temperature (Figure 4-59) and die chilling influences metal flow and may cause defects in the forged part. Die

chilling may be minimized by selecting working temperatures at which the flow stress is not sensitive to temperature and by using, whenever possible, fast-acting forging machines, such as hammers, screw presses, and mechanical presses. In these machines, the time of contact between the workpiece (the metal part being forged) and the colder dies, during which heat transfer can occur, is reduced. Moreover, with the forging of nickel, steel, and titanium alloys, glass lubricants reduce the die-chilling effect, because they also act as thermal insulators. Even with these measures, however, some die chilling is unavoidable.

In forging steels and high-temperature alloys, the need to use several preforming and blocking operations increases the forging cost. This is especially true in the

case of aerospace forgings that are produced in relatively small quantities, for which large die costs are unjustified. Thus, very often it is economical to forge a part conventionally with a large machining envelope and to reduce die costs even though the material losses and machining costs are increased. However, this alternative is undesirable in the long run for expensive materials, especially for titanium and nickel alloys. Therefore, it has been necessary to develop isothermal (dies and workpiece at the same temperature) and hot die (die temperature near that of the workpiece) forging techniques for titanium and nickel alloys.

Isothermal forging and hot die forging of high-temperature alloys offer a number of potential advantages. Elimination of die chilling allows forging to closer tolerances

TABLE 4-1. Common Cold Forged Materials

Material	Standard	Designations
Carbon steels	AISI	1010, 1015, 1020, 1036, 1045
Alloy steels	AISI	5015, 5117, 5210, 4320, 4317, 4130, 4137, 4140(H), 4197, 5132, 5140, 4340, 6150(H)
Stainless steels, ferritic	AISI	410S, 410, 430, 431, 430F
Stainless steels, austenitic	AISI	302, 304, 305, 304L, 316, 316L, 317L, 321
Aluminum (pure/ commercial quality)	AA	1050, 1100
Aluminum alloys (not hardenable)	AA	3005 (3103), 5005
Aluminum alloys (hardenable)	AA	6063, 6061, 2017, 2014, 2024, 7005, 7075, 7079
Copper and its alloys	UNS(C)	11000, 12200, 22000, 26000, 27000, 64700, 50500
Magnesium and its alloys	UNS(M)	15110, 11311, 11610, 11800, 16600
Lead and its alloys	···	Fine lead, wrought lead, 0.1% copper lead
Zinc	···	Commercially pure zinc
Tin	···	Commercially pure tin
Nickel and its alloys	···	Monel (67% Ni, 30% Cu) K-Monel (66% Ni, 29% Cu, 3% Al)
Titanium and its alloys	···	Grade 1, grade 2 titanium

Source: Ref. 36.

(a)

(b)

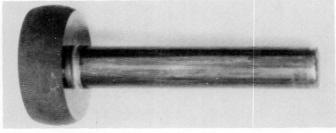

(c)

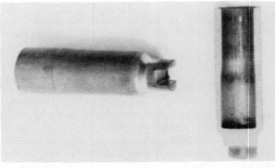

(d)

FIGURE 4-51. Examples of rotationally symmetrical cold forged parts. (a) Piston pin center weld. (b) Ball joint housing. (c) Compressor rotor. (d) Brake adjusting nut. Courtesy of Verson.

than possible with conventional forging. As a result, savings can be realized by reducing machining and material costs (Figure 4-60). Elimination of die chilling also allows a reduction in the number of preforming and blocking dies necessary for forging a given part; as a result, die costs are reduced. Because die chilling is not a problem, a slow ram speed—i.e., a hydraulic press—can be used. This lowers the strain rate and flow stress of the forged material. As a result, the forging pressure is reduced, and larger parts can be forged in existing hydraulic presses.

The main disadvantages of isothermal forging are that the process requires expensive dies made from special materials, uniform and controllable heating systems, and an inert atmosphere around the dies and the forging to avoid oxidation of the dies. Much of the early work establishing the features of the isothermal forging process was performed at the Illinois Institute of Technology Research Institute (IITRI) by Watmough, Kulkarni, and their co-workers in the mid-1960s and early 1970s. Concentrating on titanium alloys, improvement in metal flow through isother-

via conventional, nonisothermal forging by using several forging steps. In these instances, the isothermal forging had properties equal to those of conventional forgings (Table 4-4). This work also made valuable contributions in the selection of die materials that perform satisfactorily at the high temperatures (i.e., about 1650 °F, or 900 °C) used for isothermal forging of titanium alloys. These dies materials included a cast nickel-base alloy, IN-100 (Figure 4-58), which is still used widely today in isothermal forging of titanium alloys.

The isothermal and hot die forging of nickel-base alloys was pioneered by Pratt & Whitney Aircraft Company in the late 1960s and early 1970s. Located in Florida, the company called their process "Gatorizing." Developed originally for forging hard-to-work or cast superalloys such as IN-100, this process often makes use of preforms of starting shapes of fine-grained materials made via powder metallurgy techniques. The starting preform possesses a low flow stress and high ductility at isothermal forging temperatures and low strain rates, a condition often referred to as superplasticity (Figure 4-62a and b). With Gatorizing, hard-to-work nickel alloys were forged for the first time (Figure 4-63). In addition, the higher strengths obtainable in products made from these alloys resulted in high strength-to-weight ratios in parts such as jet engine disks. The invention of the process has also led to the development of techniques for isothermally forging integrally balded engine rotors.

Following the lead of IITRI and Pratt & Whitney, a number of other U.S. companies have started the production of titanium and nickel-base parts employing isothermal as well as hot die forging techniques. In developing hot die forging processes for titanium alloys, a wide range of commercial as well as experimental lubricants and various tooling concepts were evaluated. These evaluations included the critical examination of various methods of die heating, from a technical as well as an economic viewpoint.

Isothermal and hot die forging of high-temperature alloys are now well-proven and accepted production processes. In introducing and applying these processes, technical and economic factors, more than in conventional forging practice, must be evaluated in terms of:

- Workpiece material and its flow stress
- Forging equipment, load capacity, and speed control

(a)

(b)

(c)

(d)

(e)

FIGURE 4-52. Examples of axisymmetric cold forged parts. (a) Alternator claw. (b) Gear shaft. (c) Hub transmission convertor cover. (d) Flanged rotor gear. (e) Rotor gear. (a), (c) to (e) Courtesy of Verson. (b) Courtesy of National Machinery Co.

mal forging and the feasibility of isothermally forging complex titanium alloy aircraft parts in one or two operations were demonstrated (Figure 4-61). Several of these parts included intricate designs with blades or ribs that could only be produced

(a)

(b)

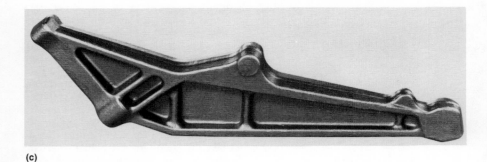

(c)

FIGURE 4-53. Examples of nonsymmetric parts. (a) Rivet head. Courtesy of Hackett Precision Co. (b) Brake cam. Courtesy of National Machinery Co. (c) Wing strut. Courtesy of Canton Drop Forge Co.

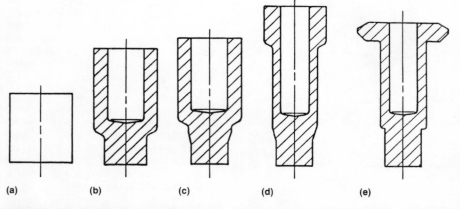

(a)   (b)   (c)   (d)   (e)

FIGURE 4-54. Schematic of forming sequences in cold forging a gear blank. (a) Sheared blank. (b) Simultaneous forward rod and backward cup extrusion. (c) Forward extrusion. (d) Hollow forward extrusion. (e) Simultaneous upset of flange and coin of shoulder.

- Die material methods and temperature control
- High-temperature lubricants

Powder Preform Forging

Forging of sintered powder preforms is a limited but important area of technology. Like isostatic pressing, powder forging produces 100% dense parts, but the similarity stops there; forging is much faster and is thus well suited for high-volume production. The beginnings of powder forging go back to the late 1940s, when the process was used to fabricate special, few-of-a-kind parts such as tungsten rocket nozzles for space vehicles. Little progress was made toward adapting powder forging for production use for the next 20 years.

One of the first production parts manufactured by powder forging was a differential pinion used in 1973 Buicks. Today, a variety of powder forged parts are being produced, principally for automotive use. These include various gears, bearing components, and transmission clutch cams. A number of powder forged parts have been produced in England, including automotive connecting rods, transmission gears, and various components for agricultural and recreational equipment. Some powder forged parts are also produced in Sweden, Germany, Japan, and Italy—also principally for automotive applications.

Federal-Mogul Corporation, one of the originators of powder metallurgy (P/M) forging, started its Precision Forged Products Division in 1972, resulting in the Sinta Forge process. In the plant, Sinta Forge production operations begin with conventional compacting of preforms that are engineered to the exact weight to produce a specific part. Preforms are compacted into cylindrical shapes using AISI 4600, modified AISI 4600, or AISI 1045 blended P/M alloys. They then are sintered at about 2000 °F (1100 °C) in a protected atmosphere furnace, cooled to room temperature, and coated with a proprietary compound to prevent oxidation during the forging process.

The actual P/M forging operation is semiautomatic, utilizing a high-tonnage press with closed dies. Preforms, manually loaded into a magazine, are fed to an induction unit that heats them to forging temperature. A single heated preform automatically transfers into the forging die, and the press is pushbutton-actuated. A ram-mounted upper die travels downward to mate with the fixed lower die while forging the preform to the finished shape. During the press upstroke, the forged part mechanically ejects to a steel belt that

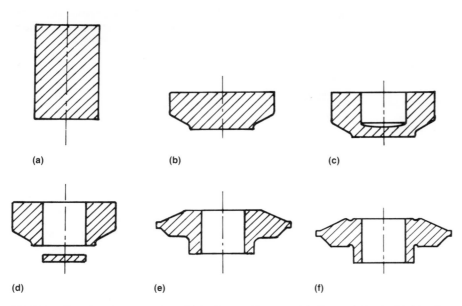

FIGURE 4-55. Forming steps in cold forging a side gear. (a) Sheared blank. (b) Heading. (c) Backward extrusion. (d) Piercing. (e) Finish forging. (f) Coining. The intermediate annealings are not shown. Source: Ref. 37.

FIGURE 4-56. Recent cold forged parts. Courtesy of Neumeyer-Fliesspressen GmbH (West Germany).

conveys it through an oil quench to discharge into a stock tub.

One problem in powder forging is the difficulty of obtaining impurity-free powders. Although cleanliness is desirable in all P/M parts, it becomes particularly important in forged pieces. By contrast, the relatively high levels of porosity in conventionally pressed and sintered parts mask the effects of impurities on properties. However, as full density is approached, the effects of impurities in reducing properties of the finished part are magnified. Impurities produce stress concentrations, which are particularly detrimental when rolling fatigue is involved, such as in bearings. Powder manufacturers are continually improving the cleanliness of powders so that they can be used with confidence in forged parts. Another problem in powder forging is cracking. Elimination of cracking and the requirements of complete densification by plastic flow are the major criteria on which preform design must be based.

Powder forgings that are most suitable for high-volume production are symmetrical shapes containing large holes (such as bearing races) and parts that would otherwise require a large amount of machining (such as gears) (Figure 4-64). Such parts made by powder forging (or by other P/M processes, for that matter) are cost-effective in material usage, and scrap is almost nonexistent.

Another major area of application has been difficult-to-work superalloys. For example, powder forgings of Astroloy, a high-strength nickel-base superalloy, have been found to have properties equal to or in excess of those specified for cast and wrought forgings (Table 4-5). The powder for these forgings was processed in an inert atmosphere of argon, canned in steel at a density of about 67% of theoretical, and then forged to 10-in. diameter disks, 1 in. thick. Oxygen contents were held to 60 ppm or less. The specimens representing the higher level of forging reduction had better properties, because the additional deformation produced a more uniform structure and eliminated evidence of particle identity. The forgings made from superalloy powder also exhibited good properties in low- and high-cycle fatigue tests at 1300 °F (700 °C) and resistance to creep at the same temperature.

Thermomechanical Treatments

An important metallurgical development is the process of thermomechanical treatment, or the combined effect of simultaneous plastic deformation and heat treatment. The combination is synergistic, producing effects not characteristic of either process alone. Specifically, it can improve toughness at a given strength level, or increase the strength level while maintaining, or even increasing, the toughness. The useful strength level in a high-strength alloy is limited by adequate notch toughness. Its widest application has been in the area of improvements in the strength levels as opposed to toughness, however.

Thermomechanical treatments, although not confined to steels, have re-

TABLE 4-2. Achievable Dimensional Tolerances in Cold Forging

Values indicated are for steel.

Process/shape	Dimension/form error	Normal quality range	Achievable either through special tooling or additional operation
Forward extrusion/rod	Diameter d_1	IT12-IT10 (rough machining)	IT10-IT6 (finish machining/grinding)
	Length l_1	IT15-IT13 (shearing, rough machining)	IT12-IT10 (rough machining)
	Straightness (bending)	Up to $0.02 \times d_1$	Up to $0.01 \times d_1$

d_1 = 50 to 120 mm
   ($\approx$2.2 to 4.8 in.)
$l_1 = 10 \, d_1$

Forward extrusion/sleeve	Outer diameter d_1	IT12-IT10 (rough machining)	IT9-IT8 (finishing machining/grinding)
	Inner diameter d_2	IT11-IT9 (finish machining/reaming	IT8-IT6 (finish grinding)
	Length l_1	IT13-IT11 (shearing, rough machining)	IT10-IT9 (finish machining)
	Concentricity of O.D. (d_1) and I.D. (d_2)	Up to $0.008 \times d_1$	Up to $0.003 \times d_1$
	Roundness of d_1 or d_2	Up to $0.005 \times$ diameter	Up to $0.001 \times$ diameter

$l_0/d_0 \geq 1.2$

Length/diameter of billet

d_1 = 10 to 120 mm
   ($\approx$0.4 to 4.8 in.)
s = 1 to 8 mm
   ($\approx$0.04 to 0.32 in.)

Backward/can	Outer diameter d_1	IT12-IT10 (rough machining)	IT9-IT7 (grinding)
	Inner diameter d_1	IT11-IT9 (finish machining)	Up to IT7 (reaming/grinding)
	Height h_1	IT15-IT13 (parting)	IT12-IT10 (rough machining)
	Bottom thickness h_b	Up to $0.01 \times h_b$	Up to $0.005 \times h_b$
	Concentricity of O.D. (d_1) and I.D. (d_1)	Up to $0.008 \times d_1$	Up to $0.003 \times d_1$
	Roundness of O.D. and I.D.	Up to $0.005 \times$ diameter	Up to $0.001 \times$ diameter

$l_0/d_0 \geq 1.2$
d_1 = 10 to 120 mm
   (0.4 to 4.8 in.)
s = 2 to 15 mm
   (0.08 to 0.6 in.)

Source: Ref. 36.

TABLE 4-3. Comparison of Hot Forging and Cold Extrusion for Manufacture of a Wheel Hub

	Hot forging	Cold extrusion
Material	8620	1010
Weight	2.85 kg (6.3 lb)	2.18 kg (4.8 lb)
Finish machining time	4.5 min	1.47 min
Finish treatment	Hardening and tempering	...
Yield strength	600 N/mm^2 (87 ksi)	550 N/mm^2 (80 ksi)

Courtesy of Neumeyer-Fliesspressen GmbH (West Germany)

ceived the greatest attention in the area of steel processing. Because of the range of treatments that the term includes for steels, it is helpful to classify them according to whether the deformation is imposed prior to, during, or following transformation of austenite, the face-centered cubic (FCC) allotrope of iron that occurs at temperatures higher than the so-called critical temperature Ac_1 and that decomposes into a variety of aggregates of body-centered cubic (BCC), or alpha, phase and cementite, Fe_3C, below the critical temperatures. These treatments are summarized in Table 4-6. In Type Ia, the main objective is to change the shape of the workpiece as in forging or rolling. Thus, the steel is typically worked above the Ac_1 temperature (where the austenite is stable and recrystallizes during or immediately following deformation).

Reduced grain size and improved mechanical properties (such as higher strength) can be achieved by careful control of these processes. In the remaining treatments in Table 4-6, processing conditions are selected to minimize recrystallization of austenite in an attempt to increase strength. For Type Ib treatments, this is done by working at low temperatures in the region in which austenite is metastable and by transformation of the austenite to martensite prior to the formation of higher temperature isothermal transformation products such as pearlite or bainite. Type Ib processes include what is variously referred to as ausforming, hot/cold working, and mar-working, all of which may be considered warm-working processes.

Type II treatments include deformation during the transformation of austenite to martensite and to isothermal-reaction products. Type IIa treatments, which involve both deforming austenite to induce martensite formation and deforming the resulting martensite, are applied commercially to nonaging and precipitation-hardenable stainless steels. The only Type IIb

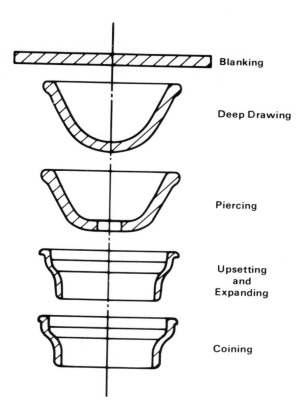

Blanking

Deep Drawing

Piercing

Upsetting and Expanding

Coining

FIGURE 4-57. Manufacturing stages for a flange piece from a thick plate. Adopted from D. Schmoeckel (West Germany).

treatment that appears to have been investigated is the flow tempering of bainite.

Type III treatments include deforming martensite followed by tempering, tempered martensite followed by retempering or aging, and isothermal reaction products such as pearlite and bainite, followed by aging. Type IIIa treatments have been investigated under the names of flow tempering and martensite cold drawing; Type IIIb treatments are also referred to as flow tempering, mar-straining, strain-tempering, tempforming, and warm working. Type IIIc treatments have been studied under the names of patenting, flow tempering, and warm working.

The above classification, based on the relative positions of deformation and transformation in the treatment cycle, has other justification in that the tensile stress-strain curves and the rate of increase in yield strength with increasing deformation have been found to be broadly similar for a variety of steels subjected to a given class of treatment and to differ for each of the classes. The effects of the three classes of treatment on stress-strain behavior are compared in Figure 4-65. When steel has been deformed before transformation (Type

I, Figure 4-65a), the general shape of the stress-strain curve remains the same, although it is raised to higher stress levels. Thus, the yield and tensile strengths are both increased similarly, and the strains at maximum load and at fracture are the same or only slightly lower. In contrast, when the steel has been deformed during transformation (Type II, Figure 4-65b), yield and tensile strengths are increased, but the shape of the curve changes, and ductility is decreased. The most striking changes are obtained when steel is deformed after transformation (Type III, Figure 4-65c). Whereas the other classes require substantial deformation to greatly affect strength, in this class, deformation of only a few percent, followed by an aging or retempering treatment, can lead to a sharp increase in yield and tensile strengths, a yield-to-tensile ratio approaching one, and markedly reduced ductility.

In a few cases, the different classes of thermomechanical treatments have been applied to the same steel, and a comparison can be made of the effectiveness of each treatment in increasing yield strength. In Figure 4-66, the increase obtained by deforming austenite followed by martens-

ite transformation and tempering (Type Ib treatment) is compared with the increase obtained by deforming the same steel in the martensitic or tempered conditions (Types IIIa and IIIb treatments). The comparison is made for a 410 stainless steel, AISI 4340, and the hot-work die steel H-11. It is evident that the rates of flow stress increase with increasing deformation fall clearly into two groups, corresponding to the two classes of treatment involved. For Type Ib treatments, the rate of flow stress increase is approximately 5 to 10% of that for Types IIIa, IIIb, and IIIc treatments.

Although the thermomechanical treatment of steels can be used to obtain high strength levels in steels, problems such as control of product uniformity throughout heavy section products, the loss of strength in welded areas of those steels, and the presence of strong directionality of properties have tended to limit the use of such treatments. Even with these drawbacks, there are a number of applications for steel products produced with these techniques (Table 4-7).

Upsetting

Forging on the horizontal forging machine (upsetter) accounts for a substantial part of forging production and can be classified from a technological point of view as impression die forging. The operational sequence is illustrated in Figure 4-67. The hot end of the bar is placed in the stationary grip die against a stop (Figure 4-67a), the moving grip die closes, and the stop retracts (Figure 4-67b), whereupon the heading tool begins compression (Figure 4-67c) and completes it at the end of its stroke (Figure 4-67d). The fact that deformation is basically upsetting imposes some limits on production—limits that must be observed if a sound forging is to be obtained.

Forging of Solid Bodies. The limits on the maximum length that can be upset successfully in a single stroke are set by the instability, or buckling, of the unsupported workpiece. In general, the unsupported length is not more than three times the diameter (or square) of the bar. This applies whether the stock overhangs the face of the grip dies (Figure 4-68a) or part of it is formed in cavities of grip dies or a heading tool (Figure 4-68b through d). The permissible length is further reduced if the end of the bar is sheared nonperpendicular to its axis, or if an impression is to be formed in the surface. Table 4-8 provides representative examples.

Greater amounts of material can be gathered only if the bar is supported to

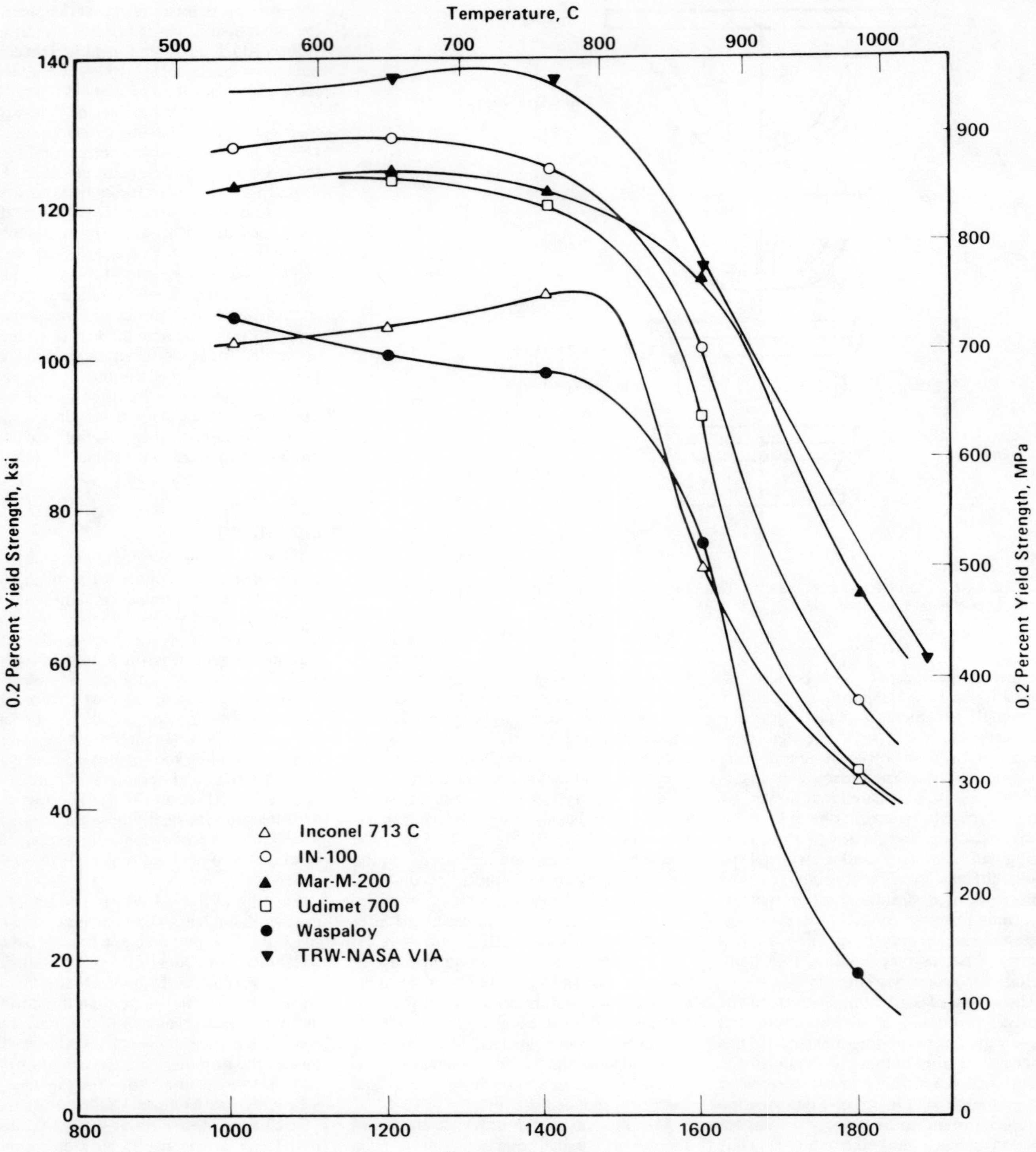

FIGURE 4-58. Yield strength of several cast and wrought nickel-base superalloys. Source: Ref. 38.

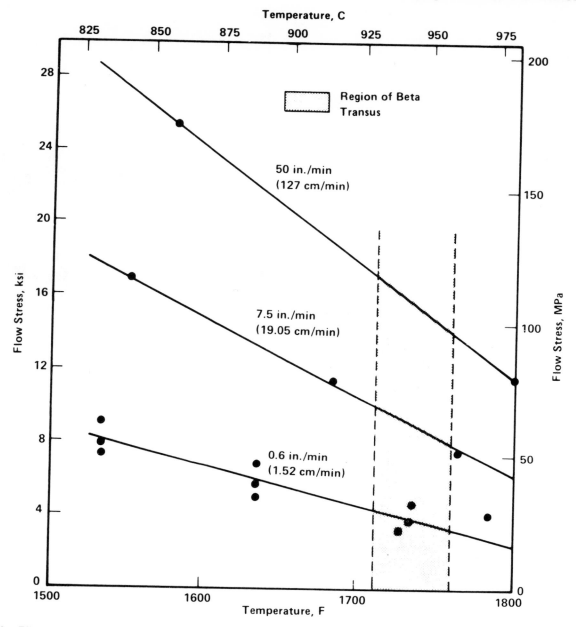

FIGURE 4-59. Effects of deformation rate and temperature on flow stress of Ti-6Al-6V-2Sn under isothermal forging conditions. Source: Ref. 40.

prevent buckling. Thus, in die where the diameter of the hole is not more than $1\frac{1}{2}$ times the diameter of the stock, the length upset can, in principle, be infinitely long, as the slight initial buckling is eliminated in further compression without leading to forging defects (Figure 4-69a). Again, the limitation applies whether the metal is gathered in a cavity of the heading tool, the grip die, or both, and whether the shape is cylindrical or conical (Figure 4-69b).

If the shape of the forging requires a greater accumulation of metal in the head,

several subsequent operations may be necessary, with a consequent increase in forging and die costs. In experience, multiple conical upsetting has been found most economical and gives the best material flow.

The process is not limited to forging heads at the end of the bar. Material can be gathered at any part of its length, but in this instance the die set will be somewhat more complicated, because sliding dies must be inserted in the grip die frames (Figure 4-70). The same limitations apply

as in upsetting the end of the piece, and care must be taken not to exceed the permissible unsupported length. Otherwise, a one-sided flash may be formed when the kinked bar is caught between the faces of the sliding and stationary grip tools.

Forging of Hollow Bodies. The horizontal forging machine is particularly suited for producing hollow and pierced components, because hollows can be made with very small or no taper and because piercing can be performed with little or no material loss. In the simplest case, the ends

CONVENTIONAL FORGING

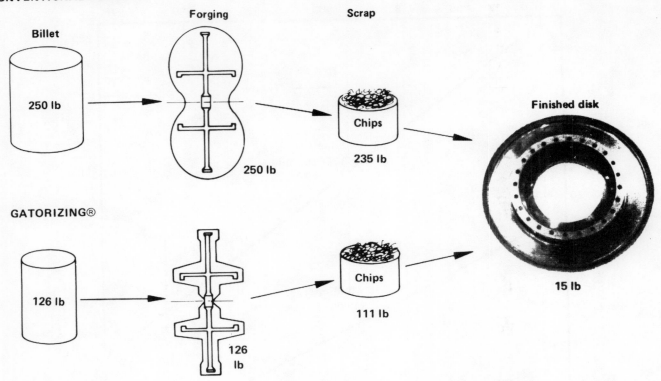

FIGURE 4-60. Isothermal forging process cost reduction. Gatorizing is a Pratt & Whitney process for isothermal and hot-die forging of nickel- and titanium-base alloys. Data and photograph courtesy of Pratt & Whitney Aircraft Co.

TABLE 4-4. Tensile Properties of Bulkhead Forgings

Specimen type	Tensile strength, ksi (MPa)	Yield strength, ksi (MPa)	Elongation, %	Reduction in area, %
MCAIR isothermal forging data				
Round specimens (22) from thicker portions	max 148 (1020)	140 (965)	19	49
	min 140 (965)	126 (870)	13	38
	avg 144 (995)	132 (910)	15	45
Flat specimens (9) from thin web	max 146 (1005)	134 (925)	16	...
	min 138 (950)	125 (860)	8	...
	avg 142 (980)	129 (890)	12	...
Alcoa isothermal forging data				
Specimens (22) from thicker portions	max 154 (1060)	138 (950)	19	48
	min 140 (965)	128 (885)	15	39
	avg 145 (1000)	132 (910)	17	43
Specimens (9) from thin web	max 146 (1005)	134 (925)	19	49
	min 143 (985)	130 (895)	17	38
	avg 144 (995)	132 (910)	17	44
MCAIR conventional forging data				
Unspecified	max 146 (1005)	138 (950)	17	40
	min 137 (945)	128 (885)	13	25
	avg 141 (970)	132 (910)	15	34
MIL-F-83142 minimum properties				
	130 (895)	120 (825)	10	25

Source: Ref. 42.

of a tube are upset. If the material is gathered so as to increase the outer diameter, the same rules apply as for the upsetting of solid bars, except that the wall thickness of the tube is now substituted for the diameter of the bar (Figure 4-71a). Heavier deformation is permissible if the outer diameter is kept constant and the inner diameter is reduced, because the natural flow of the material tends to give sound flow patterns even at higher reductions (Figure 4-71b).

Deep holes are pierced progressively, using relatively sharp punches of approximately 60° to 70° included angle. If the hole is to have a square bottom, the next to last piercer may have an included angle of 120° maximum, followed in the final stroke by a flat punch. Normally, however, the flat punch is used directly after a 75° piercer. The penetration of the punch into a long, unsupported bar would lead to buckling and eccentric piercing. A rim initially forged will prevent this, as shown by the forging of a drag link (Figure 4-72). If it is not desired in the final product, the rim may be subsequently trimmed off.

Rings of a relatively low weight-to-diameter ratio are easily produced in an upsetting and piercing operation, starting from stock that is the size of the hole diameter (Figure 4-73). When the diameter of the hole desired in a ring is larger or smaller than the diameter of the stock most suitable for upsetting, special techniques are used to modify the stock diameter adjacent to the ring before punching the ring from the stock. A particular advantage of this operation is the savings of material through elimination of internal flash.

Roll Forging

The properties of roll forged components are very satisfactory. In most instances, forging is carried out without flash formation, and the fiber structure is favorable and continuous in all sections. The rolls perform a certain amount of descaling. Therefore, the surface of the rolled product is generally smooth and free of scale pockets. Dimensional tolerances are similar to those obtained in conventional hot rolling. Some of the newer processes are continuous and therefore suitable for mass production purposes.

In its classical form, roll forging is carried out on two-high rolling mills. There are, however, basic differences compared to the conventional rolling process. The

Blank Temperature: 1650 °F (900 °C)

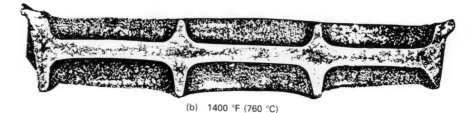

(a) 1650 °F (900 °C)

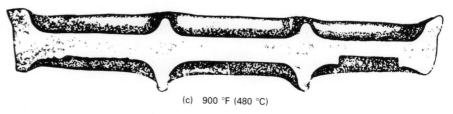

(b) 1400 °F (760 °C)

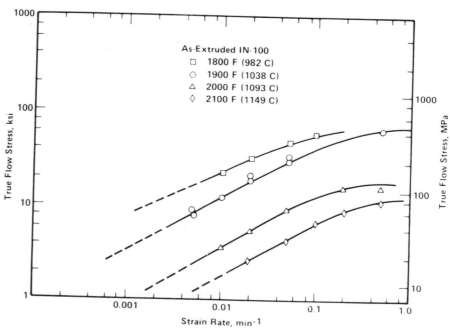

(c) 900 °F (480 °C)

FIGURE 4-61. Cross sections of structural forgings showing effect of die temperature on die filling. Source: Ref. 41.

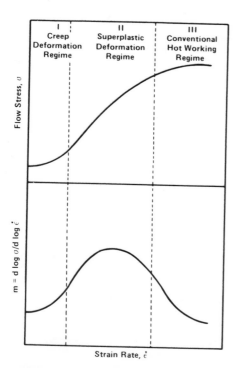

FIGURE 4-62a. Deformation regimes for superplastic materials. Source: Ref. 43.

FIGURE 4-62b. Superplastic deformation of as-extruded IN-100 at various temperatures plotted as log true flow stress versus log strain rate. Source: Ref. 44.

FIGURE 4-63. Typical Gatorized disks made from IN-100 alloy. Top photograph shows a disk and the starting billet preform. Courtesy of Pratt & Whitney Aircraft Co.

TABLE 4-5. Mechanical Properties of Astroloy Powder Metal Forgings

	Specification value(a)	Powder forging(b)	Reforged powder forging(c)
Tensile properties at 70 °F (20 °C)			
0.2% yield strength, ksi (MN/m²)	140 (965)	148.6 (1025)	156.4 (1078)
Ultimate strength, ksi (MN/m²)	195 (1345)	196.5 (1389)	214.5 (1513)
Elongation, % ...	16.0	11.5	20
Reduction in area, %	18.0	10.0	24.3
Tensile properties at 1400 °F (760 °C)			
0.2% yield strength, ksi (MN/m²)	125 (862)	132.5 (914)	140.4 (968)
Ultimate strength, ksi (MN/m²)	150 (1034)	161.3 (1112)	161.7 (1096)
Elongation, % ...	20	13.0	18.7
Reduction in area, %	30	10.4	38.1
Stress-rupture properties at 1400 °F (760 °C)			
Stress, ksi (MN/m²)	85 (586)	85 (586)	85 (586)
Life, h	30	47, 59	66.1
Elongation, % ...	17	11.9, 10.8	21.1

(a) Pratt & Whitney Specification 1013 E. Heat treatment consists of heating at 1600 °F (870 °C) for 8 h. AC; 1800 °F (870 °C) 4 h. AC; 1200 °F (650 °C) 24 h. AC; 1400 °F (760 °C) 8 h. AC. (b) Forged from canned powder. Solution treated at 2200 °F (1205 °C) then heat treated as in (a). Grain size ASTM 4 to 5. (c) Reforged to an upset reduction of 50%. Solution treated at 2075 °F (1135 °C), then heat treated as in (a). Grain size ASTM 4 to 5. Courtesy of Pratt & Whitney Aircraft Co.

TABLE 4-6. Classification of Thermomechanical Treatments

I. Deformation before austenite transformation
  a. Normal hot-working processes
  b. Deformation before transformation to martensite
II. Deformation during austenite transformation
  a. Deformation during transformation to martensite
  b. Deformation during transformation to ferrite-carbide aggregates
III. Deformation after austenite transformation
  a. Deformation of martensite followed by tempering
  b. Deformation of tempered martensite followed by aging
  c. Deformation of isothermal transformation products

Source: Ref. 45.

rolls are of relatively small diameter and serve as arbors onto which the forging tools are secured. The active surface of the tool occupies only a portion (usually half) of the roll circumference. Thus, there is a big enough gap left on the rest of the circumference to accommodate the full cross section of the stock. The stock is introduced from the delivery side of the rolls. Forming begins when the tools grip the stock and push it back toward the operator (Figure 4-74). The tools may take as little as 120° of the circumference on fast, small rolls to allow more time for handling and transporting the stock to the next cavity. On large rolls, as much as 220° of the circumference may be taken up by the tools, but the rolls are stopped and started by the operator to allow more time for handling.

The reduction in cross section obtainable in one pass is limited by the tendency of the material to spread and to form an undesirable flash, which may be forged into the surface as a forging defect in subsequent operations. The workpiece is introduced repeatedly with a 90° rotation between subsequent passes. In most applications, however, several roll passes (dies) are required to finish the component, and die design is based primarily on the general principles discussed previously.

FIGURE 4-64. Parts containing large center openings and parts that would otherwise require extensive machining. The four transmission components (left) and the three bearing parts (right) were produced by Federal-Mogul Corp.

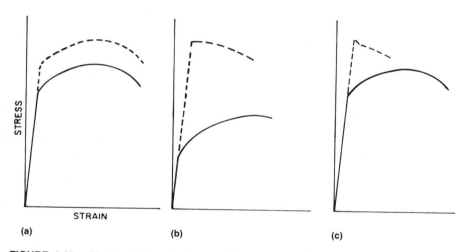

FIGURE 4-65. Effects of different classes of thermomechanical treatment on the shape of the tensile stress-strain curve. (a) Type I. (b) Type II. (c) Type III. Source: Ref. 45.

Radial Forging

In manufacturing gun barrels through conventional methods, it is necessary to bore a large-diameter forged billet. The rifling is then introduced into the barrel by conventional broaching or electrochemical machining. This method of manufacture is expensive, and it does not offer the means for obtaining the best possible fatigue and wear properties at the I.D. of the gun barrel.

The radial forging process was developed in Europe, and it is routinely used for forging of gun barrels and other axisymmetric parts. This method of manufacture offers the advantage of lower per unit cost compared to the conventional method. Part of this savings for gun barrels results from the fact that bore rifling can be forged simultaneously with working of the outer diameter of the workpiece. In addition, gun barrels and other products produced by radial forging show improved mechanical and metallurgical properties.

Radial forging machines use the radial hot or cold forging principle, with three, four, or six hammers to produce solid or hollow, round, square, rectangular, or profiled sections. The principle of these machines is shown in Figure 4-75 for forging of rods with four hammers. The machines used for forging large gun barrels are of a horizontal type and can size the bore of the gun barrel to the exact rifling machined on the mandrel. The horizontal radial forging machine consists of a forging box with gear drive, two chuck heads to manipulate the preform, centering devices, and necessary hydraulic and electronic control components (Figure 4-76).

Forging Box and Stroke Adjustment. The core of the machine is a robust cast steel forging box that absorbs all forging forces. It is mounted with the gear box on a support bolted to the foundation frame. The forging box (Figure 4-77) contains four rotatable adjustment housings in which the eccentric shafts are mounted. The eccentric shafts, which are driven by an electric motor through a gear system, actuate the connecting rods and the forging dies, arranged at right angles to each other, at the rate of 200 strokes/min.

The stroke position of the connecting rods (or the dies) is adjusted in unison by rotating each adjustment housing through a link, screw, adjustment nut, and worm gear drive powered by one or two hydraulic motors. Each adjustment nut rests, through a piston, on an oil cushion of a hydraulic cylinder. During operation, the forging pressure generates in this oil cushion a pressure proportional (about 20%) to the forging pressure. The pressure in the oil cushion is monitored continuously, and if it exceeds a certain limit, the adjustment housings are immediately rotated to bring the dies to open position while the movement of the chuck heads is stopped simultaneously. This system protects the machine from overloading.

The stroke of the four connecting rods is constant and very short compared to conventional crank or eccentric presses. This design permits very high stroking rates, up to 220 strokes/min in a large machine having a 1000-ton load capacity per forging tool. Despite the short stroke length, the deformation speed remains within the range usual for conventional presses because of these high stroking rates. Thus, the radial forging process is not a hammering process but a pressing process in which the material is forged by the simultaneous action of the four forging tools.

Chuck Heads and Mandrel Forging. For holding the material at the proper lo-

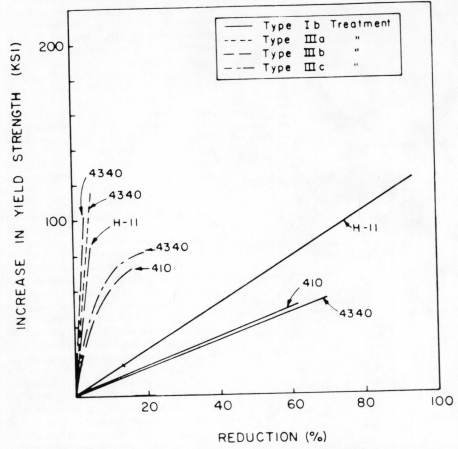

FIGURE 4-66. Comparison of increase in yield strength obtained by type I and type III treatments of three steels: type 410 stainless, AISI 4340, and H-11. Source: Ref. 45.

TABLE 4-7. Applications of Thermomechanical Treatments

Type of process	Simple shapes	Finished components
Type I		
Deformation before transformation	Wire cables, tires, springs	Rocket motor cases, mortar tubes, torsion bars, armor, damascus swords
Type II		
Deformation during transformation	Strip springs, wrap-type composites	Rocket motor cases
	Sheet for aircraft and missile hardware (300 series and semi-austenitic PH stainless steel)	
Type III		
Deformation after transformation	Sheet honey-comb and wrap-type composites	Rocket motor cases, armor plate, autofrettage of gun tubes
	Patented wire "Martensite" wire (Japan)	
	Rod and bar drawn at elevated temperatures for vehicle hardware	

Source: Ref. 45.

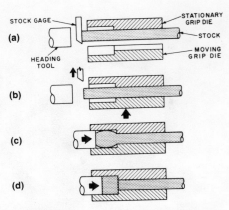

FIGURE 4-67. Schematic of typical operating sequence in upsetting on a forging machine. See text for discussion of (a) to (d).

cation and for feeding it to the forging box, one or two chuck heads are provided, as shown in Figure 4-76. Two chuck heads are used when components must be forged in one heat over the entire length, as is the case in radial forging of gun barrels. At each forging blow, the forged material moves axially toward the chuck heads. These intermittent blow-like motions press against the springs located in the chuckheads. In forging round components, the material is rotated. To avoid twisting the forging, the rotating chuck heads are stopped during each blow.

In hot or cold forging of a gun barrel, an axially stationary mandrel is used (Figure 4-78). The smooth (or rifled) cylindrical mandrel is mounted in a holder at the rear of the machine. The chuck head holds the hollow stock, which has a bore diameter larger than that of the mandrel. The head advances at the beginning of the operation to the furthermost position, pushing the tube through the open tools over the mandrel. The tools then close, and the tube is forged radially over the mandrel to obtain the required O.D. (and/or the internal rifling in case of cold forging). During hot forging, the mandrel is exposed to intense heat and therefore is provided with internal cooling. The feed of the chuck heads and the penetration of the hammers, i.e., the reduction, are regulated by numerical control. Thus, it is possible to obtain a highly reproducible accuracy that ensures uniform quality of the forged components.

Special Forging Techniques

There are several forging operations that do not fit easily into any of the previously discussed categories. Some of these are described below.

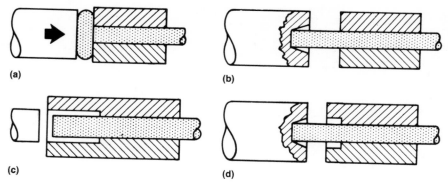

FIGURE 4-68. Schematic of the upsetting of an unsupported bar on a horizontal forging machine. See text for discussion of (a) to (d).

TABLE 4-8. Permissible Stock Length in Free Upsetting of Bars

Condition of the bar end	Angle of cut	Shape of the heading tool end face	Permissible length, expressed as multiple of given bar diameter:		
			1 in.	2 in.	4 in.
Sawed or sheared (even)	<1°	Hollow	2.4	. . .	. . .
		Flat	2.2	2.5	3.0
		With preforming punch	1.8	2.0	2.0
	>3°	Flat	1.8	2.0	2.5
		With preforming punch	1.25	1.5	1.5
Uneven	<1°	Flat	1.8	. . .	. . .

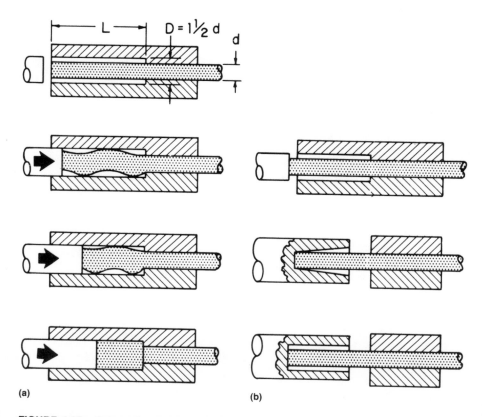

FIGURE 4-69. Schematic of typical operations involving upsetting to a limited diameter. See text for discussion of (a) and (b).

Bending is often employed as part of the operational sequence in hammer or press forging. Because bending of larger parts requires a machine of long stroke, special mechanical or hydraulic presses (bulldozers) are built for this purpose. Simple shapes can be bent in one operation, but more complicated contours require successive steps. If complex shapes are to be formed in a single operation, the tool must contain moving elements.

Swaging is much like the process of drawing out between flat, narrow dies. The difference is that two hammers, instead of the stock, are rotated. It is a useful method of primary working, although in industrial production its role is normally that of finishing.

The principle of operation is rather simple. The machine is housed in a steel frame. Rollers spaced by a race are supported on a forged steel ring. The anvils slide freely in a radial direction in slots of the rotor. On spinning the rotor, the centrifugal force throws the anvils outward. However, the end faces of the anvils encounter the rollers during rotation and are forced inward. Thus, the tools secured to the anvils rain fast blows on the workpiece. At a typical speed of 200 rpm, machines can provide 2400 strokes/min. Tubes, as well as solid bars, can be reduced in diameter. The outer surface is always circular in cross section, but the inner surface can take any form from which the mandrel can be removed.

Punching is very similar to trimming, except that the cut surface is an internal one. A rigid or spring-loaded stripper plate helps to remove the punched forging from the punch. As in trimming, removal of draft may frequently be accomplished by punching, but necessitates more elaborate and expensive tooling and occasionally requires an added operation. Trimming and punching carried out simultaneously (Figure 4-79) increase the accuracy of the cuts relative to each other.

Coining and ironing are essentially sizing operations performed in dies where pressure is applied to all or some portion of a forging surface to obtain closer tolerances and smoother surfaces or to eliminate draft. Both operations normally are performed on a press and may be done while forgings are hot or cold.

Coining is usually done on surfaces of the forging parallel to the parting line, while ironing is usually typified by the forcing of a cup-shaped component through a ring to size its O.D. Little metal flow is involved in either operation, and flash is not formed.

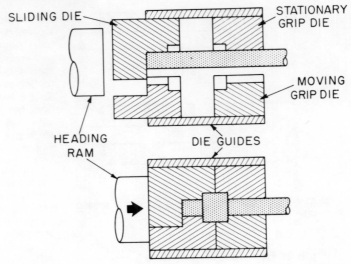

FIGURE 4-70. Schematic of the use of sliding dies on a forging machine.

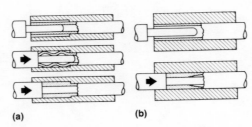

(a)   (b)

FIGURE 4-71. Schematic of the upsetting of tube ends on a forging machine. See text for discussion of (a) and (b).

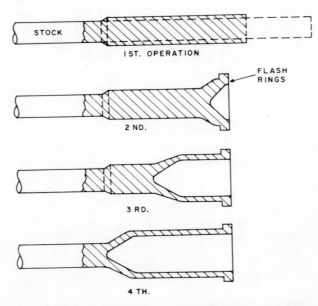

FIGURE 4-72. Schematic showing forging of a drag link on a forging machine.

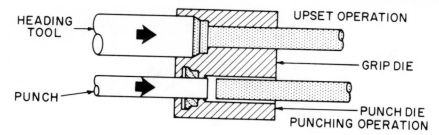

FIGURE 4-73. Schematic showing typical operations for forging a ring on a forging machine. Diameter of original bar stock and I.D. of forged ring are identical.

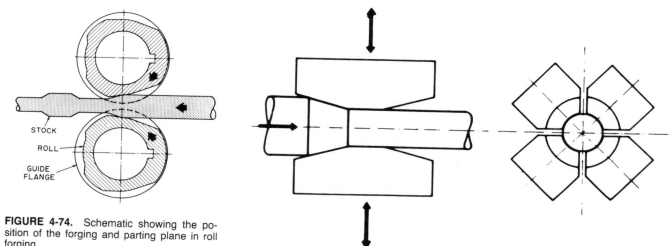

FIGURE 4-74. Schematic showing the position of the forging and parting plane in roll forging.

FIGURE 4-75. Principle of the radial forging machine. Source: Ref. 46.

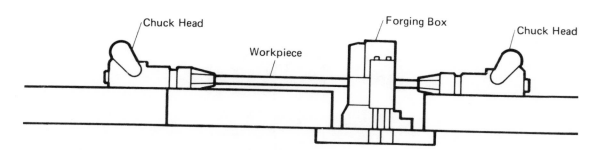

FIGURE 4-76. Schematic of a radial precision forging machine with two chuck heads. Source: Ref. 46.

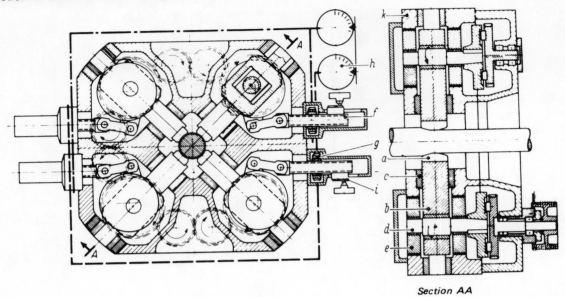

FIGURE 4-77. Forging box of a radial precision forging machine illustrating the tool function and adjustment. a, dies; b, pitman arm; c, guides; d, eccentric shaft; e, adjustment housing; f, adjustment screw; g, worm gear drive; h, adjustment input; i, adjustable cam; k, forging box. Source: Ref. 46.

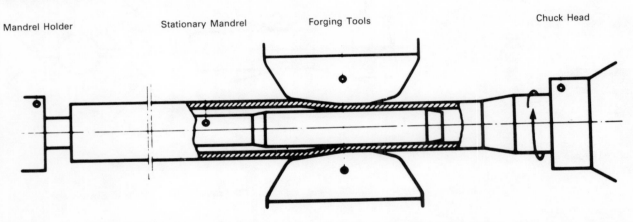

FIGURE 4-78. Forging of a tubular product over a stationary mandrel. Source: Ref. 46.

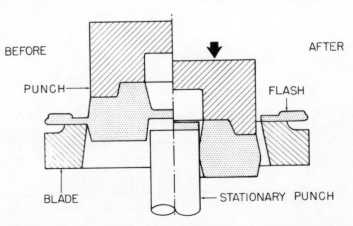

FIGURE 4-79. Schematic of combined trimming and punching operation.

Equipment for Forging

Contributing Author: Ken J. Sorace, General Manager,
Ajax Manufacturing Company, Cleveland, Ohio

The commercial manufacture of impression die forgings requires a wide variety of heavy industrial machines and equipment. Thus, even a forge plant of modest size represents facilities requiring a substantial capital investment. Certain machine tools used in the forging industry, such as lathes, drill presses, grinders, and milling machines, are common to many metalworking industries. However, equipment for various forging operations—particularly those involving deformation by impact—have no counterpart. This subsection provides a brief survey of typical forging facilities, highlighting the principal types of forging equipment and auxiliary machines and equipment in current use.

By definition, forging involves the shaping of metal by the application of impact or pressure. A fundamental difference among various forging methods is the rate at which energy is applied to the workpiece. In many cases, each primary type of forging equipment is useful in performing several different types of forging operations, and rates of deformation vary accordingly. For example, hammers, mechanical and hydraulic presses, and horizontal forging machines (upsetters) are all useful in impression die forging operations that involve compression of a workpiece between dies containing cavities. Upsetting (compression of the workpiece parallel to its longitudinal axis) is also commonly performed by equipment in each of these categories. The same is true, to a lesser degree, of other forging operations such as ring rolling, swaging, extrusion and drawing out, in which specialized forging equipment is often used. The decision as to whether a particular forging operation will be performed at relatively rapid rates of deformation (in a hammer) or at much slower rates (hydraulic press) depends on a variety of factors, including forging characteristics of the workpiece material, shape, size and quantity of the desired forging, and the availability of equipment.

Hammers

In the past, forging hammers have been the most widely used type of equipment for impression die forging and are generally considered the most flexible in the variety of forging operations they can perform. The three principal types of forging hammers in common use operate on the same basic principle: a heavy ram containing the upper die is raised and is driven or allowed to fall on the workpiece, which is placed on the bottom die. These hammers may be classified by the method used to raise the ram: board hammers, air-lift hammers, and steam hammers. Other types include counterblow equipment, and helve and trip hammers.

Board hammers, or "board drop hammers," operate by gravity and are responsible for the term "drop forging." The ram is attached to a set of hardwood boards. The boards (and ram) are raised vertically by action of contrarotating rolls, then released. Energy for forging is obtained by the mass and velocity of the freely falling ram and attached die.

The operating cycle begins as the rolls engage the boards and draw them upward. The boards (and ram) are then suspended at the top of the stroke by a set of clamps. The operator releases the clamps by depressing a foot treadle, and the hammer delivers a blow. The cycle is repeated without interruption (up to 70 times per minute) as long as the treadle is depressed. Each successive blow contains a like amount of energy, because the length of stroke is constant and can be varied only by adjustments made between jobs.

Board hammers are rated by their falling weight and range in size from 100 to 7500 lb, although hammers smaller than 500 or larger than 6500 lb are seldom found. The anvil, upon which the hammer rests, extends underground and serves as a massive inertia block. The usual ratio of anvil-to-ram weight is 20:1.

Air-lift hammers are similar to board hammers in that energy for forging is supplied by the falling weight of the ram and upper die. However, they differ from the board hammer in two important respects: the ram is raised by action of an air cylinder and, on some hammers, the length of each stroke can be varied. Ram velocity and thus energy delivered to the workpiece can be altered to predetermined levels with each blow of the hammer.

The ram is joined to a piston rod and raises directly upon the introduction of compressed air beneath the piston in the cylinder at the top of the unit. The oper-

ator's treadle controls a mechanical clamp on the piston rod. By depressing the treadle, the operator releases the clamp, thus allowing the ram to fall. As the ram nears the bottom of its stroke, it activates the air valve to begin the next cycle. While ascending, it once again triggers the valve, exhausting air from beneath the piston in readiness for the next blow. The hammer may be adjusted in advance to provide strokes of intermediate length, as well as the full stroke.

During operation, the operator selects the appropriate stroke using a device on the treadle. The opportunity of varying the intensity of each work stroke allows the operator to exercise greater skill and increases somewhat the range of preliminary forging operations possible. Air-lift hammers range from 500 to 10,000 lb in falling weight and are adaptable to programmed blow control. An air-lift gravity-drop hammer is shown in Figure 4-80.

Steam Hammers. The general characteristics of steam hammers are somewhat

FIGURE 4-80. Air-lift gravity-drop hammer. Courtesy of Chambersburg.

FIGURE 4-81. Double-acting power-drop hammer. Courtesy of Chambersburg.

similar to those of the air-lift hammer, with several important exceptions. The ram is raised by a piston and is driven down on the workpiece by steam or air pressure in addition to gravity. Striking force can be varied over the entire range from a light tap to full power, and hammer construction is much sturdier to accommodate higher energy levels. In practice, either steam or compressed air can be used as the power source. A double-acting power drop hammer is shown in Figure 4-81.

Movement of the piston is controlled by the operator's foot treadle, which activates a valve to admit steam to either the upper or lower side of the piston in the desired amounts. This complete control of each work stroke places higher requirements on operator skills than other types of hammers.

Largest and most powerful of conventional forging hammers, steam hammers range in falling weight from 1000 to 50,000 lb. Anvils are between 10 and 25 times heavier than rams, requiring tremendous underground installations.

Other Hammers. Several additional types of hammers are less widely used. These include vertical counterblow hammers, impacters, helve hammers, and trip hammers. Of these, impacters are most popular.

Vertical Counterblow Hammers. The steam-driven vertical counterblow hammer is characterized by mutually opposed rams that are activated simultaneously and stroke the workpiece repeated blows at a midway point. Counterblow hammers develop combined ram velocities of $1^1/_2$ times those for conventional hammers. For this reason, ratings are not directly comparable, but it is probable that the largest counterblow unit develops greater energy than the largest conventional hammers now available. A large (125-mkg rating) counterblow hammer is shown in Figure 4-82.

Impacters. Although impacters are commonly regarded as special forging equipment, they are fundamentally air-operated horizontal counterblow hammers. Mutually opposed rams, called "impellers," simultaneously move in from the sides to strike successive blows on the workpiece, which is held in position by a programmed manipulator. Impacters may be combined with automatic heating equipment and work-handling equipment to achieve a system of continuous, automatic impression die forging.

The types of forgings produced on the impacter are generally comparable to those produced on gravity hammers. Impacters deliver blows at velocities equal to those of gravity hammers and are available in energy ratings covering the complete range of gravity hammers and up to 5000-lb steam hammers. A relatively recent innovation, impacters are finding application in the commercial impression die forging industry. A 31-in. impacter is shown in Figure 4-83.

FIGURE 4-82. 125-Mkg counterblow hammer. Courtesy of Ladish Co.

FIGURE 4-83. Three-inch impacter. Courtesy of Ajax Manufacturing Co.

Helve and trip hammers are rapid-cycling, light-hitting hammers compared to conventional types. In the commercial forging industry, they are used principally for preforming operations such as fullering, rolling, and drawing, or for post-forging operations such as planishing and coining. They are also used in the manufacture of small forgings such as knives, chisels, and other edged tools and to provide hammered finishes on ornamental iron products.

In the trip hammer, the guided sliding ram is activated by a toggle connection to a rotating shaft at the top of the hammer. A conical clutch connected to the shaft is controlled by the operator's foot pedal so that continuously variable speeds may be transmitted to the ram, regulating the frequency and intensity of the blows. Trip hammers range in size from about 15 to 500 lb.

Helve hammers are designed for the same type of work performed by the trip hammer, but the method of transmitting power to the ram is very different. The ram is located on the end of a wooden helve (or beam) and is caused to strike the workpiece in a manner similar to the action of the blacksmith's hand sledge. Several models are in use, ranging in size up to 500 lb.

Forging Presses

Forging presses comprise the second type of primary forging equipment regularly employed in impression die forging and are commonly classified acording to the basic means used to deliver energy to the workpiece. Mechanical forging presses provide a fixed stroke; hydraulic presses have a variable stroke that can be adjusted to predetermined speeds, pressures, and dwell times.

Mechanical Forging Presses

Fundamentally, all mechanical forging presses are characterized by a ram (or slide) that moves in a vertical direction by mechanical means. Several basic types of eccentric drives—crankshaft, eccentric shafts, eccentric gears, and knuckle levers—can be used to convert the rotary motion of the drive to the reciprocating motion of the ram. However, the typical forging press comprises a heavy, rigid frame, in which energy from an electric motor is intensified in a flywheel and transmitted through an air-operated clutch to an eccentric shaft.

The press is activated by a foot pedal, and after each cycle the ram is stopped at the top of the stroke by a brake on the eccentric shaft, unless the cycle is to be repeated. Usually, mechanical ejectors or "knockouts" can be provided in top and bottom dies and are operated by cams working off the eccentric shaft.

As with a hammer, the upper die is attached to the ram, and the downward stroke of the ram exerts force on the workpiece beneath. However, the stroke of the press is of set speed, length, and duration, and energy is delivered to the workpiece at a slower rate than with the hammer. Usually, the forging is struck only once in each die impression. Maximum pressure is built up at the bottom of the work stroke, and the estimated load at this point is the basis for rating press capacity. Capacities of typical forging presses range from approximately 100 to 10,000 tons. Mechanical presses of 2500 tons and 10,000 tons are shown in Figures 4-84 and 4-85, respectively.

Some mechanical presses are double-acting in that a second power motion, in addition to the stroke, can be used. Dies can be made in two parts and moved together in a horizontal direction as the ram descends in the vertical direction. Because of its unvarying stroke, the mechanical forging press provides opportunity for consistent forging results and offers high productivity and accuracy without requirements for special operator skills.

Special Press Designs. A type of me-

FIGURE 4-84. 2500-ton mechanical press. Courtesy of National Machinery Co.

chanical press called the wedge-type press is claimed to reduce tilting under off-center loading in both directions (front to back and left to right) and to offer increased overall stiffness. In this press, the load acting upon the ram is supported by the wedge, which is driven by a two-point crank mechanism. This design greatly reduces the deflection of the drive mechanism (to about one-quarter that of a single-point drive) so that the total deflection of the press is only about 60% of that for a one-point eccentric press. The eccentric mechanism driving the wedge is provided with an eccentric bushing that can be rotated through a worm gear. Thus, the shut height, or the forging thickness, can be adjusted by using this mechanism instead of the more commonly used wedge ad-

justment at the press bed. More than 50 forging presses of this type are operating; the largest known wedge press is 12,000 tons. A wedge press with an 8000-ton capacity is in operation in a West German forge plant.

In conventional mechanical forging presses, the eccentric shaft is oriented from left to right of the press. This is not so for some sheet metal and trimming presses. A novel forging press design uses a front-to-back shaft orientation and claims to offer a number of advantages. The shaft is shorter and deflects less. During off-center forging, both eccentric shaft bearings, being located in front and back, are loaded evenly. The ram guides are built in one piece, and they are longer than in a conventional design. The pitman is connected

to the ram through an eccentric pin. By slight swiveling of the eccentric pin, the ram-to-bolster distance can be finely adjusted without modifying the lateral positioning of the dies. The adjustable pin can also be rotated hydraulically and used to free the press if it is overloaded and blocked.

The "Scotch-yoke" design represents a well-established and proven drive mechanism for forging presses. In this design, the ram contains top and bottom eccentric blocks that retain the eccentric shaft. As the shaft rotates, the eccentric blocks move in both horizontal and vertical directions, while the ram is actuated by the eccentric blocks only in the vertical direction. This design compares favorably with the wedge-type press drive and offers rigid guiding and good off-center loading capacity.

Recent developments in mechanical press design indicate a strong emphasis on large capacities, increased stiffness to provide improved forging accuracy, mechanization and automation, and high forging speed, not necessarily in terms of production rate, but rather in terms of contact time. Mechanical presses are replacing hammers in ever-increasing numbers, not only because of environmental problems, but also because, in most circumstances, mechanical press forging is less costly than hammer forging.

Large-Capacity Mechanical Presses. In recent years, presses over 8000-ton capacity have been installed or have been ordered. These giant machines are used for forging of large components for the trucking industry or for energy applications. Among the large mechanical presses are a 12,000-metric-ton German-built wedge press, an American-built 12,000-ton Scotch-yoke-design eccentric press, a Japanese-built 11,000-ton eccentric press, a Japanese-built 16,000-ton eccentric press, and a German-built 16,000-ton eccentric press.

Hydraulic Presses

Hydraulic forging presses are operated by large pistons driven by high-pressure hydraulic or hydropneumatic systems. They are usually slow moving (up to 3 in./s) under pressure. The basic difference between hydraulic press forging and other methods is that pressure is applied in a squeezing manner rather than by impact.

Among the wide variety of hydraulic presses available in the forging industry, there are two basic types: (1) direct-drive hydraulic presses, which operate with hydraulic fluid (oil or water) pressurized directly by high-pressure pumps; and (2) hy-

FIGURE 4-85. 10,000-ton mechanical press. Courtesy of Erie Press Systems.

coining operations. Because screw presses offer advantages in certain types of forming operations, more forgers have become interested in screw presses as information becomes available and more experience is accumulated. The screw press uses a friction, gear, electric, or hydraulic drive to accelerate the flywheel and the screw assembly, and it converts the angular kinetic energy into the linear energy of the slide or ram.

In the friction drive press, the driving disks are mounted on a horizontal shaft and are rotated continuously. For a downstroke, one of the driving disks is pressed against the flywheel by a servomotor. The flywheel, which is connected to the screw either positively or by a friction slip clutch, is accelerated by this driving disk through friction. The flywheel energy and the ram speed continue to increase until the ram hits the workpiece. Thus, the load necessary for forging is built up and transmitted through the slide, the screw, and the bed to the press frame. When the entire energy in the flywheel has been used in deforming the workpiece and elastically deflecting the press, the flywheel, the screw, and the slide stop. At that moment, the servometer activates the horizontal shaft and presses the upstroke driving disk wheel against the flywheel. Thus, the flywheel and the screw are accelerated in the reverse direction, and the slide is lifted to its top position.

In the direct electric drive press, a reversible electric motor is built directly on the screw and on the frame, above the flywheel. The screw is threaded into the ram or slide and does not move vertically. To reverse the direction of flywheel rotation, the electric motor is reversed after each downstroke and upstroke.

A variation of the direct electric drive is the gear drive with slip clutch used in large-capacity presses. The flysheel is in two parts: the smaller inner wheel is positively connected to the screw shaft; the outer ring, in which the larger portion of the energy is stored, is connected to the inner flywheel by a slipping clutch. Thus, the total torque is limited and, during operation, the drive gears and the screw are protected from overloading.

In a screw press, the forging load is transmitted through the slide, screw, and bed to the press frame. The available load at a given stroke position is supplied by the energy stored in the flywheel. At the end of a forging stroke, the flywheel and the screw come to a standstill before reversing the direction of rotation. Thus, the total flywheel energy is transformed into (1) available energy for the process, (2)

dropneumatic presses, which operate with hydraulic fluid supplied from accumulators that are in turn pressurized by high-pressure pumps. The large presses are of this type.

In operation, hydraulic pressure is applied to the top of the piston, moving the ram downward. When forging is completed, pressure is applied to the underside of the piston to raise the ram. Speeds and pressures can be closely controlled.

In many hydraulic presses, circuits are available that provide for a rapid advance stroke, followed by a predetermined initial pressing speed, in turn followed by a controlled, preselected second pressing speed. The press can be regulated to dwell at the bottom of the stroke for the desired time, then raised at a slow predetermined release speed, and accelerated to its original position to complete the cycle. In forging certain deformation-rate-sensitive

materials in dies with intricate contours, the slower speeds maintained under constant pressure can be advantageous. However, when needed, hydraulic press speeds can be increased considerably.

In the United States, hydraulic forging presses range up to 50,000 tons, and larger sizes are contemplated. Figure 4-86 shows a 50,000-ton hydraulic press. The 35,000-ton press shown in Figure 4-87 is extruding hollow tube vertically through the top of the press. Figure 4-88 shows a 4000-ton/23,000-ton press complex used for upsetting and impression die forging.

Screw Presses

The screw press (Figure 4-89), along with hammers and mechanical presses, is a widely used type of equipment for die forging in Europe. A few U.S. companies are using screw presses for forging or

FIGURE 4-86. 50,000-ton hydraulic press. Courtesy of Wyman-Gordon Co.

operation of all presses, whether hydraulic, mechanical, or screw. The off-center loading capacity of the press influences the parallelism of forged surfaces. This capacity is increased in modern presses by long gibs and by finish forging at the center whenever possible. The average off-center loading capacity of screw presses is much less than that of mechanical presses and hammers. Consequently, the screw press is not suitable for performing several operations in the same heat such as descaling, preforming, and trimming. Thus, when the screw press is used for finish forging, additional machines are necessary to perform these auxiliary operations.

A screw press is operated like a hammer, i.e., the top and bottom dies "kiss" at each blow. Therefore, the stiffness of the press, which affects the load and energy and characteristics, does not influence the thickness tolerances in the forged part. Due to this mode of operation, die setup is also relatively simple, because, as in hammer and hydraulic press dies, no adjustment for flash is necessary. In screw presses, in addition to static or quasistatic deflections, the dynamic behavior of the slide is also of importance. In this respect, the screw press behaves dynamically similar to the drop hammer.

Forging Machines (Upsetters)

Forging machines (upsetters) were originally developed to upset metal for bolt heads and similar shapes. For this reason, they are sometimes referred to as "headers." Modern forging machines are used to gather or upset metal preparatory to forging operations on other equipment, but are used primarily to produce a variety of complex, finished configurations with precision, such as gear blanks, bearing races, shafts, and spindles.

Forging machines are basically double-acting mechanical presses operating in a horizontal plane. Like the press, they employ a flywheel, air clutch, and eccentric shaft to operate the slide (or heading ram). In operation, the bar stock is placed against the stationary die, and the moving grip die moves laterally against the stationary die, gripping the stock tightly. As the heading ram with its attached heading tool (die) moves forward, the tool is forced against the end of the workpiece, thus displacing the stock into the die impressions. As the ram recedes, the moving die returns to its original position, releasing the workpiece. The stock is then ready for subsequent forging operations in additional die impressions, as required. In some cases, the forging is punched or sheared off the bar stock in the final step. Figure 4-90

energy to overcome machine friction, and (3) energy to elastically deflect the machine. If the total flywheel energy is larger than necessary for overcoming machine losses and for carrying out the forging process, the excess energy is transformed into additional deflection energy, and both the die and press are subjected to an unnecessarily high load. To prevent the excess energy, along with its resultant increased die wear and noise, a modern screw press is equipped with an energy-metering device that controls the flywheel velocity and regulates the total flywheel energy. The energy metering can also be programmed so that the machine supplies different amounts of energy during successive blows.

A screw press designed for forging operations in which large energies are needed can also be used for coining where smaller energies are required. Here, however, a friction clutch is installed between the flywheel and the screw. When the ram load reaches the nominal load, this clutch starts slipping and uses part of the flywheel energy as frictional heat energy at the clutch. Consequently, the maximum load at the end of downstroke is reduced, and the press is protected from overloading. Screw presses used for coining are designed for hard blows (i.e., die-to-die blows without any workpiece) and do not have a friction-slip-clutch on the flywheel.

In general, the dimensional accuracies of press components under unloaded conditions, such as parallelism of slide and bed surfaces, clearances in the gibs, etc., have basically the same significance in the

shows stock in the three stages of upsetting.

Reminiscent of previous days, forging machines are rated for size according to the maximum bolt size for which they can provide an upset head. Thus, a 2-in. upsetter could theoretically head bolts or sizes up to 2-in. in stem diameter. The rating has nothing whatsoever to do with the size of stock that can be accommodated or the machine's forging ability. A rough measure of hot upsetter tonnage can be made by multiplying the "inch" rating by 200 tons, i.e., 2-in. upsetter = 400 tons. Forging machines range in rating from $1/4$ to 10 in. and produce upset forgings of comparable size to all but the largest impression die forgings made on gravity hammers. A quick forging machine with automatic feed is shown in Figure 4-91.

Special hot upsetting machines have been developed to be used in upsetting the ends of tubes. These machines normally have a much longer die length to provide an increase in the grip length to hold the tubes from slipping. A double knuckle system is employed in the die slide closing mech-anism to help the grip dies to close with both ends parallel.

Various small forging machines, equipped with automatic work-handling equipment and stock transfer mechanisms, use multiple-station dies in the rapid production of finished products, such as fasteners complete with upset and trimmed heads, cold rolled threads, and chamfered ends. In these applications, coiled stock is fed into the machine from a reel. These are usually called "cold headers" or "bolt-making machines." They are currently made in sizes up to $1^7/_8$-in.-diameter cold stock.

When a large volume of parts is to be produced, e.g., automotive and bearing manufacturers, automatic, multiple-station hot forging machines are increasingly being used. They are fed with bars that are heated to forging temperatures by induction heating, resistance heating, or gas furnaces. The stock is sheared hot and transferred in a similar manner to cold transfer headers. Multiple-station forging can produce complex designs and configurations at high production rates.

FIGURE 4-87. 35,000-ton press extruding hollow tube. Courtesy of Cameron Iron Works.

FIGURE 4-88. 4000-ton/23,000-ton hydraulic press complex for upsetting and impression die forging. Courtesy of Cameron Iron Works.

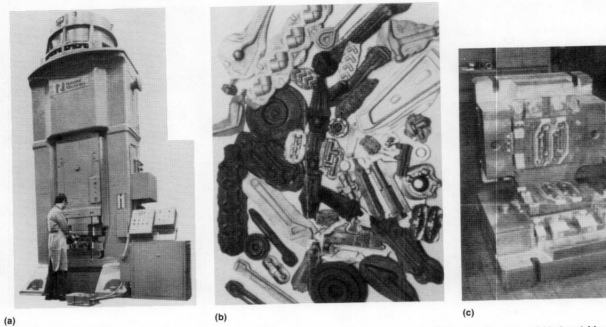

(a)   (b)   (c)

FIGURE 4-89. (a) Screw press in operation with (b) typical forged parts and (c) bolster set. Courtesy of National Machinery Co.

FIGURE 4-90. Upsetter showing three stages in upsetting sequence. Courtesy of Ajax Manufacturing Co.

FIGURE 4-91. Nine-inch automatic-feed forging machine. Courtesy of National Machinery Co.

Selection of Forging Equipment

Contributing Author: Dr. Taylan Altan, Senior Research Leader, Battelle's Columbus Laboratories, Columbus, Ohio

Forging equipment influences the forging process, because it affects deformation rate, temperature, and rate of production. The forging engineer must have sound knowledge of different forging machines to:

- Use existing machinery more efficiently

- Define the existing plant capacity with accuracy

- Better communicate with, and at times request improved performance from, the machine builder

- If necessary, develop in-house proprietary machines and processes not available in the machine-tool market

Interaction Between Process Requirements and Forging Machines

The interaction between the principal machine and process variables is illustrated in Figure 4-92 for hot forging conducted in presses. As shown at the left in Figure 4-92, the flow stress, $\bar{\sigma}$, the interface friction conditions, and the forging geometry (dimensions and shape) determine the load, L_p, at each position of the stroke and the energy, E_p, required by the forming process. The flow stress, $\bar{\sigma}$, increases with increasing deformation rate, $\dot{\varepsilon}$, and with decreasing temperature, θ. The magnitudes of these variations depend on

the specific workpiece material. The frictional conditions deteriorate with increasing die chilling.

As indicated by the lines connected to the temperature block, for a given initial stock temperature, the temperature variations in the part are largely influenced by the surface area of contact between the dies and the part, the part thickness or volume, the die temperature, the amount of heat generated by deformation and friction, and the contact time under pressure.

The velocity of the slide under pressure, V_p, determines mainly the contact time under pressure, t_p, and the deformation rate, $\dot{\varepsilon}$. The number of strokes per minute under no-load conditions, n_0, the machine energy, E_M, and the deformation energy, E_p, required by the process influence the slide velocity under load, V_p, and the number of strokes under load, n_p; n_p determines the maximum number of parts formed per minute (i.e., the production rate) provided the feed and unloading of the machine can be carried out at that speed. The relation-

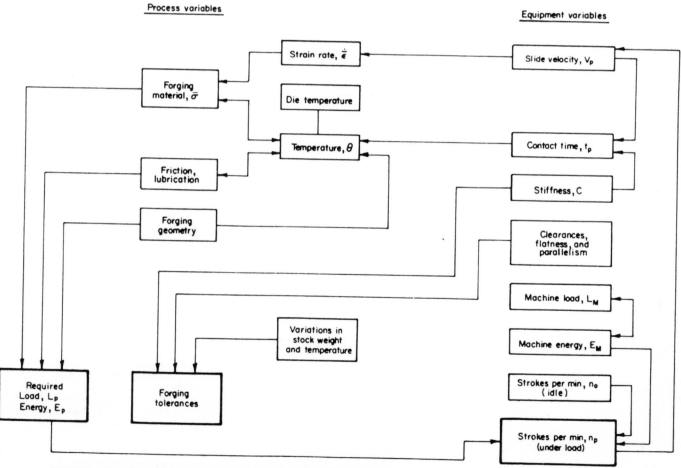

FIGURE 4-92. Relationship between process and machine variables in hot forging processes conducted in presses.

ships illustrated in Figure 4-92 apply directly to hot forging in hydraulic, mechanical, and screw presses.

For a given material, a specific forging operation (such as closed die forging with flash, forward or backward extrusion, upset forging, bending, etc.) requires a certain variation of the load over the slide displacement (or stroke). This fact is illustrated qualitatively in Figure 4-93, which shows load versus displacement curves characteristic of various forming operations. For a given part geometry, the absolute load values will vary with the flow stress of the material as well as with frictional conditions. In the forming operation, the equipment must supply the maximum load as well as the energy required by the process.

Classification and Characterization of Forging Machines

Forging machines can be classified into three types:

- Load-restricted machines (hydraulic presses)
- Stroke-restricted machines (crank and eccentric mechanical presses)
- Energy-restricted machines (hammers and screw presses)

The significant characteristics of these machines comprise all machine design and performance data that are pertinent to the machine's economic use, including characteristics of load and energy, time-related characteristics, and characteristics of accuracy.

Hydraulic Presses

The operation of hydraulic presses is relatively simple and is based on the motion of a hydraulic piston guided in a cylinder. Hydraulic presses are essentially load-restricted machines; i.e., their capability for carrying out a forming operation is limited mainly by the maximum available load.

The operational characteristics of a hydraulic press are essentially determined by the type and design of its hydraulic drive system. The two types of hydraulic drive systems shown in Figure 4-94 provide different time-dependent characteristic data. Direct-driven presses usually employ hydraulic oil as the working medium. Accumulator-driven presses usually employ a water-oil emulsion as the working medium and use nitrogen, steam, or air-loaded accumulators to keep the medium under pressure. The sequence of operations is essentially similar to that for the direct-driven press, except that the pressure is built up by means of the pressurized water-oil emulsion in the accumulators.

In both direct and accumulator drives, as the pressure builds up and the working medium is compressed, a slowdown in penetration rate occurs. This slowdown is larger in direct oil-driven presses, mainly because oil is more compressible than a water emulsion.

The approach and initial deformation speeds are higher in accumulator-driven presses. This improves the hot forging conditions by reducing contact times, but wear in hydraulic elements of the system also increases. Sealing problems are somewhat less severe in direct-oil drives, and control and accuracy in manual operation are in general about the same for both types of drives.

From a practical point of view, in a new installation the choice between direct and accumulator drive is decided by the economics of operation. Usually, the accumulator drive is more economical if one accumulator system can be used by several presses, or if very large press capacities (10,000 to 50,000 tons) are considered. In direct-driven hydraulic presses, the maximum press load is established by the pressure capability of the pumping system and is available throughout the entire press stroke. Thus, hydraulic presses

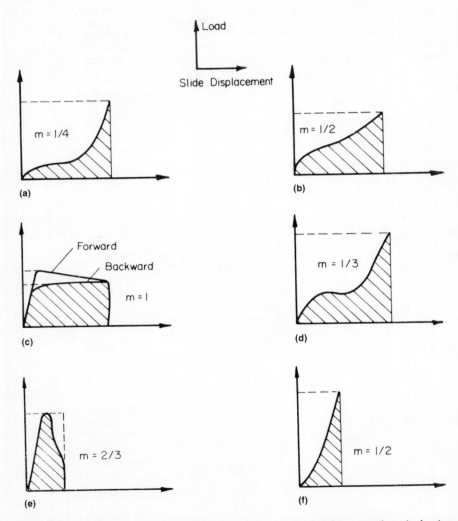

FIGURE 4-93. Load-versus-displacement curves for various forming operations in forging. Energy = load × displacement × m, where m is a factor characteristic of the specific forming operation. (a) Closed die forging (with flash). (b) Upset forging (without flash). (c) Forward and backward extrusion. (d) Bending. (e) Blanking. (f) Coining. Source: Ref. 48, 49.

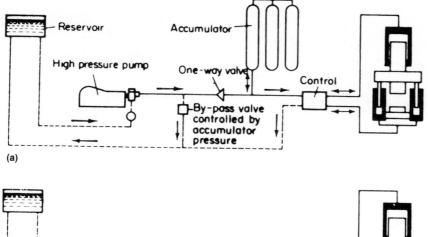

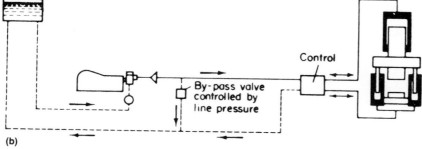

FIGURE 4-94. Schematic of drives for hydraulic presses. (a) Accumulator drive. (b) Direct drive. Source: Ref. 50.

are ideally suited for extrusion-type operations requiring very large amounts of energy. With adequate dimensioning of the pressure system, an accumulator-driven press exhibits only a slight reduction in available press load as the forming operation proceeds.

In comparison with direct drive, the accumulator drive usually offers higher approach and penetration speeds and a shorter dwell time prior to forging. However, the dwell at the end of processing and prior to unloading is larger in accumulator drives. This is illustrated in Figures 4-95 and 4-96, in which the load and displacement variations are given for a forming process using a 2500-ton hydraulic press equipped with either accumulator- or direct-driven systems.

Mechanical Crank and Eccentric Presses

The drive system used in most mechanical presses (crank or eccentric) is based on a slider-crank mechanism that translates rotary motion into reciprocating linear motion. The eccentric shaft is connected through a clutch and brake system directly to the flywheel (Figure 4-97). In

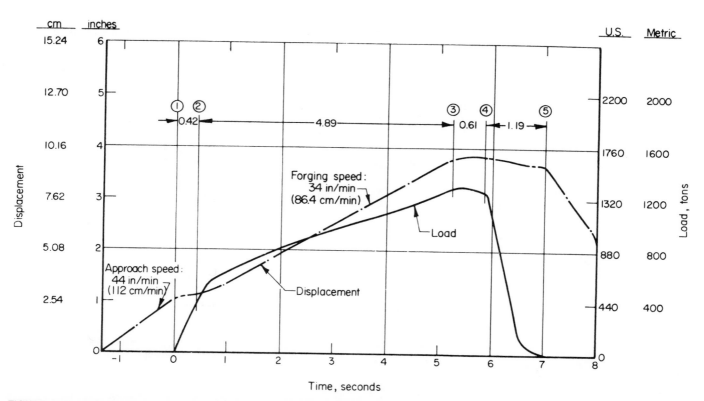

FIGURE 4-95. Load- and displacement-versus-time curves obtained on a 2500-ton hydraulic press in upsetting with direct drive. 1, start of deformation; 2, initial dwell; 3, end of deformation; 4, dwell before pressure release; 5, ram lift. Source: Ref. 51.

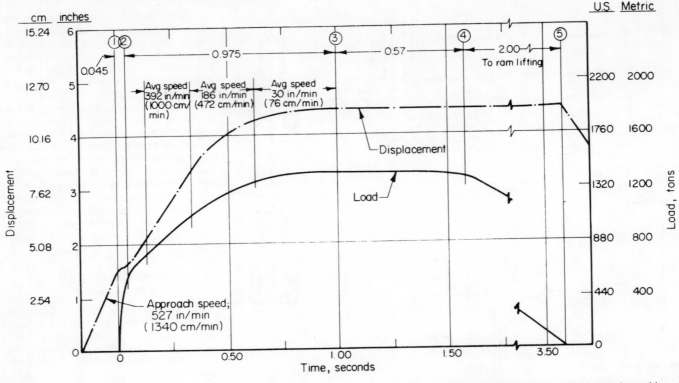

FIGURE 4-96. Load- and displacement-versus-time curves obtained on a 2500-ton hydraulic press in upsetting with accumulator drive. 1, start forming; 2, initial dwell; 3, end of forming; 4, dwell before pressure release; 5, ram lift. Source: Ref. 51.

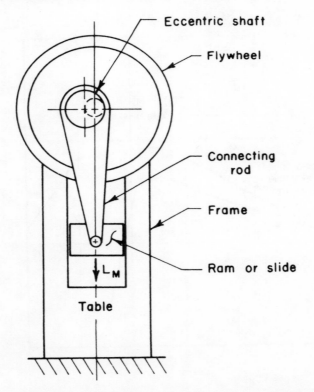

FIGURE 4-97. Schematic of a mechanical press with eccentric drive (clutch and brake on eccentric shaft). Source: Ref. 51.

designs for larger capacities, the flywheel is located on the pinion shaft, which drives the eccentric shaft (Figure 4-98).

Kinematics of the Slider-Crank Mechanism. The slider-crank mechanism is presented schematically in Figure 4-99. The following valid relationships can be derived from the geometry illustrated.

The distance (w) of the slide from the lowest possible ram position, the "bottom dead center" (BDC) position, can be expressed in terms of r, l, s, and α, where (from Figure 4-99) r is the radius of the crank or half of the total stroke (2r = S), l is the length of the pitman arm, and α is the crank angle before BDC.

Because the ratio of r/l is usually small, a close approximation is:

$$w = \frac{S}{2}(1 - \cos\alpha) \quad (Eq\ 4\text{-}1)$$

Equation 4-1 gives the location of the slide at a crank angle (α) before BDC. This curve is plotted in Figure 4-99(b), along with the slide velocity (V), which is given by the close approximation:

$$V = \frac{S\pi n}{60}\sin\alpha \quad (Eq\ 4\text{-}2)$$

where n is the number of strokes per minute.

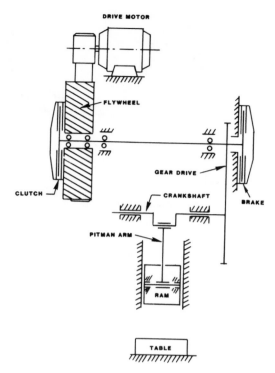

DRIVE MOTOR

FLYWHEEL

CLUTCH

GEAR DRIVE

CRANKSHAFT

BRAKE

PITMAN ARM

RAM

TABLE

FIGURE 4-98. Schematic of crank press with pinion gear drive. Clutch and brake are on pinion shaft; for large capacities, this design is more stable and provides high flywheel energy. Source: Ref. 33.

The slide velocity (V) with respect to slide location (w) before BDC is given by:

$$V = 0.105 \; wn \sqrt{\frac{S}{w} - 1} \quad \text{(Eq 4-3)}$$

Thus, Equations 4-1 and 4-2 give the slide position and the slide velocity at an angle α above BDC. Equation 4-3 gives the slide velocity for a given position (w) above BDC, if the number of strokes per minutes (n) and the press stroke (S) are known.

Load and Energy Characteristics. An exact relation exists between the torque (M) of the crankshaft and the available load (L) at the slide (Figure 4-99a and c). The torque (M) is constant, and for all practical purposes, angle β is small enough to be ignored (Figure 4-99a). A very close approximation then is given by:

$$L = \frac{2M}{S \; \sin\alpha} \quad \text{(Eq 4-4)}$$

Equation 4-4 gives the variation of the available slide load (L) with respect to the crank angle (α) above BDC, as shown in Figure 4-99(c). From Equation 4-4, it is apparent that as the slide approaches the BDC—i.e., as angle α approaches zero—the available load (L) may become infi-

nitely large without exceeding the constant clutch torque (M), or without causing the friction clutch to slip.

From the observations that have been made thus far, the following conclusions are drawn. Crank and the eccentric presses are displacement-restricted machines. The slide velocity (V) and the available slide load (L) vary accordingly with the position of the slide before BDC. Most manufacturers in the United States and the United Kingdom rate their presses by specifying the nominal load at 1/2 in. before BDC. For different applications, the nominal load may be specified at different positions before BDC, according to the standards established by the American Joint Industry Conference. If the load required by the forming process is smaller than the load available at the press (i.e., if the curve EFG in Figure 4-99(c) remains below the curve BCD), the process can be carried out, provided the flywheel can supply the necessary energy per stroke.

For small angles (α) above BDC, within the CD portion of the curve BCD in Figure 4-99(c), the slide load (L) can become larger than the nominal press load if no overload safety (hydraulic or mechanical) is available on the press. In this case, the press stalls, the flywheel stops, and the

entire flywheel energy is transformed into deflection energy by straining the press frame, the pitman arm, and the drive mechanism. The press can be freed in most cases only by burning out the tooling.

If the applied load curve EFG exceeds the press load curve BCD (Figure 4-99c) before point C is reached; then, the friction clutch slides, the press slide stops, but the flywheel continues to turn. In this case, the press can be freed by increasing the clutch pressure and by reversing the flywheel rotation if the slide has stopped before BDC.

The energy needed for the forming process during each stroke is supplied by the flywheel, which slows to a permissible percentage, usually 10 to 20% of its idle speed. The total energy stored in a flywheel is:

$$E_{FT} = \frac{I\omega^2}{2} = \frac{I}{2} \; \frac{\pi n}{30} \quad \text{(Eq 4-5)}$$

where I is the moment of inertia of the flywheel, ω is the angular velocity in radians per second, and n is the rotation speed of the flywheel.

The total energy (E_s) used during one stroke is:

$$E_s = 2\frac{1}{I} \; \omega_0^2 - \omega_1^2 = \frac{I}{2} \; \frac{\pi^2}{30} \; a_0^2 - n_1^2 \quad \text{(Eq 4-6)}$$

where ω_0 is the initial angular velocity, ω_1 is the angular velocity after the work is done, n_0 is the initial flywheel speed, and n_1 is the flywheel speed after the work is done.

Note that the total energy (E_s) also includes the friction and elastic deflection losses. The electric motor must bring the flywheel from its slowed speed (n_1) to its idle speed (n_0) before the next stroke for forging starts. The time available between two strokes depends on the mode of operation—namely, continuous or intermittent. In a continuously operating mechanical press, less time is available to bring the flywheel to its idle speed; consequently, a larger horsepower motor is necessary.

Frequently, the allowable slowdown of the flywheel is given in percentage of the nominal speed. For instance, if 13% slowdown is permissible, then:

$$\frac{n_0 - n_1}{n_0} = \frac{13}{100} \text{ or } n_1 = 0.87 \; n_0 \quad \text{(Eq 4-7)}$$

The percentage energy supplied by the flywheel is obtained by using Equations 4-5 and 4-6 to give:

$$\frac{E_s}{E_{FT}} = \frac{n_0^2 - n_1^2}{n_0^2}$$
$$= 1 - (0.87)^2 = 0.25 \quad \text{(Eq 4-8)}$$

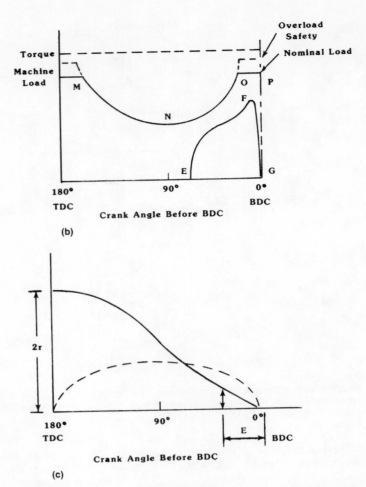

FIGURE 4-99. Load, displacement, velocity, and torque in a simple slider-crank mechanism. (a) Slider-crank mechanism. (b) Torque and load. (c) Displacement and velocity. Source: Ref. 51.

Equations 4-7 and 4-8 illustrate that for a 13% slowdown of the flywheel, 25% of the flywheel energy will be used during one stroke.

Time-Dependent Characteristics. The number of strokes per minute (n) has been discussed previously as an energy consideration. For a given idle flywheel speed, the contact time under pressure (t_p) and the velocity under pressure (V_p) depend primarily on the dimensions of the slide-crank mechanism and on the total stiffness (C) of the press. The effect of press stiffness on contact time under pressure (t_p) is shown in Figure 4-100. As the load builds, the press deflects elastically. A stiffer press (larger C) requires less time (t_{p1}) for pressure to build and also less time (t_{p2}) for pressure release, as shown in Figure 4-100a. Consequently, the total contact time under pressure ($t_p = t_{p1} + t_{p2}$) is less for a stiffer press.

Characteristics for Accuracy. The working accuracy of a forging press is substantially characterized by two features: the tilting angle of the ram under off-center loading and the total deflection

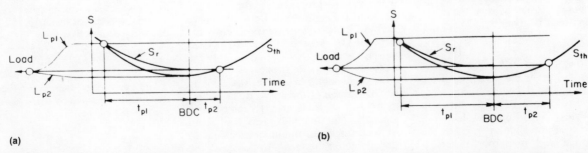

FIGURE 4-100. Effect of press stiffness on contact time under pressure. (a) Stiffer press. (b) Less stiff press. Source: Ref. 52.

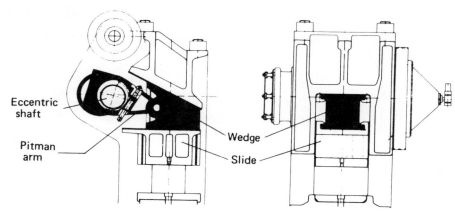

FIGURE 4-101. Design principle of the wedge-type press. Source: Ref. 53.

TABLE 4-9. Distribution of Total Deflection

	Eccentric press (one point), %	Eccentric press (two point), %	Wedge press, %
Slide and Pitman arm 30	21	21 (includes wedge)	
Frame 33	31	29	
Drive shaft and bearings 37	33	10	
Total deflection 100	85	60	

Source: Ref. 53.

under load or stiffness of the press. The tilting of the ram produces skewed surface and an offset on the forging; the stiffness influences the thickness tolerance.

Under off-center loading conditions, two- or four-point presses perform better than single-point presses, because the tilting of the ram and the reaction forces into gibways are minimized. A relatively new type of mechanical press, called the wedge-type press, has been claimed to reduce tilting under off-center stiffness. The design principle of the wedge-type press is shown in Figure 4-101. In this press, the load acting on the ram is supported by the wedge, which is driven by a two-point crank mechanism.

Assuming the total deflection under load for a one-point eccentric press to be 100%, the distribution of the total deflections was obtained after measurement under nominal load on equal-capacity forging presses (Table 4-9). It is interesting to note that a large percentage of the total deflection is in the drive mechanism, i.e., slide, pitman arm, drive shaft, and bearings.

For the same presses discussed above, Figure 4-102 illustrates table-load diagrams, which show, in the percentage of the nominal load, the amount and location of off-center load that causes the tilting of the ram. The wedge-type press has advantages, particularly in front-to-back off-center loading. In this respect, it performs like a four-point press.

Crank Presses With Modified Drives. The velocity versus stroke and the load versus stroke characteristics of crank presses can be modified by using different press drives. A well-known variation of the crank press is the knuckle joint design (Figure 4-103). This design is capable of generating high forces with a relatively small crank drive. In the knuckle joint drive, the ram velocity slows much more rapidly toward the BDC than the regular crank drive. This machine is successfully used for cold forming and coining applications.

A recently developed mechanical press drive uses a four-bar linkage mechanism, as shown in Figure 4-104. In this mechanism, the load-stroke and velocity-stroke behavior of the slide can be established at the design stage by adjusting the length of one of the four links, or by varying the connection point of the slider link with the drag link. Thus, with this press, it is possible to maintain the maximum load, as specified by press capacity, over a relatively long deformation stroke. Using a conventional slider-crank-type press, this capability can be achieved only by using a much larger capacity press. A comparison is illustrated in Figure 4-105, in which the load-stroke curves for a four-bar linkage press and a conventional slider-crank press are shown. It is apparent that a slider-crank press equipped with a 1700 ton-in. torque drive can generate a force of about 1500 tons at 1/32 in. above BDC. The four-bar press equipped with a 600 ton-in. drive generates a force of about 750

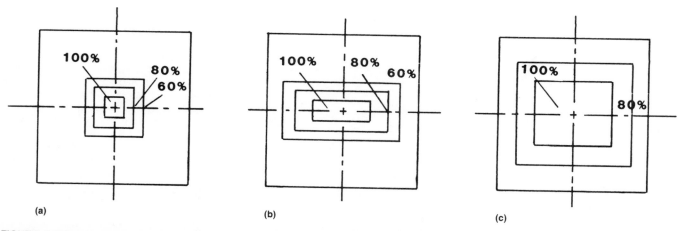

FIGURE 4-102. Amount and location of off-center load that causes tilting of the ram. (a) Eccentric, one point. (b) Eccentric, two point. (c) Crank wedge. Source: Ref. 53.

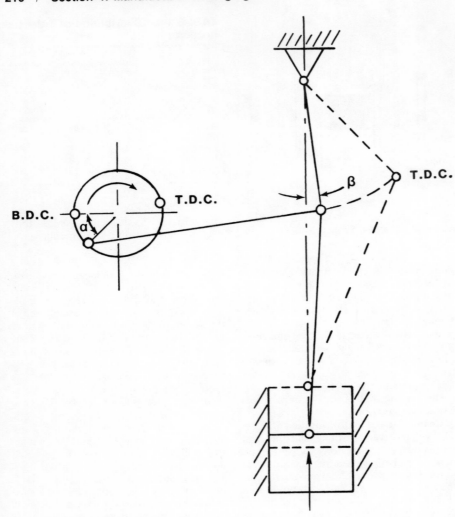

FIGURE 4-103. Schematic of a toggle (or knuckle) joint mechanical press. After Ref. 54.

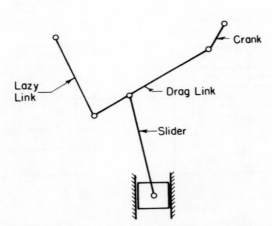

FIGURE 4-104. Four-bar linkage mechanism for mechanical press drives. Source: Ref. 55.

tons at the same location. However, in both machines, a 200-ton force is available at 6 in. above BDC. Thus, a 750-ton, four-bar press could perform the same forming operation, requiring 200 tons over 6 in., as a 1500-ton eccentric press. The four-bar press, which was originally developed for sheet metal forming and cold extrusion, is well suited for extrusion-type forming operations, in which a nearly constant load is required over a long stroke.

Screw Presses

As mentioned above, the screw press uses a friction, gear, electric, or hydraulic drive to accelerate the flywheel and the screw assembly, and it converts the angular kinetic energy into the linear energy of the slide or ram. Figure 4-106 shows two basic designs of screw presses.

Load and Energy. In screw presses, the forging load is transmitted through the slide, screw, and bed to the press frame. The available load at a given stroke position is supplied by the stored energy in the flywheel. At the end of the downstroke after the forging blow, the flywheel comes to a standstill and starts its reversed rotation. During the standstill, the flywheel no longer contains any energy. Thus, the total flywheel energy (E_{FT}) has been transformed into:

- Available energy for deformation (E_M) to carry out the forging process
- Friction energy (E_f) to overcome frictional resistance in the screw and in the gibs
- Deflection energy (E_d) to elastically deflect various parts of the press

At the end of a downstroke, the deflection energy (E_d) is stored in the machine and can be released only during the upward stroke.

The load versus displacement diagrams of a forging operation are illustrated in Figure 4-107. The flywheel in Figure 4-107(a) is accelerated to such a velocity that at the end of downstroke the deformation is carried out, and no unnecessary energy is left in the flywheel. This is done by using an energy-metering device that controls flywheel velocity. The flywheel in Figure 4-107(b) has excess energy at the end of the downstroke. The excess energy causes additional unnecessary elastic straining of the press by transformation into additional deflection energy.

From the above discussion, it is apparent that, in screw presses, the load and energy are in direct relation with each other. For given friction losses, elastic deflection properties, and available flywheel energy,

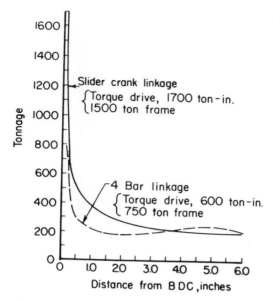

FIGURE 4-105. Load-stroke curves for a 750-ton four-bar linkage press and a 1500-ton slider-crank press. Source: Ref. 55.

heat energy (E_e) at the clutch. Consequently the maximum load at the end of downstroke is reduced from L to L_{max}.

The energy versus load curve has a parabolic shape so that energy decreases with increasing load. This is due to the fact that the deflection energy (E_d) is given by a second-order equation:

$$E_d = \frac{1}{2}\,dL = \frac{L^2}{2C} \quad \text{(Eq 4-9)}$$

where d is the elastic deflection of the press, L is the load, and C is the total stiffness of the press.

A screw press can be designed so that it can sustain die-to-die blows without any workpiece for maximum energy of the flywheel. In this case, a friction clutch between the flywheel and the screw is not required. It is important to note that a screw press can be designed and used for forging operations in which large deformation energies are required, or for coining operations in which small energies but high loads are required. Another interesting feature of screw presses is that they cannot be loaded beyond the calculated overload limit of the press.

Time-Dependent Characteristics. For a screw press, the number of strokes per minute (n) is a dependent characteristic. Because modern screw presses are equipped with energy-metering devices, the number of strokes per minute depends on the energy required by the process. In general, however, the production rate of screw

the load available at the end of the stroke depends mainly on the deformation energy required by the process. Thus, for a constant flywheel energy, low deformation energy (E_p) results in high end load (L_M), and high deformation energy (E_p) results in low end load (L_M). These relations are illustrated in Figure 4-108, a load-energy diagram of a screw press.

The screw press generally can sustain

maximum loads (L_{max}) up to 160 to 200% of its nominal load (L_M). Hence, the nominal load of a screw press is set rather arbitrarily. The significant information about the press load is obtained from its load-energy diagram (Figure 4-108). Many screw presses have a friction clutch between the flywheel and the screw. At a preset load, this clutch starts to slip and uses part of the flywheel energy as friction

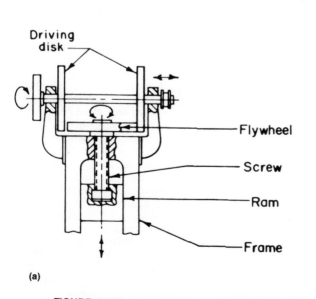

(a)

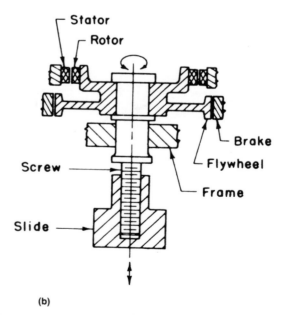

(b)

FIGURE 4-106. Two common screw press drives. (a) Friction drive. (b) Direct electric drive. Source: Ref. 56, 57.

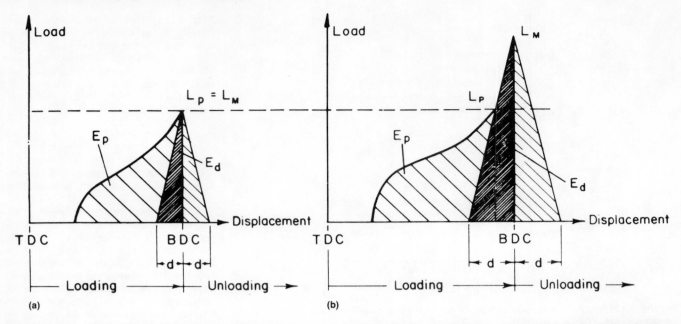

FIGURE 4-107. Load and energies in die forging under a screw press. (a) With energy or load metering. (b) Without energy or load metering. E_p, energy required by process; L_p, load required by process; L_M, maximum machine load; E_d, elastic deflection energy; d, elastic deflection of the press. Source: Ref. 58.

presses compares with that of mechanical presses.

The velocity under pressure (V_p) is generally higher than in mechanical presses, but lower than in hammers. This is because the slide velocity of a mechanical press slows toward the BDC, and the velocity of the slide in a screw press accelerates until deformation starts and the load builds. This fact is more pronounced in forging thin parts such as airfoils or in coining and sizing operations.

The contact time under pressure (L_p) is related directly to the ram velocity and to the stiffness of the press. In this respect, the screw press ranks between the hammer and the mechanical press. The contact times (t_p) for screw presses are 20 to 30 times longer than for hammers. A similar comparison with mechanical presses cannot be made without specifying the thickness of the forged part. In forging turbine blades, which require small displacement but large loads, the contact times for screw presses have been estimated to be one fourth to one tenth of that for mechanical presses.

Variations in Screw Press Drives. In addition to direct friction and electric drives, there are several other types of mechanical, electric, and hydraulic drives that are commonly used in screw presses. A relatively new screw press drive is shown in Figure 4-109. A flywheel (1), supported on the press frame, is driven by one or several electric motors and rotates at a constant speed. When the stroke is initiated, an air-operated clutch (2) engages the rotating flywheel against the stationary screw (3). This feature is similar to that used to initiate the stroke of an eccentric forging press. On engagement of the clutch, the screw is accelerated rapidly and reaches the speed of the flywheel. As a result, the ram (4), which acts like a giant nut, moves downward. During this downstroke, the air is compressed in the pneumatic lift-up cylinders. The downstroke is terminated by controlling either the ram position, by means of a position switch, or the maximum load on the ram, by disengaging the clutch and the flywheel from the screw when the preset forming load is reached. The ram is then lifted by the lift-up cylinders, releasing the elastic energy stored in the press frame, the screw and, the lift-

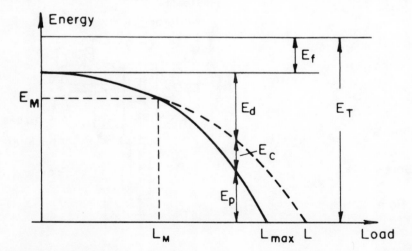

FIGURE 4-108. Energy-load curve of a screw press. ---, without friction clutch at flywheel; —, with slipping friction clutch at flywheel; E_T, total flywheel energy; E_f, friction energy; E_d, deflection energy; E_p, energy required by process; E_c, energy lost in slipping clutch; L_M, nominal machine load; L_{max}, maximum load; E_M, nominal machine energy available for forging. Source: Ref. 59.

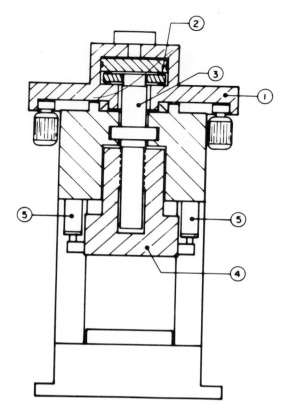

FIGURE 4-109. New screw press drive that combines the characteristics of mechanical and screw presses. 1, flywheel; 2, air-operated clutch; 3, screw; 4, ram; 5, lift-up cylinders. Courtesy of G. Siempelkamp GmbH & Co. (Krefeld, West Germany).

up cylinders. At the end of the upstroke, the ram is stopped and held in position by a hydraulic brake.

This press provides several distinct benefits: a high and nearly constant ram speed throughout the stroke, full press load at any position of the stroke, high deformation energy, overload protection, and short contact time between the workpiece and the tools. The press can also be equipped with variable-speed motors so that different flywheel and ram speeds are available. Thus, it offers considerable flexibility and can be used for hot as well as cold forming operations.

Hammers

The hammer is the least expensive and most versatile type of equipment for generating load and energy to carry out a forming process. Hammers are used primarily for hot forging, for coining, and, to a limited extent, for sheet metal forming of parts manufactured in small quantities—for example, in the aircraft/airframe industry. The hammer is an energy-restricted machine. During a work-

ing stroke, the deformation proceeds until the total kinetic energy is dissipated by plastic deformation of the material and by elastic deformation of the ram and anvil when the die faces contact each other. Therefore, it is necessary to rate the capacities of these machines in terms of energy, i.e., foot-pounds, meter-kilograms, or meter-tons. The practice of specifying a hammer by its ram weight is not useful for the user. Ram weight can be regarded only as model or specification number.

There are basically two types of anvil hammers: gravity-drop and power-drop types. In a simple gravity-drop hammer, the upper ram is positively connected to a board (board-drop hammer), a belt (belt-drop hammer), a chain (chain-drop hammer), or a piston (oil-, air-, or steam-lift drop hammer) (Figure 4-110). The ram is lifted to a certain height and then dropped on the stock placed on the anvil. During the downstroke, the ram is accelerated by gravity and builds up the blow energy. The upstroke takes place immediately after the blow; the force necessary to ensure quick lift-up of the ram can be three to five times the ram weight. The operation principle of

a power-drop hammer is similar to that of an air-drop hammer (Figure 4-110d). In the downstroke, in addition to gravity, the ram is accelerated by steam, cold air, or hot air pressure. Electrohydraulic gravity-drop hammers, introduced in the United States in recent years, are used more commonly in Europe. In this hammer, the ram is lifted with oil pressure against an air cushion. The compressed air slows the upstroke of the ram and contributes to its acceleration during the downstroke. Thus, the electrohydraulic hammer also has a minor power hammer action.

Counterblow hammers are used widely in Europe; their use in the United States is limited to a relatively small number of companies. The principles of two types of counterblow hammers are illustrated in Figure 4-111. In both designs, the upper ram is accelerated downward by steam, cold air, or hot air. At the same time, the lower ram is accelerated by a steel band (for smaller capacities) or by a hydraulic coupling system (for larger capacities). The lower ram, including the die assembly, is approximately 10% heavier than the upper ram. Therefore, after the blow, the lower ram accelerates downward and pulls the upper ram back up to its starting position. The combined speed of the rams is about 25 ft/s; both rams move with exactly half of the total closure speed. Due to the counterblow effect, relatively little energy is lost through vibration in the foundation and environment. Therefore, for comparable capacities, a counterblow hammer requires a smaller foundation than an anvil hammer.

Characteristics of Hammers. In a gravity-drop hammer, the total blow energy is equal to the kinetic energy of the ram and is generated solely through free-fall velocity, or:

$$E_T = \frac{1}{2} m_1 V_1^2 = \frac{1}{2} \frac{G_1}{g} V_1^2 = G_1 H$$

where m_1 is the mass of the dropping ram, V_1 is the velocity of the ram at the start of deformation, G_1 is the weight of the ram, g is the acceleration of gravity, and H is the height of the ram drop.

In a power-drop hammer, the total blow energy is generated by the free fall of the ram and by the pressure acting on the ram cylinder, or:

$$E_T = \frac{1}{2} m_1 V_1^2 = (G_1 + pA)H$$

where, in addition to the symbols given above, p is the air, stream, or oil pressure acting on the ram cylinder in the down-

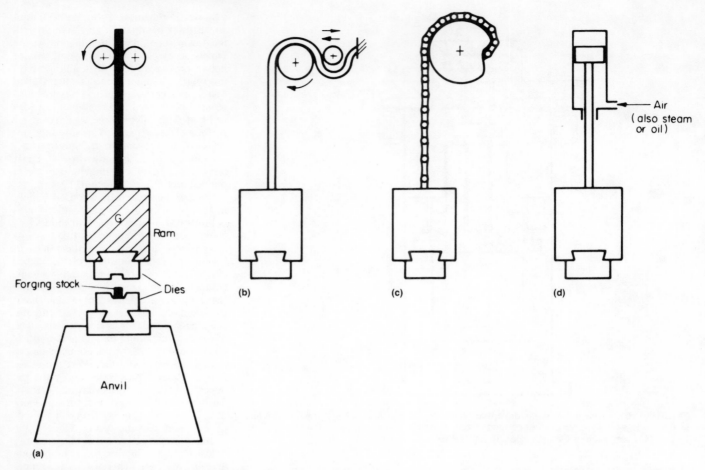

FIGURE 4-110. Principles of various types of gravity-drop hammers. (a) Board drop. (b) Belt drop. (c) Chain drop. (d) Air drop. Source: Ref. 51.

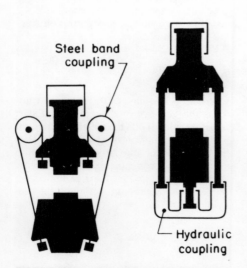

FIGURE 4-111. Principles of operation of two types of counterblow hammers. Courtesy of Beche.

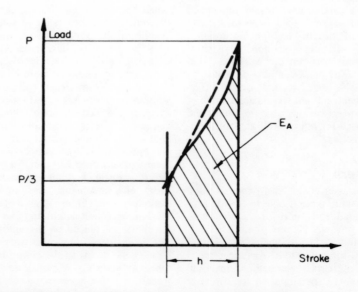

FIGURE 4-112. Example of a load-stroke curve in a hammer blow. E_A = energy available for forging = $E_{T\eta}$. Source: Ref. 60.

stroke, and A is the surface area of the ram cylinder.

In counterblow hammers, when both rams have approximately the same weight, the total energy per blow is given by:

$$E_T = 2\left(\frac{m_1 V_1^2}{2}\right) = \frac{m_1 V_1^2}{4} = \frac{G_1 V_1^2}{4g}$$

where m_1 is the mass of one ram; V_1 is the velocity of one ram; V_t is the actual velocity of the blow of the two rams, which is equal to $2V_1$; and G_1 is the weight of one ram.

During a working stroke, the total nominal energy, E_T, of a hammer is not entirely transformed into useful energy available for deformation, E_A. A small amount of energy is lost in the form of noise and vibration to the environment. Thus, the blow efficiency, $n = E_A E_T$, of hammers varies from 0.8 to 0.9 for soft blows (small load and large displacement) and from 0.2 to 0.5 for hard blows (high load and small displacement).

The transformation of kinetic energy into deformation energy during a working blow can develop considerable force. For example, consider a deformation blow where the load, P, increases from P/3 at the start to P at the end of the stroke, h. The available energy, E_A, is the surface area under the curve in Figure 4-112. Therefore:

$$E_A = \frac{P/3 + P}{2} h = \frac{4Ph}{6}$$

Consider a hammer with a total nominal energy, E_T, of 35,000 ft·lb and a blow efficiency, η, of 0.4; here, $E_A = E_{T\eta} = 14,000$ ft·lb. With this value, for a working stroke, h, of 0.2 in., Equation 4-10 gives:

$$P = \frac{6E_A}{4h} = 1,260,000 \text{ lb} = 630 \text{ tons}$$

If the same energy were dissipated over a stroke, h, of 0.1-in., the load, P, would reach approximately double the calculated value. The simple hypothetical calculations given above illustrate the capabilities of relatively inexpensive hammers in exerting high forming loads.

Tool and Die Manufacture

Contributing Authors: M. "Bubs" Wells, Consultant, Downey, California; and Dr. S. L. Semiatin, Principal Research Scientist, Battelle's Columbus Laboratories, Columbus, Ohio

In the final analysis, it is the design and fabrication of dies and tooling that hold the key to the manufacture of any forged part. With good forging design, many parts can be produced before the dies must be replaced or reworked; poor design may result in the production of only a few parts before catastrophic die failure via gross fracture occurs. In addition to proper design, material selection is important, followed by multistep processes to facilitate a set of dies that will produce the part to proper dimensions and in sufficient quantities to justify the expense required to make the tooling. Aspects of tooling manufacture are addressed in this section.

Material Considerations

Die life and die failure are affected significantly by the mechanical properties of the die materials (usually steels) under the conditions that exist in performing a given forging operation. Generally, the properties that are most significant depend on process temperature. Thus, die steels used in hot forging processes are quite different from those used in cold forging.

Die Steels for Hot Forging

Die steels commonly used for hot forging can be grouped in terms of alloy content (Table 4-10). Low-alloy steels with ASM designations such as 6G, 6F2, and 6F3 possess good toughness and shock resistance qualities, with reasonable resistance to abrasion and heat checking. However, these steels are tempered at relatively low temperatures, usually 840 to 930° F (450 to 500° C); therefore, they are suited for applications that do not entail higher die surface temperatures, such as die holders for press forging or as hammer die blocks.

Low-alloy steels with higher (2 to 4%) nickel content, with ASM designation 6F5 and 6F7, have higher hardenability and toughness and can be used in more severe applications than the steels 6G, 6F2, and 6F3. The precipitation-hardening steel 6F4 can be hardened by a simple aging operation without cracking or distortion. In hot forming in presses, heat transfer from the hot stock to the dies causes this steel to harden and to become more abrasion resistant.

Hot-work die steels are used at temperatures of 600 to 1200 °F (315 to 650 °C) and contain chromium, tungsten, and, in some cases, vanadium or molybdenum, or both. These elements induce deep hardening characteristics and resistance to abrasion and softening. Hence, these die steels are often used for press forging dies for which surface temperatures may be substantially higher than those experienced by hammer dies. Generally, these steels are hardened by quenching in air or molten salt baths.

The chromium-base steels have about 5% chromium (Table 4-10). High molybdenum content gives these steels high resistance to softening; vanadium content increases resistance to heat checking and abrasion. Tungsten improves toughness and hot hardness; however, steels containing tungsten are not thermal shock resistant and cannot be cooled intermittently by water. The tungsten-base hot-work die steels have 9 to 18% tungsten. They also contain 2 to 12% chromium and may have a small amount of vanadium.

High-speed steels, originally developed for metal cutting, can also be used in warm or hot forging applications. There are two types of high-speed steels—molybdenum types designated by the letter M, and tungsten types designated by the letter T. These steels offer a good combination of hardness, strength, and toughness at elevated temperatures.

Selection of steels for hot forging depends not only on forging temperature and equipment, but also on the ability to harden the die block, machinability, and resistance to abrasion, to thermal and mechanical fatigue plastic deformation, and to catastrophic crack growth. Die life is often controlled by abrasive wear. Thus, commercially, die wear and die life are often thought to be synonymous. However, wear is but one of several mechanisms that may render hot forging dies unusable. Another common mechanism in hot forging is thermal fatigue or thermal cycling, which gives rise to superficial cracks, often known as heat checks. In mechanical fatigue, cracking results from the cyclic application of forged loads. If the loads are very high or the dies relatively soft, plastic deformation of the dies may occur, making it impossible to impart the desired shape to the workpiece. It is not uncommon for one or more of these mechanisms to contribute to

TABLE 4-10. Classification and Compositions of Principal Types of Tool Steels

AISI-SAE except for the first group of steels.

	C	Mn	Si or Ni	Cr	V	W	Mo	Co
Low-alloy tool steels								
6G	0.55	0.80	0.25 Si	1.00	0.10		0.45	
6F2	0.55	0.75	0.25 Si	1.00	0.10 opt		0.30	
			1.00 Ni					
6F3	0.55	0.60	0.85 Si	1.00	0.10 opt		0.75	
			1.80 Ni					
6F4	0.20	0.70	0.25 Si				3.25	
			3.00 Ni					
6F5	0.55	1.00	1.00 Si	0.50	0.10		0.50	
			2.70 Ni					
6F6	0.50		1.50 Si	1.50			0.20	
6F7	0.40	0.35	4.25 Ni	1.50			0.75	
6H1	0.55			4.00	0.85		0.45	
6H2	0.55	0.40	1.10 Si	5.00	1.00		1.50	
Chromium hot-work tool steels								
H10	0.40			3.25	0.40		2.50	
H11	0.35			5.00	0.40		1.50	
H12	0.35			5.00	0.40	1.50	1.50	
H13(a)	0.35			5.00	1.00		1.50	
H14	0.40			5.00		5.00		
H19	0.40			4.25	2.00	4.25		4.25
Tungsten hot-work tool steels								
H21	0.35			3.50		9.50		
H22	0.35			2.00		11.00		
H23	0.30			12.00		12.00		
H24	0.45			3.00		15.00		
H25	0.25			4.00		15.00		
H26	0.50			4.00	1.00	18.00		
Molybdenum hot-work tool steels								
H41	0.65			4.00	1.00	1.50	8.00	
H42	0.60			4.00	2.00	6.00	5.00	
H43	0.55			4.00	2.00		8.00	

(a) Also available as free cutting grade. Source: Ref. 61.

die failure. Each of these mechanisms is discussed below.

Abrasive wear of the dies is probably the most common form of die failure, resulting in the inability to produce parts to the desired shape. In general, abrasive wear can be defined as material removal due to sliding contact between the two surfaces, or between one surface and a number of abrasive particles. This should not be confused with erosive wear, in which metal removal occurs as a result of particles impinging on the surface. Unfortunately, the terms abrasive wear and erosive wear often are used interchangeably in the literature.

Most investigations of abrasive wear of forging dies and machinery parts have shown that the volume of material removed is directly proportional to the interface pressure and distance traveled and inversely proportional to the hardness of the metal surfaces. Because of this, the major factors affecting failure of forging dies by abrasive wear are the die material composition and hardness, the ambient temperature of the die surface, the deformation resistance of the workpiece, and the interface conditions (which depend on lubrication and the presence or absence of scale).

Forging dies are most likely to experience abrasive wear at locations where pressures are highest and metal sliding greatest. For impression dies, these positions include the flash lands and the other regions in which reductions in vertical cross section (leading to large amounts of sliding) are greatest. Such positions for this, as well as for the other die failure mechanisms, are illustrated schematically in Figure 4-113.

There are several solutions to the problem of abrasive wear. Perhaps the most direct is to use a die steel that is more resistant to wear, i.e., one that is harder and retains its hardness at high die tempera-

tures. This could mean changing from a low-alloy die steel to a chromium hot-work die steel. Tradeoffs between expected increases in die life must be weighed against increases in material and machining costs, however.

Coatings (e.g., chromium or cobalt), hardfacing (e.g., nickel alloy weld-fusion deposits), and surface treatment (e.g., nitriding) of hot forging dies also improve wear resistance. Another means of reducing wear in forging of steel parts involves reducing the scale on heated billets, which acts as an abrasive between the dies and workpiece. Methods of reducing scale include using a reducing, or inert, furnace atmosphere; applying a billet coating to prevent oxidation; and minimizing the time at temperature in the furnace, or using induction heating.

Thermal Fatigue. The second most common reason for discarding hot forging dies is the development of heat cracks (or "craze" cracks) during thermally induced fatigue deformation. This arises during conventional forging when a hot workpiece contacts cooler dies, leading to intermittent heat transfer between the two. As the die surfaces are heated, compressive stresses are developed because of the constraint imposed by the interior of the die blocks. Subsequently, after a given forging cycle and before the next cycle, heat is conducted away from the surface to the interior of the dies via conduction and to the surrounding air via convection, resulting in tensile loading and tensile straining of the surface layers. Thus, the surface is subjected to a type of deformation analogous to low-cycle fatigue.

As expected, the tendency for failure because of thermal fatigue is related to a number of factors, the most important of which is the strain amplitude imposed during each forging cycle. In turn, this depends on the magnitude of the temperature change during each cycle and the thermal properties of the die material. Other factors include the composition and purity of the die material. Composition is important because of its effect on metallurgical phase transformations, which may lead to additional large increments of straining above that which is due solely to thermal effects. Purity is important because of the effects of inclusions on fatigue crack initiation, with dirty die steels typically having much poorer thermal crack resistance.

Regions of forging die cavities most prone to thermal fatigue and heat checking are those in contact with the hot workpiece for the longest period of time and thus most likely to experience temperature extremes (Figure 4-113). In addition, in-

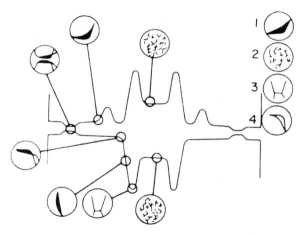

FIGURE 4-113. Common failure mechanisms for forging dies. 1, abrasive wear; 2, thermal fatigue; 3, mechanical fatigue; 4, plastic deformation. Source: Ref. 62.

creasing dwell time under load raises the average temperature of the dies and leads to an increase in thermal fatigue.

Methods of alleviating thermal fatigue problems include using a die steel with a higher yield strength (which has a larger elastic strain at yield and thus exhibits a smaller plastic strain amplitude during low-cycle fatigue), or decreasing the maximum die surface temperature variation (by lowering the maximum die surface temperature experienced or by raising the bulk die temperature by preheating the die blocks). In addition, using high-quality, clean die steels (in which fatigue cracks are initiated with difficulty) or special surface finishes (removing stress raisers such as machining marks, for example) can often alleviate persistent heat checking problems.

Mechanical fatigue of forging dies is affected by the magnitude of the applied loads, the average die temperature, and the condition of the surface of the dies. Fatigue cracks usually initiate at points where the stresses are highest, such as cavities with sharp radii of curvature whose effects on the fatigue process are similar to notches (Figure 4-113). Other regions where cracks may initiate include holes, keyways, and deep stamp markings used to identify die sets.

Redesign to lower the stresses is probably the best way to minimize fatigue crack initiation and growth. Redesigning may include changes in the die impression itself or modification of the flash configuration to lower the overall stresses. Surface treatments may also be beneficial in reducing fatigue-related problems. These treatments include nitriding, mechanical polishing, and shot peening and are effec-

tive because they induce surface residual (compressive) stresses or eliminate notch effects, both of which delay fatigue crack initiation. On the other hand, surface treatments such as nickel, chromium, and zinc plating, which may be beneficial with regard to abrasive wear, have been found to be deleterious to fatigue properties.

Catastrophic Die Failure. The fracture of hot forging dies after only one or a few forging cycles is a special case of mechanical fatigue. In these cases, failure occurs quickly because of high applied stresses or low toughness of the die material. As with mechanical fatigue, high applied stresses may result from poor die design, improper press or shrink fitting of dies and die inserts into containers, or lack of control of the forging load and energy.

Die steel toughness depends on chemical composition, grain size, hardness, and operating temperature. Because of the problems of energy control and impact loading, die steels of generally low alloy content and low hardness are employed primarily for the manufacture of forging dies for hammers to avoid the problem of catastrophic failure. Moreover, hydraulic and mechanical press dies, usually not subject to such problems, are typically made of more highly alloyed die steels (such as the 5% chromium hot-work die steels).

Plastic Deformation. Failures of forging dies by plastic deformation results from excessive pressure, and solutions to the problem involve either reducing the pressure or selecting a die material with higher strength at elevated temperatures. With some die materials, it may also be possible to obtain the desired effect by cooling the die more thoroughly. Such a reduction

in temperature is beneficial in that tempering or overaging and thus softening is lessened. However, the use of such techniques must be considered carefully to avoid surface temperature cycles that lead to deformation and phase transformations that promote heat checking.

Die Materials for Cold Forging

Because of the generally higher flow stresses of workpiece materials at cold working temperatures, tooling for cold forging and extrusion forging must be fabricated from tool materials that are harder than those used in hot forging. These materials frequently must have good resistance to wear and fatigue strength. Several tool steels that meet these requirements are shown in Table 4-11, which also lists several grades of tungsten carbide that are useful for cold forging applications.

Fabrication of Impression Forging Dies

Die sinking is a machine trade whereby a craftsman known as a die sinker performs certain steps to produce a forging die. In addition to personal skills, the die sinker needs the appropriate machines and hand tools. As the forging industry has demanded more complex forgings, the machine tool industry has developed more sophisticated machine tools to facilitate the production of these complex dies. The die sinker still uses the same basic steps that have been used for years, but with new machine tools and refined techniques so that a die can be fabricated to furnish extremely complex and close-tolerance forgings. The die-making process includes (1) selection of materials for the die, (2) die preparation, taking into consideration the forging machine that will produce that particular forging, (3) preparing a design,

TABLE 4-11. Materials for Die Inserts

AISI	Hardened to HRC
Steels	
D2	60–62
M2	60–64

Co, wt%	Density, g/cm³	Hardness, DPH
Tungsten carbides		
25–30	13.1–12.5	950–750
19–24	13.6–13.2	1050–950
15–18	14.0–13.7	1200–1100

(4) machining the dies, (5) benching the dies, and (6) taking a cast of the dies.

Quality forging dies are achieved through a blending of the skill and knowledge of both the forging engineer and the die sinker. When the forging design has been completed and approved, the die sinker, after consulting with the designer on any special details of the job, begins the process of sinking the desired impression in the die blocks of alloy steel. Rough die blocks, carefully forged and heat treated, usually are obtained from firms that specialize in their manufacture. Blocks may be purchased in a variety of shapes, sizes, and tempers, depending on the type and size of forging intended and, accordingly, the type and size of equipment to be used. They may range from a few hundred pounds to several tons in weight.

The die shop begins its work by generally following this sequence of operations: top, bottom, one side, and one end need to be finish surfaced either on a planer, a milling machine, and/or a surface grinder. All surfaces must be flat, parallel, and 90° to each other. Because of the size and weight, handling holes are drilled in the ends or sides so that the dies can be handled more easily. The rough blocks are then moved to a planer or planer mill, where they are paired as upper and lower die blocks of a die set. "Dovetail" shanks for holding the blocks in the forging equipment are cut. Die faces are often ground to a fine finish to obtain a smooth surface for layout work (Figure 4-114).

After the material has been selected and prepared, the die sinker is given a print of the customer's forging and a die design. He is now ready to sink the die. So the layout lines on the die steel are more easily visible, a solution of copper sulfate or dye blue is applied to the face of each die. The outline of the forging is scribed on the face of the dies to the exact dimensions dictated by the drawing. Mold lines are identified first, and the draft lines are added (3°, 5°, 7°, etc.). Dimensions for the draft are determined by the depths of the impressions. To ensure that impressions in each die match, the layout is located on the dies in relation to the side and end match edges. Special shrink scales are used that are based on the shrink factor of the material to be forged (1/10 in., 3/16 in., etc.). The design will dictate how many impressions—roller, fuller, edger, cutoff, and gate—are in each set of dies.

Layout lines are scribed on each die using a square and a blade protractor, dividers, and a hardened scriber (Figure 4-115). If it is possible to stand the dies on end or side on a surface plate, a height gage can

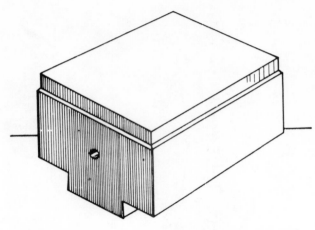

FIGURE 4-114. Machined and processed die block.

be used to scribe lines that are parallel to the match edges. This method is very accurate; some tools have digital readouts and a programmable shrink factor. The finishing impression usually is positioned so that its weight center will be aligned as nearly as possible with the center of the hammer or press ram, as measured from all sides. This helps ensure perfect balance in the forging equipment, permits full utilization of maximum ram impact as the forging is in the finishing impression, and eliminates wear-causing side thrusts and pressures during forging. After the layout is finished and checked, the dies are ready for machining of the impression.

The machine tools for die shrinking have changed dramatically over the years. The simple vertical milling machine has developed into a very sophisticated machine tool with hydraulic movement of ram, table, and spindle, with the ability to trace from a template or tracing mold. The impression (cavity) is sunk to within a few thousandths of an inch of its finished part size (Figures 4-116 and 4-117).

The cutting tools used are fabricated from high-speed tool steel and have two, three, or four flutes (straight or spiral). They may also have angles to produce 3°, 5°, 7°, etc., drafts. For heavy flat cutting, a carbide insert cutter is used. As the die sinking be-

FIGURE 4-115. Die layout prior to machining the impression.

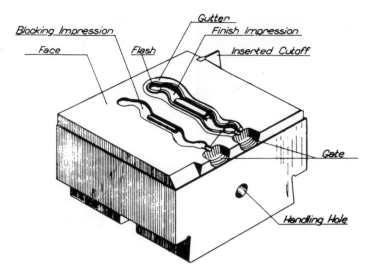

FIGURE 4-116. Typical finish machined die cavity with gate and flash gutter included.

gins, the deepest section will be cut first with the largest cutter, working up to the shallowest section, until all vertical walls are machined. The webs and radii are machined last. The X and Y dimensions are machined according to the scribed lines on the face, with control of the Z dimensions or depth via a depth gage or profile template. If the design calls for more than one impression, only the first impression is made until it has been benched and a cast

has been submitted for approval. Regardless of when the rest of the operations are completed, the same procedure is used. Flashing and guttering of the dies can be done at either time.

The complexity of some forgings may dictate that a better die be fabricated by using a wooden pattern of the forging. The pattern is then used to construct a plaster mold that is used to trace the impression into the die. This method requires mini-

mal layout. The dimensions of the impression are determined by the mold.

Finishing of impressions is primarily done by hand with the aid of power hand grinders. All tool marks and sharp corners must be removed, and all vertical and horizontal radii made true to size. The surfaces are then polished. Most of the surfaces have been machined within a few thousandths of the finish dimensions; consequent benching is not done to remove too much stock, but only to polish the surfaces to ensure that they are true in every dimension and free of tool marks, blemishes, and sharp corners. These hand operations help ensure filling of the impressions with the least resistance to metal flow during forging. Likewise, it minimizes abrasive wear on the impressions (Figure 4-118).

When the bench work on the finishing impression is completed, a parting agent is applied to the surface of the impression to prepare it for proofing of the impression. The pair of dies is clamped together in exact alignment, using the matched edges as guides, and the cavity formed by the finishing impression is filled with molten lead, plaster, or special nonshrinking compounds to obtain a die proof. The die proof is then checked for dimensional accuracy. When all dimensions are correct, it is submitted to the customer for approval if requested.

Other die impressions may then be sunk (to preform edging, fullering, and bending operations), depending on the complexity of the forging. These impressions for preliminary forging operations may also be sunk in a separate set of dies. The arrangement and sequence of preliminary operations differ widely according to variations in practice throughout the forging industry (Figure 4-119).

Ordinarily, the final machining operations on the faces of a set of dies are performed on the flash gutter. When a forging is worked to its final dimensions in the finishing impressions, there is always some metal in excess of that required to completely fill the cavity. This metal, called flash, is squeezed out of the impression between the die blocks. Flash may begin to appear during the blocking operation; however, at the finishing impression, where the full impact of the hammer is utilized, a gutter of carefully calculated dimensions must be provided around the entire shape of the finishing impression to accommodate the excess metal and to ensure proper die closure.

After guttering of dies, dowel pockets are ordinarily milled into one side of the shank of each die block. The dowel pocket

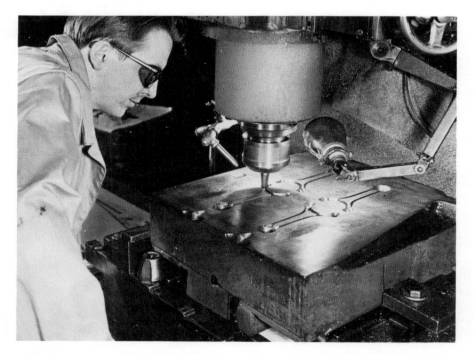

FIGURE 4-117. Machining of blocking and finishing impressions for automotive connecting rods.

FIGURE 4-118. Hand finishing of die impressions.

converts AC current to DC current through the use of diodes. It also converts DC current to an on-and-off current. The rate of burn is designated by length of on-time, amount of current, and off-time. The longer the on-time and the higher the current, the faster the metal is removed. Also, this roughing operation leaves a very rough surface. A shorter on-time and lower current will cause less metal removal, but a much better surface for finish polishing.

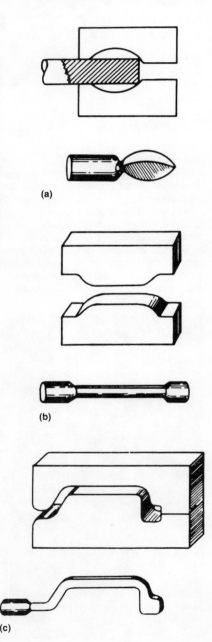

FIGURE 4-119. Dies for (a) edging, (b) fullering, and (c) bending operations. Adapted from Ref. 63.

accommodates the dowel key, which is inserted by the hammer or press operator to maintain die alignment in the equipment from front to back.

Another close inspection of the dies is generally scheduled as a final precaution. All dimensions of blocking, as well as finishing impressions, are carefully compared with the blueprint dimensions and specifications again. Figure 4-120 shows the bottom half of a finish die set.

Extreme care is required in bringing the dies into exact alignment as they are placed in the forging equipment so that forgings will be on match and there will be a minimum of strain on the equipment and wear on the dies. Dies correctly and properly handled are normally capable of producing thousands of uniform forgings of identical shape and size. Figure 4-121 shows a typical die manufacturing facility.

An alternative method has been developed for sinking dies using electric discharge machining (EDM) in place of a vertical mill. This method is used when minimal draft angles and very narrow ribs are required, and it has the ability to pro-

duce dies accurately. Also, if several of the same cavities are to be sunk in one die, use of EDM ensures reproducibility.

The machine tool for this method of die fabrication has a hydraulic-powered ram and table. The table is a large tank that is open at the top. All metal removal is done with the die block submerged in a dielectric solution, which is used as a flushing agent to keep the burning area clean. Also, it acts as the carrier for electric current between the electrode and the die block. The solution is constantly circulated through a separate filter system to keep it clean and free of contaminants from the burning operation. A clean solution is necessary for an efficient burn. The electrode never makes contact with the die block as the electric current passes through the dielectric solution to the die block and erodes the die steel to create the impression.

Power for the burning operation is supplied by a power source that is separate from the machine tool. The power source has a wide range and is usually matched to the machine tool for the type and size of work to be done. The power source

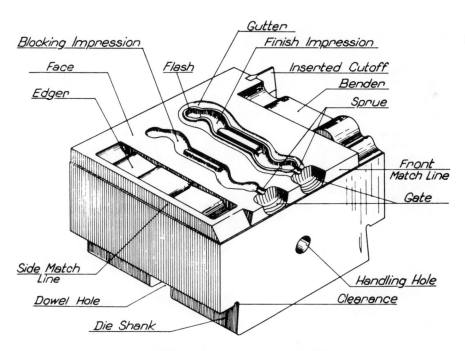

FIGURE 4-120. Elements of a drop forging die block.

This combination is used for the finish burn. Very little metal will be removed, but the impression made will be accurate and will have a suitable surface for benching.

Electrode material is a medium-grade carbon (graphite) that has been developed for EDM. Other materials that have been used are brass, aluminum, copper, silver, gold, and epoxy plates with copper or brass. Most materials that will conduct electricity can be used for electrodes, but carbon is most widely used. It is easily fabricated, relatively inexpensive, and has acceptable wear characteristics.

The most common machine tool for fabricating electrodes is a vertical mill with three heads and a syncrotrace control. This allows three electrodes to be cut simultaneously. By controlling the cutter size in relation to the tracing stylus, one rough and two finish electrodes can be machined from the same pattern. Electrodes may be full impressions or a partial of an area of the die that is not practical to machine conventionally. For the full impression, a pattern is fabricated from wood or another suitable material to the exact shape of the electrode. Partial electrodes can be laid out and machined on a single-spindle vertical mill. The graphite is cut to a rough shape and mounted onto a precision plate that is compatible with both the machine tool for tracing and EDM. Another method of fabricating an electrode is to fabricate a fe-

male model impregnated with an abrasive material using an abrader machine to create an electrode.

An electrode for burning can also be fabricated using an N/C machine tool. This system is very good for developing airfoils and complex contours. It is also used for high-production forgings and when several dies are needed. After machining, electrodes are benched to remove any tool marks, and flush holes are drilled. Because the electrodes never make contact with the die block, allowance must be made for the gap between the electrode and the die block. The gap is determined by the amount of power used. The higher the power used, the greater the gap will be. The rough and finish electrodes are sized by the power to be used. The gap can be from a few thousandths of an inch to 0.070 in. or more.

The first burning operation is done with the rougher electrode, followed by the first finisher and second finisher. The electrodes are remounted on to the plates and attached to the ram of the EDM machine tool. A positive and negative power source is attached to the electrode and die block. After the die block is positioned properly in relation to an electrode and a layout on the die block, the tank is filled with dielectric solution. The operator presets the desired depth and starts the burning operation. As previously mentioned, the rate and amount of burn can be controlled. For

roughing, a long burning time and high current are used. After roughing, the first finisher electrode is mounted on the same plate, and the power source is adjusted for a shorter burn time and less current. The final finish is accomplished in the same manner and provides an impression that is true dimensionally and has a relatively smooth surface finish for final benching.

Usually after the burning operation is completed, the die block is transferred to a conventional machine tool for flashing, guttering, or any other machine operation that is required. The die is now ready for benching. For a vertical wall and narrow ribs, special tools and techniques have been developed. For the majority of impressions, conventional tools and practices are used. After finishing benching, dies are prepared, and a proof cast is taken and submitted for inspection.

Dies for Precision Forging

The airframe industry requires forgings that undergo a minimum of machining. The forging industry has responded by developing precision, or no-draft, dies that produce forgings that require little or no machining before assembly.

Dies are being designed and fabricated with not only zero draft, but also with an undercut and closer tolerances. These dies consist of several pieces of steel that lock together to form a single unit. The simplest precision die has only a top and bottom die with a knockout pin to help remove the forging during the forging operation. As the complexity of a forging increases, the design of the die requires more pieces to form the part. The die may consist of two or more pieces to form the outside of the forging (wraps), and a bottom and top punch to form the inside configuration. All of these pieces must fit together—the wraps and bottom punch, which fits into the wraps to make a bottom die, and top punch, which then fits into the bottom assembly to make a complete set of forging dies. For the forging operation, the dies are contained in a holder or ring die designed to accept several different precision dies. During the forging operation, the bottom assembly has to separate so that the forging can be removed.

The fabrication of precision dies is different from that of conventional dies. All pieces are rough cut to shape and surface ground on top, bottom, and two sides. Layout is usually done on a surface table using a height gage for applying the lines of the layout. The same machine tools are used, but the machining is more open with less cavity work. Outside ribs can be machined as a wall. The inserts will be positioned on a machine for the easiest machining operation. Inside ribs with low draft angles will be burned in with an EDM tool. Wood patterns and tracing molds are also used in sinking precision dies. Usually, the different pieces are fitted together before the impression is sunk. After machining, the dies are benched using the same equipment as for conventional dies. More draw filing and polishing is needed to ensure that vertical walls and ribs are flat and true. Undercuts are unacceptable, because they would prevent the dies from releasing the forging. Plaster is not always used for a cast; sometimes, an epoxy will be used.

After all impression work is completed, the outside of the assembled inserts must be sized to fit either a holder or ring die. For a holder die, usually two sides are 90°, and two sides have a 3° taper. For a ring die, usually all four sides have a 3° taper.

In a precision forging operation, forging frequently can be made with only a finish die. If a block operation is required, the blocker die will be separate from the finish die and will resemble a conventional die. Figures 4-122 through 4-125 illustrate several precision die configurations.

Use of Computerized Systems

Developments in the computer field have also influenced die sinking. Equipment and programs have been developed for fabrication of all types of forging dies. With computerized methods, machining of a complete die, sections of a die, or even electrodes for EDM can be programmed, and a die can be fabricated.

The equipment is entirely different from any other type of die sinking. The machine tool can have either a horizontal or a vertical spindle. All movements are controlled automatically by the program designed for a specific impression. An automatic tool changer is not required, but can save downtime. In place of layout equipment, a computer with proper software is used. The programmer must have knowledge of machining of die steel, die sinking practices, and mathematics, particularly geometry and trigonometry. An accurate cutter tool grinder should be available as the program is written. This program should include the exact sizes for the cutting tools to be used in production.

Computerized die sinking requires an approach different from other methods. The programmer takes the design and part print and develops an N/C tape for fabricating the die. He assigns cutting tools, numbers

FIGURE 4-121. Typical die manufacturing facility.

of each size, and feeds the speed of each. He decides how much of the die is to be sunk by tape and develops the cutter path and sequence.

After a program is fully developed, a tape is made that controls all movements of the machine tool. All cutting tools are sized and set in the tool chamber at proper depth. Before machining of the die, the tape is tested by cutting the impression in styrofoam. This allows the operator to check that all movements are correct and that the dimensions of the impression are accurate.

When the program is accepted, the die block is set up in the machine tool, and the die is sunk. Because there are no lay-out lines for the operator to follow, all machining is performed with the cutting tool flooded with a coolant. This not only keeps steel chips out of the cavity, but also keeps the cutting tools cool, thereby increasing tool life. Flashing, guttering, or other operations can be included in the program to be completed following sinking of the cavity. The programmer will decide whether it is faster or easier to accomplish these operations conventionally or by tape.

When all machining is completed, the dies are then benched using the same equipment and methods as for conventionally sunk dies. Casts are then taken and submitted for inspection.

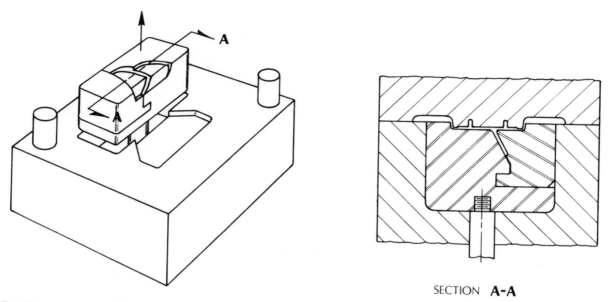

FIGURE 4-122. Conventional die with two-piece knockout in up position. Only bottom die is shown. Courtesy of Aluminum Precision Products, Inc.

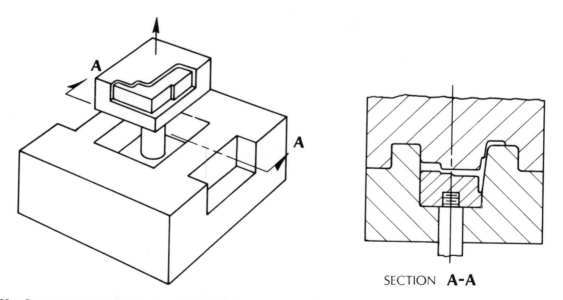

FIGURE 4-123. Conventional die with one-piece knockout in up position. Only bottom die is shown. Courtesy of Aluminum Precision Products, Inc.

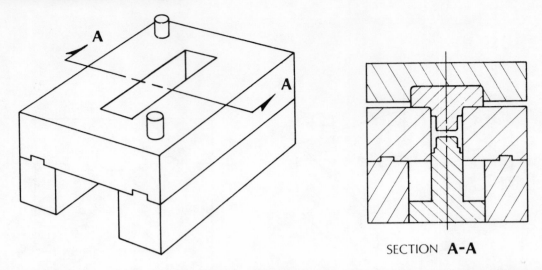

FIGURE 4-124. Through die with top and bottom punches. Only bottom die is shown. Courtesy of Aluminum Precision Products, Inc.

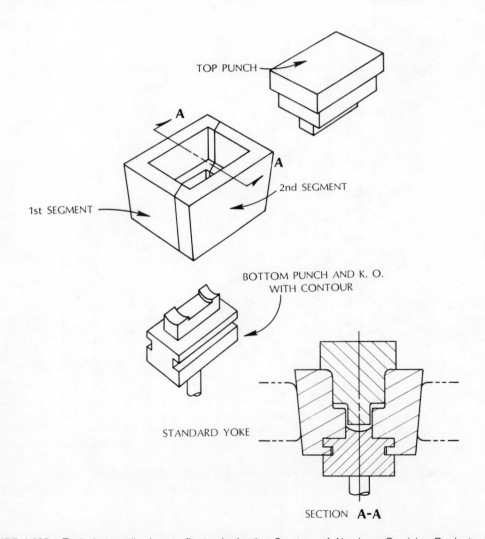

FIGURE 4-125. Typical wrap die. Inserts fit standard yoke. Courtesy of Aluminum Precision Products, Inc.

Lubrication in Forging

Contributing Authors: Dr. S. L. Semiatin, Principal Research Scientist, Battelle's Columbus Laboratories, Columbus, Ohio; Robert Stansbury, Industry Manager, Pennwalt Corporation, Philadelphia, Pennsylvania; and Walter Levine, Manager, DAG Application Equipment, Acheson Colloids Company, Port Huron, Michigan

Conventional Hot Forging

During the deformation phase of a conventional hot forging operation in which the die temperature is much less than the billet temperature, the lubricant provides three basic effects:

- Lubricity, ensuring that the correct coefficient of friction is maintained between the hot billet and the tooling
- Physical barrier, preventing physical contact between billet and dies
- Thermal insulation, retarding the rate of heat transfer between the hot metal and the dies

All of these effects are equally important in any forging operation, but become more critical as the degree of metal flow increases. Simple forgings for hand tools, for example, place fewer demands on a lubricant than do severe extrusions for automotive and ordnance purposes. As the demand for more complex forgings increases, so will the need for lubricants that can be relied on to move with the metal and help it fill the die (Figure 4-126).

The manner in which a part is forged can also affect the proper functioning of a lubrication system. It is not uncommon for a billet to be held in a set of dies for 5 or 10 s or even longer. The resulting die temperature creates lubrication requirements that are different from those in high-production hot forging operations.

Because each forge shop and process characteristic requires a different composition of lubricant, a wide variety of compounds have been developed. Lubricants can be classified into four groups, although a lubricant compound can be a blend of the first two (or possibly the first three) categories to obtain the desired performance or to improve performance:

- Solids, including graphites, clays, mica, and talc
- Meltable pigments, principally salts and glass-type materials
- Organic chemicals, including mineral oil, animal fats, and polymers
- Water solubles, comprising dispersions, suspensions, emulsions, or solutions of one of the three above types

Solid lubricants contain inert substances with melting points, if appropriate, that are considerably above the temperature of the metal being formed. They function mainly as physical-barrier lubricants, although graphite and some other solid compounds also provide lubricity. Furthermore, such materials may serve as thermal barriers.

Meltable pigments function differently in the physical process of providing lubrication. The concept of this class of compounds is to use a material with a predictable melting point within the temperature range of a particular forging process. As the lubricant is heated, the pigment liquefies and provides a hydrodynamic film for lubricating.

Organic chemicals generally function by reliance on type, rate, and amount of oxidation and/or polymerization residue formed when they contact hot metal. Silicon polymers are inert and provide fluid-film lubrication much the same as meltable pigments. Organic chemicals may ultimately liberate gases that help to eject the forged part from the die.

Water-soluble compounds are considered the primary trend of advanced lubrication technology. Water-soluble compounds appear to have outdistanced conventional dispersions, suspensions, and emulsions.

The characteristics of two hot forging lubricant systems are:

Oil-based/graphite

- Adaptable to a wide range of application methods, which can be very simple
- Environmental pollution usually occurs. Smoke can contain graphite dust, which can settle on equipment and cause problems. However, smoke may prevent "stickers" (parts that hang up in the dies).
- Many of these lubricants are conducive to long die life.

Oil-free/graphite-free

- Spray equipment is required for application. "Spreading" may be incomplete

FIGURE 4-126. Multiple-weld yokes for car drive shafts emerge from a Ford Motor Co. forging press as steam cools the die. The lubricant for this operation minimizes smoke, flash, and odor. Courtesy of American Machinist.

if other methods are used. Complete wetting is usually attainable.

- No graphite dust is generated and virtually no smoke. Housekeeping is thus easier.
- Flexibility in the manufacture of forgings with a variety of shapes and compositions can be a challenge.

Selecting the Proper Lubricant

A plant survey by potential suppliers of lubricants is recommended as the best way to evaluate performance criteria versus product characteristics. The survey should seek specific information in these areas:

- *Tooling:* What is the die alloy? How hot will it get? How complex is the die?
- *Workpiece:* What is its composition? What is its proper forging temperature?
- *Forging equipment:* Press? Hammer? Type and size/capacity?
- *Forging sequence:* Number and type of die stations? Function of each?
- *Cycle times*
- *Lubricant presently used:* Perceived advantages? Disadvantages?
- *Application methods presently used*

Although accurate answers to these survey questions may lead to the proper choice, minor changes in the makeup of the lubricant may be required:

- Proportion of graphite
- Size of graphite particles (customarily expressed in microns)
- Addition of other melting pigments
- Viscosity modifications to aid in lubricant application
- Flashpoint adjustment

Oil/graphite-type lubricants can be applied in a number of ways. The oil base is the medium that holds and distributes the lubricating additives across the die surface. Because it is quite fluid and readily coats the entire die surface, it can be swabbed as well as sprayed and is somewhat forgiving in terms of incomplete or spotty application. Although sophisticated application equipment can also be used for these lubricants, the basic application method is quite simple.

Smoke and fire and their consequences are the most dramatic drawbacks of oil-based lubricants. The fact that smoke escaping between the surfaces of the dies and the forgings helps to prevent "stickers" is overshadowed by the air pollution problem. Oil/graphite lubricants also generate graphite dust, which creates a haze and settles throughout the area. Penetrating electrical equipment, it can cause serious problems.

Water-soluble forging lubricants—oil emulsions containing graphite—were hailed as solutions to the pollution problem and a means of placating government agencies seeking to enforce worker health and safety standards. This group of lubricants has scored high in forging productivity, and water dilution has cut lubricant consumption. However, smoking was not eliminated, the graphite settling problems continued, die life improvement—together with lubricity and metal flow—was marginal, and application became an uncertain operation.

Ordinary salt water was probably the first water-soluble lubricant used. A meltable pigment, the salt provided a hydrodynamic lubrication, but it also built up in the dies. When dies became increasingly complex, graphite replaced salt. In general, water-soluble lubricants followed the same developmental path as oil/graphite compounds. Various agents and additives were incorporated to enhance suspension of solids, wetting, gassing, and lubricity.

Forging companies considering graphite-free synthetic lubricants must understand both the benefits and the drawbacks. In addition, for manufacturers of synthetic lubricants, one basic problem is developing greater flexibility to handle a wide variety of forged parts.

Forging companies that evaluate currently available lubricants will learn that a conversion from oil-based to water-based compounds usually reduces costs and smoke levels. The use of a graphite-free synthetic will virtually eliminate the smoking problem. Beyond these two advantages, no blanket statement can completely satisfy the requirements of lubricity, die release, die life, and die buildup, all of which are related to the sophistication of the lubricant formulation. If lubricity is inadequate and metal flow is incomplete, the other factors are of no consequence.

Lubricant Application

Lubricants for conventional hot forging usually are applied to dies by a variety of swabbing or spraying techniques. In all cases, care should be exercised to avoid lubricant buildup. For precise control of lubrication, automatic spray units are desirable. In these units, air and lubricant mixtures typically are mixed under pressure in metering blocks, and the air-atomized liquid is supplied to the spray nozzle or nozzles, which direct it onto the dies. Such units often include a timer to adjust the time of spray, and the amount of lubricant can be adjusted by metering screws at the metering blocks. The amount of lubricant sprayed in a given time is also influenced by the lubricant and air pressure in the lines connected to the metering blocks. Figure 4-127 illustrates a commercial spray system.

Isothermal and Hot Die Forging

The development of suitable systems of lubrication and part separation for isothermal and hot die forging has been one of the most difficult tasks during the commercial application of these processes. These systems must provide low friction for good metal flow, ease of release from the dies, and good surface finish on the forgings, but should not lead to lubricant buildup and difficulty in obtaining desired finished tolerances, particularly in parts such as blades and structural forgings. Also, lubricants that are applied to the forging billet (by dipping, swabbing, or spraying) rather than the dies themselves often must serve as coatings to protect the billet surface from oxidation during furnace heating and forging. However, they should not contain substances that react with either the workpiece or the dies at elevated temperatures. An example of this is sulfur and sulfur-containing compounds, which cause rapid intergranular attack on nickel alloy dies (used for isothermal forging of titanium alloys) at high temperatures.

The characteristics of lubricants used for isothermal forging are somewhat different from those for conventional nonisothermal forging. Unlike conventional forging, a lubricant for isothermal forging does not operate over a range of temperatures and therefore can be selected to have an optimal composition for a given temperature or narrow temperature range. Furthermore, the danger of a lubricant changing to an abrasive as the die temperature changes during a forging run, particularly at thin localized areas, is avoided. However, because the dies and lubricant are maintained at elevated temperatures over extended periods of time, poor lubricant selection may lead to undesirable reactions between the two. Also, petroleum-base, graphite-base, and other lubricants such as molybdenum disulfide and tungstic oxide rapidly decompose and burn off at high temperatures and thus are unsuitable for isothermal and hot die forging. Therefore, totally different lubrication/

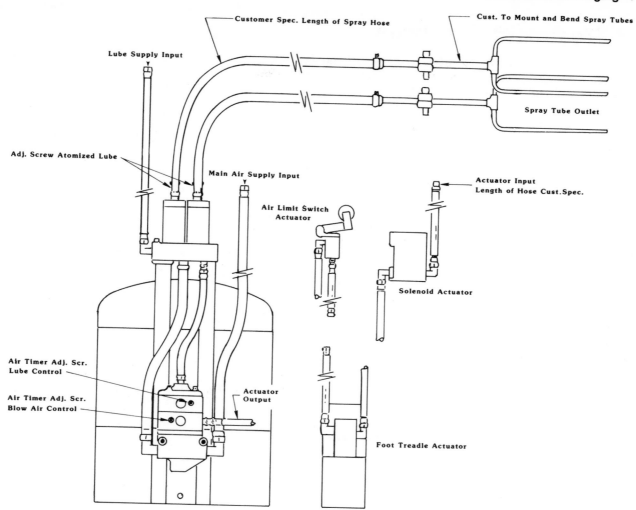

FIGURE 4-127. Commercial automatic lubricant spray system. Courtesy of Acheson Colloids Co.

separation systems are needed for this new process.

Various glass mixtures provide the best lubrication for isothermal and hot die forging. The glasses may come as frits (composed of a variety of glass-forming oxides) or as premixed compounds with their own aqueous or organic solvent carriers. In the former, the frits are ground into a fine powder. When this is completed, a slurry of frits is made (usually in an alcohol bath), and the forging billets are dipped into it. The alcohol then evaporates in air prior to forging, which may be speeded up by a low-temperature bake. This leaves a powder-like coating on the billets that, during heating to the forging temperature, becomes a viscous, glassy layer suitable for lubrication and oxidation

protection. Using viscosity charts, glass frit selection may be based on the desired forging temperature. Experience has shown that a glass with a viscosity of 200 to 1000 poises at the forging temperature provides optimal lubricity and a good continuous film characteristic that is required to prevent galling (metal pickup), which may occur when bare metal surfaces come in contact under high pressures.

In production isothermal or hot die forging, the use of glass frits may be inconvenient; the frits tend to provide poor separation of adhesion properties. In this case, premixed or commercial isothermal forging lubricants are available. They come with their own carriers (such as water, xylene, alcohol, and isopropanol). Besides vitreous and carrier components, these lu-

bricants usually contain various particulate phases that aid in part separation from the dies and control lubricant buildup.

Particulates include lamellar solids, nucleating agents, and semiabrasive particles. With lamellar or layered solids, such as boron nitride and graphite, a network of easily sheared particles effect separation. Nucleating agents, including titanium dioxide and cerium oxide, act to weaken the affinity of the lubricant to the die material. Semiabrasive particles, such as titanium carbide and tantulum carbide, decrease the amount of contact between the dies and lubricant, thereby enhancing separation. Care must be exercised, however, to keep the size of this type of particle small to avoid abrasion of the forgings or the dies.

Heat Treatment Practices

Contributing Authors: Robert T. Morelli, Consultant, Pittsburgh, Pennsylvania; and Dr. S. L. Semiatin, Principal Research Scientist, Battelle's Columbus Laboratories, Columbus, Ohio

In general, heat treatment involves specific controlled thermal cycles of heating and cooling to improve one or more of the properties of the forged part. The primary objectives of such treatment may be to relieve internal stresses (as in tempering and stress relieving), control distortion, optimize the depth of hardening (hardenability), refine grain size (by normalizing), change the microstructure to improve machinability, or develop the final specification for mechanical or physical properties. Most important, of course is achievement of mechanical or physical properties, because the forged component must meet performance engineering design requirements and ensure safety and reliability in service.

Underlying the selection of the proper treatment are cost considerations that must factor in the variables of forging shape and size, metal composition, machining, and final properties. For example, it would be wasteful of time and energy to reheat a forging when cooling from forging heat would satisfy the specification, or to add a temperature cycle when normalizing suffices. An examination of the practical aspects of material changes that occur and how the correct heat treatment can be applied to meet the end use requirements of a component is presented below.

Heat Treatment of Steel Forgings

Steel forgings generally are specified by the purchaser in one of four principal conditions: as forged with no further thermal processing, heat treated for machinability, heat treated for final mechanical/physical properties, or specially heat treated to enhance dimensional stability, particularly in more complex part configurations.

As Forged. Although the vast majority of steel forgings are heat treated prior to use, a large tonnage of low-carbon steel (0.10 to 0.25% carbon) is used in the as-forged condition. In such forgings, machinability is good, and little is gained in terms of strength by heat treatment. In fact, a number of widely used ASTM and federal specifications permit this economic option. It is also interesting to note that, compared to normalizing, strength and machinability are slightly better, which is most likely attributable to the fact that grain size is somewhat coarser than in normalizing.

Heat Treated for Machinability. When the forging purchaser must produce a finished machined component from a roughly dimensioned forging, machinability becomes a vital consideration to optimize tool life, increase productivity, or both. The purchase specification or forging drawing may specify the heat treatment, in which case the forger must follow instructions. However, when specifications give only maximum hardness or microstructural specifications, the forger must select the most economical and effective thermal cycle. Choices of full anneal, spheroidize anneal, subcritical anneal, normalize, or normalize and temper depend on the steel composition and the machining operations to be performed. Some steel grades are inherently soft and gummy, others become quite hard in cooling from the finishing temperature after hot forging, and still other grades of high carbon and high alloy content, such as stainless and tool steels, are difficult to machine, because they may contain higher levels of undissolved hard, brittle carbides. Some type of annealing is usually required or specified to improve machinability.

Heat Treated to Final Physical Properties. Normalizing or normalizing and tempering may produce the required minimum hardness and by a direct correlation give the minimum ultimate tensile strength. However, for most steels, a hardening (austenitize) and quenching (in oil or water, depending on section size and hardenability) cycle is employed, followed by tempering to produce the proper hardness,

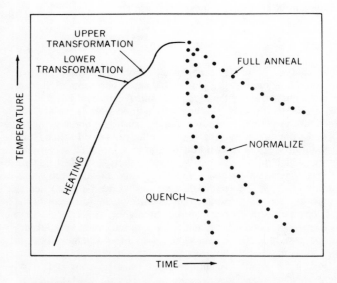

FIGURE 4-128. Typical steps in the heat treatment of steel forgings.

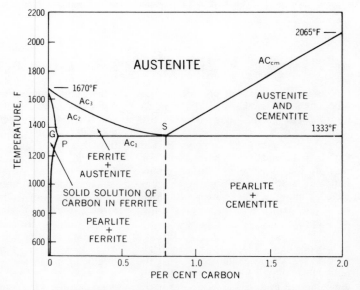

FIGURE 4-129. Iron-carbon equilibrium diagram.

strength, ductility, and impact properties. For steels to be heat treated above the 150-ksi strength level, it is general practice to normalize prior to austenitizing to produce a uniform grain size and to minimize internal residual stresses. In some instances, it is common practice to use the heat for forging as the austenitizing cycle and to quench at the forge unit. The forging is then tempered to complete the heat treat cycle. Although there are obvious limitations to this procedure, definite economies are possible when the procedure is applicable (usually symmetrical shapes of carbon steels).

Special heat treatments provide control of dimensional distortion, relief of residual stresses before or after machining operations, avoidance of quench cracking, prevention of thermal shock, surface (case) hardening, and stabilization of microstructure in certain high-alloy steels. Although most of the cycles discussed above can apply, very specific cycles may be required. Such treatments usually apply to complex forging configurations with adjacent differences in section thickness, or to very high hardenability steels and alloys. When stability of critically dimensioned finished parts permits only light machining of the forging after final heat treatment to final properties, special treatments are available, including marquenching (martempering), stress relieving, and multiple tempering.

Many applications, such as crankshafts, camshafts, gears, forged rolls, rings, certain bearings, and other machinery components, require increased surface hardness for wear resistance. The important surfaces usually are hardened after machining by flame or induction hardening, carburizing, carbonitriding, or nitriding. These processes are listed in the approximate order of increasing cost and decreasing maximum temperature. The latter consideration is important in that dimensional distortion usually decreases with decreasing temperature. This is particularly true of nitriding, which usually is performed below the tempering temperature for the steel used in the forging.

Most of these special treatments are performed by the forging customer or on their behalf. Although most forge plants have in-house heat treating equipment to improve machinability or to develop required minimum mechanical properties, outside facilities for certain special treatments exist. These treatments are discussed in subsequent sections. Heating for cold shearing of forging stock is of special interest to all forgers and can properly be regarded as a special heating operation.

TABLE 4-12. Heat Treatment of Steel Forgings: Approximate Critical Points for Carbon and Alloy Steels

AISI/ SAE No.	Upon heating at 50 °F/h (10 °C/h)		Upon cooling at 50 °F/h (10 °C/h)			AISI/ SAE No.	Upon heating at 50 °F/h (10 °C/h)		Upon cooling at 50 °F/h (10 °C/h)		
	Ac_1, °F	Ac_3, °F	Ar_3, °F	Ar_1, °F	M_s, °F		Ac_1, °F	Ac_3, °F	Ar_3, °F	Ar_1, °F	M_s, °F
1010	... 1335	1610	1560	1260	904	5045	... 1300	1450	1370	1290	600
1015	... 1335	1580	1525	1250	871	5046	... 1320	1420	1350	1260	620
1020	... 1335	1555	1500	1260	838	5120	... 1410	1540	1470	1290	760
1025	... 1340	1545	1440	1265	805	5130	... 1370	1490	1370	1280	680
1030	... 1340	1495	1450	1250	752	5140	... 1360	1450	1340	1280	620
1035	... 1340	1475	1425	1255	720	5150	... 1330	1420	1330	1290	555
1040	... 1340	1450	1395	1240	690	5160	... 1310	1410	1320	1250	490
1045	... 1340	1435	1385	1260	655	52100	... 1340	1415	1320	1270	485
1050	... 1340	1415	1365	1260	610	6120	... 1410	1530	1420	1300	760
1055	... 1340	1390	1350	1260	590	6145	... 1380	1460	1370	1280	580
1060	... 1340	1375	1340	1265	555	6150	... 1380	1450	1370	1280	545
1066	... 1340	1350	1325	1270	501	8620	... 1350	1525	1415	1220	745
1070	... 1340	1350	1310	1275	490	8625	... 1350	1485	1390	1220	710
1080	... 1345	1355	1290	1280	415	8630	... 1355	1460	1370	1220	680
1090	... 1345	1370	1290	1270	365	8635	... 1350	1450	1345	1225	640
1095	... 1350	1415	1340	1290	351	8640	... 1350	1435	1340	1230	610
1112	... 1355	1660	1560	1250	862	8645	... 1350	1430	1310	1230	575
1117	... 1350	1550	1450	1245	809	8650	... 1350	1420	1295	1210	545
1118	... 1345	1520	1495	1245	782	8655	... 1345	1410	1270	1220	515
1137	... 1315	1420	1360	1220	654	8660	... 1345	1410	1270	1230	485
1141	... 1310	1400	1340	1210	628	8719	... 1350	1540	1430	1220	745
1320	... 1335	1510	1375	1190	740	8720	... 1350	1530	1420	1220	740
1330	... 1325	1470	1350	1170	675	8735	... 1355	1455	1350	1220	640
1335	... 1330	1440	1340	1160	640	8740	... 1345	1435	1330	1220	605
1340	... 1320	1430	1330	1150	610	8745	... 1345	1425	1315	1215	575
2330	... 1280	1375	1190	1040	625	8750	... 1340	1415	1310	1210	540
2340	... 1275	1355	1190	1050	555	8822	... 1330	1540	1445	1195	725
2345	... 1275	1345	1190	1060	525	9255	... 1400	1500	1380	1320	585
3310	... 1330	1435	1240	1160	655	9260	... 1370	1500	1380	1315	550
4023	... 1350	1540	1440	1240	775	9310	... 1320	1510	1230	1080	685
4027	... 1340	1485	1400	1240	755						
4037	... 1340	1495	1390	1210	690	**Standard boron steels**					
4042	... 1340	1460	1350	1210	650						
4047	... 1340	1440	1330	1200	615	50B46	... 1330	1440	1340	1210	620
4063	... 1340	1400	1275	1185	515	50B60	... 1345	1420	1345	1250	513
4130	... 1395	1490	1390	1280	685	51B60	... 1335	1420	1345	1250	490
4135	... 1390	1485	1380	1280	640	81B45	... 1310	1450	1325	1215	598
4140	... 1350	1480	1370	1255	595	86B45	... 1330	1420	1280	1200	582
4145	... 1340	1470	1380	1250	569	94B17	... 1300	1540	1420	1180	780
4147	... 1355	1455	1350	1240	556	94B30	... 1330	1485	1380	1210	695
4150	... 1370	1410	1345	1240	530	94B40	... 1335	1455	1350	1220	629
4320	... 1335	1490	1365	1170	720						
4340	... 1335	1425	1310	1210	545	**Selected grades**					
4422	... 1350	1550	1490	1195	780						
4427	... 1330	1540	1425	1200	750	6407	... 1335	1515	900	615	615
4615	... 1340	1490	1400	1200	780	6417					
4620	... 1330	1475	1380	1190	755	6418					
4640	... 1315	1430	1300	1150	605	6419					
4718	... 1285	1510	1395	1175	740	6427	... 1320	1485	915	610	595
4815	... 1275	1450	1370	800	725	6429					
4820	... 1270	1440	1245	780	695	6431					

Note: Standard AISI/SAE H steels have wider chemical ranges than the comparable standard. However, critical temperature points are approximate and actual grade compositions virtually overlap; thus, critical temperatures for 8620H will be the same as for the 8620, for example. The M_s temperatures were calculated for the mean of the specified chemical composition ranges, according to the following formula (R. A. Grange and H. M. Stewart, *Metals Technology*, June 1946): M_s (°F) = 1000 − (650 × %C) − (70 × %Mn) − (35 × %Ni) − (70 × %Cr) − (50 × %Mo).

Heat Treatment of Carbon and Alloy Steels and Martensitic Stainless Steels

The following principles, primarily applicable to carbon and alloy steel forgings, also apply to martensitic stainless steels and certain other iron-base high-temperature alloys. Aluminum, titanium, and other alloys have unique characteristics, and their heat treatments are discussed in separate sections.

Fundamental Considerations. Heat treating processes involve four fundamental considerations: heating, cooling, the element of time at one or more temperatures, and the rate at which temperature changes occur in heating or cooling a particular part. Heating rates must be considered to avoid undue thermal stresses in highly alloyed materials and in forgings of complex design. Cooling rates are most important in producing the desired microstructural (phase) changes, which in turn

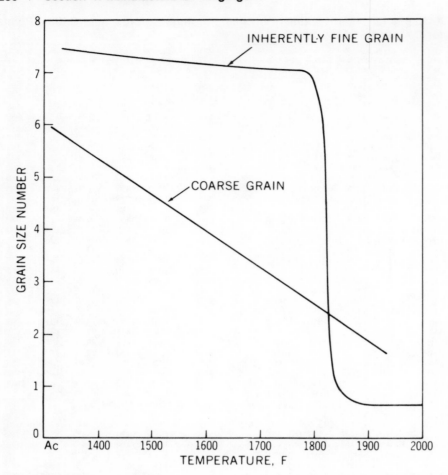

FIGURE 4-130. General pattern of grain coarsening in inherently fine-grained and coarse-grained steels.

develop the required hardness or other mechanical properties. In general, temperatures published in graphs or tables refer to the actual steel temperature unless otherwise specified.

Usually, the first operation in the heat treatment of carbon and alloy steel forgings is to raise the temperature of the steel (Figure 4-128) from room temperature to a temperature either in or above the transformation range, usually well above a lower limit of 1335 °F (725 °C). This causes ferrite, cementite, and other constituents that exist at room temperature to dissolve and form a homogeneous solid solution called austenite. The metal is then in the austenitic (gamma) condition, or "austenitized." Austenite may be regarded as the raw material at high temperatures out of which is generated the final metal structure upon subsequent cooling, or cooling and tempering, operations. Temperature regimes in which various crystal structures or phases exist under equilibrium conditions are given

in representations known as phase diagrams. Such a diagram for steels containing only iron and carbon (plain-carbon steels) is shown in Figure 4-129.

Selection of Austenitizing Temperature. The "transformation" or "critical" temperature (Ac_3) represents the minimum temperature to which the steel must be heated to austenitize it. An intermediate temperature, above which ferrite and austenite or cementite and austenite may be present at the same time, is also of importance. This critical point is known as the eutectoid temperature (Ac_1). The approximate critical points for the more commonly specified carbon and low-alloy steels are shown in Table 4-12. In carbon and alloy steels, the temperature used for austenitizing must be high enough to ensure that all carbides are in solution to take full advantage of the hardenability effects of the alloying elements, but not so high that extensive austenite grain growth occurs. In some instances, particularly with

high-carbon steels, the austenitizing temperature is kept intentionally below the Ac_3 temperature to retain carbides (previously spheroidized by heat treatment) intact and thereby promote wear resistance.

The general pattern of grain coarsening when steel is heated to above the critical temperature is shown in Figure 4-130. Coarse-grained steel increases its grain size gradually and consistently as the temperature is increased, but fine-grained steel remains so until a certain critical temperature is attained, above which grain coarsening occurs abruptly.

Time at Maximum Temperature. The length of time that steel forgings are held at temperature exerts an influence similar to that of the temperature employed. In any heat treating process, temperature and time must be considered simultaneously, because nearly all constitutional changes in metals require time and usually occur more rapidly at higher temperatures.

Because the changes sought at the maximum temperature are relatively rapid, the factor of time is less important than temperature. In general, a comparatively small increase in temperature will have a greater effect in accomplishing the desired change than a longer time at a lower temperature. In addition, decarburization, scaling, and distortion are increased with time at temperature. To reduce or avoid these effects, the time factor must be carefully controlled, and it is generally advisable to use the shortest soaking times possible. However, uniform heating throughout a cross section is important, and the use of pyrometers or thermocouples is advisable for optimal results.

Cooling Reactions

After the material has been heated to the predetermined temperature for an appropriate duration, it is cooled at the rate required to develop the desired structure. The precise temperature at which the transformation of austenite to ferrite and cementite takes place is dependent on the chemical composition and rate of cooling. Thus, the cooling rate helps determine the resultant structure and therefore the mechanical properties. The transformation of austenite takes place slightly below the lower critical temperature (Ar_1) on slow cooling. The resulting structure features coarsely laminated pearlite of relatively low hardness, low strength, and high ductility. Slow cooling also causes precipitation of ferrite in plain-carbon steels with less than approximately 0.80% carbon.

Faster rates of cooling, such as air cooling, cause the pearlite to become progres-

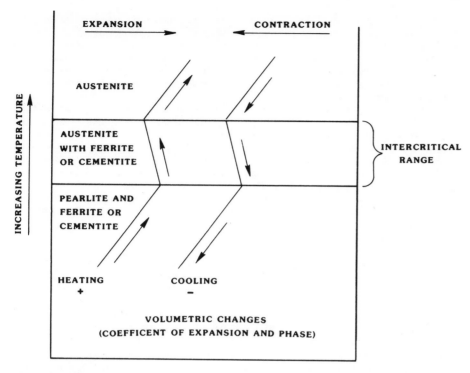

FIGURE 4-131. Schematic of density changes associated with phase transformations in steel. Adapted from Ref. 64.

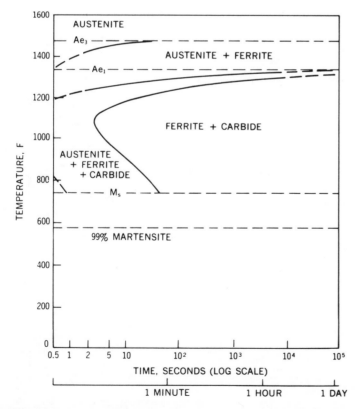

FIGURE 4-132. Schematic of the isothermal transformation diagram for 0.35% carbon and 0.37% manganese steel.

sively finer. Also, the precipitation of the ferrite becomes finer and more widely dispersed throughout the matrix. This results in a stronger, less ductile, and considerably harder steel.

By increasing the cooling rate still further, such as in liquid quenching, a new series of hard constituents, called bainite or martensite, are formed instead of pearlite and ferrite. These products are characteristic of fully hardened steel. The exact rate of cooling that causes the steel to harden fully to a martensitic condition is known as the "critical cooling rate" and depends primarily on the composition and section thickness of the steel. Martensite formed by quenching is very hard and brittle in the untempered condition.

As mentioned previously, the alloy content of steel has an important effect on the metallurgical structure obtained during cooling. Increased alloying retards the transformation rate of the austenite, causing the finer and slightly harder lamellar structures to be formed even at relatively slow cooling rates. Additional alloying causes bainite to form upon slow cooling. The air-hardening steels are those in which the transformation rate has been so extremely retarded by alloying that martensite or bainite structures are formed even as a result of air cooling from the austenitic condition.

Section thickness and its influence on the structures and properties finally achieved must also be considered. In general, the interior of a heavy section cools more slowly than the surface, and, accordingly, transformation in the interior regions is delayed. This results in a relatively softer internal structure, as compared with the more rapidly cooled exterior, and can produce a difference in properties between the interior and exterior, as well as internal stresses, possible distortion, and mixed microstructures.

Furthermore, the formation of pearlite, bainite, or martensite during cooling (or their reversion during heating) involves changes in density and coefficients of thermal expansion (Figure 4-131). The expansions or contractions associated with these changes are an important consideration in avoiding quench cracking or severe distortion, particularly in parts of large cross sections in which large temperature gradients are likely during rapid heating or cooling.

The operation known as tempering is performed to relieve internal stresses, to obtain the required hardness, and to recover toughness and ductility after heat treating for hardness. Tempering consists of heating to temperatures below the Ac_1

transformation temperature (Figure 4-129). The relieving of stresses and recovery of ductility are achieved mainly by precipitation of cementite during tempering.

Transformation Diagrams. The iron-carbon phase diagram (Figure 4-129) shows that steel at high temperatures is a single-phase solid solution called austenite. In the austenitic condition, it is in a state from which it may be either hardened or full annealed (softened). These changes are determined by the rate of cooling. To understand the effect of different rates of cooling, it is necessary to observe the time required for decomposition of austenite as it transforms to other structures at specific, lower temperatures.

The pattern of transformation is determined experimentally by quenching several specimens from above the transformation temperature in a bath held at a constant elevated temperature. Individual specimens are then removed from the bath at specified time intervals and quenched in brine. After microscopic examination, the amount of decomposition of the austenite that has occurred during the immersion time at the given temperature can be determined. A series of these tests conducted at a number of temperature levels provides data for the isothermal transformation diagram (also termed time-temperature-transformation, or TTT-curves), such as the one for a 1035 steel shown in Figure 4-132.

The contours of the curves and their locations, relative to the time scale, are influenced by the composition and grain size of the austenite. Most alloying elements that dissolve in austenite change the shape of the curves. Thus, the curves become somewhat characteristic of the alloying elements, and it is possible to classify various steels according to their isothermal transformation diagrams. In general, an increase in alloy content or in grain size of the austenite results in a retardation of the transformation of the austenite (the curve in Figure 4-132 is moved to the right in relation to the time axis). Steels whose curves lie farther to the right have greater hardenability.

To develop high strength and toughness properties, a fully martensitic structure is initially desired for quenched steel forgings. Such a structure may be obtained through use of a cooling rate that would appear to the left of the "nose" of the TTT-curve. Martensite does not begin to form until the temperature is below a critical temperature known as the M_s temperature. In addition, the exact amount of martensite that is formed from austenite is determined by the temperature below the M_s to which the steel is taken. The transforma-

tion is complete once the M_f temperature, which depends on alloy composition, is reached. Sometimes, the M_f temperature is below room temperature, necessitating cryogenic treatment to avoid the retention of austenite. Thus, an understanding of the shape of transformation diagrams is essential to control microstructure and properties.

The isothermal transformation diagrams of most low-alloy steels are not as simple in shape as the one shown in Figure 4-132. Some diagrams show two noses, an upper and a lower, on the left curve (Figure 4-133). When this occurs, the nose representing the shorter of the two time intervals for the beginning of transformation offers the key indication for hardening.

The area labeled "Austenite + ferrite" shows a region in which austenite and ferrite exist together. The longer the time the steel is held at a given temperature, the greater the transformation and, thus, the percentage of ferrite.

Heat treatment can be determined for various steels by using isothermal transformation diagrams. Treatments that can be established with precision if TTT-curves of the steels are available include normalizing, spheroidizing, annealing, harden-

ing, and martempering. It must be realized, of course, that quenching steel in water or oil at room temperature does not result in transformation at a constant temperature, but rather in transformations that occur during continuous cooling. In such cases, the transformation begins at points located somewhat below and to the right of the isothermal transformation diagrams.

In subsequent sections, specific heat treatments such as annealing, normalizing, and hardening and tempering will be discussed in the context of the basic transformation principles outlined.

Annealing

Annealing of steel is a heat treatment operation that can be performed to attain a variety of results, including the softening and alteration of the steel structure to develop formability, machinability, and the required mechanical properties. and to relieve cooling stress or stresses induced by cold or hot working. The complete annealing process may involve a heating cycle, a holding period, and a controlled cooling cycle. Each part of the process can be varied to suit the desired end result;

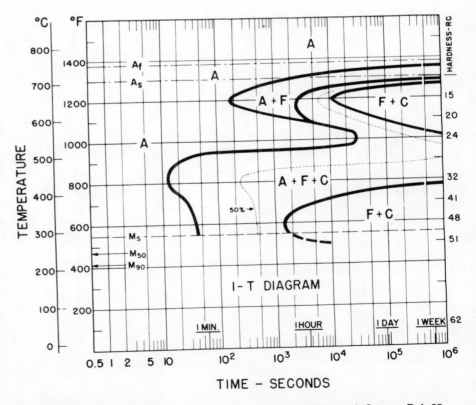

FIGURE 4-133. Isothermal transformation diagram for 4340 steel. Source: Ref. 65.

TABLE 4-13. Recommended Temperatures and Time Cycles for Annealing Alloy Steels to Obtain a Predominantly Pearlitic Structure(a)

Steel	Austenitizing temperature, °F (°C)	Full anneal(b) From °F (°C)	Full anneal(b) To °F (°C)	Full anneal(b) Rate °F/h (°C/h)	Isothermal anneal(c) Cool to °F (°C)	Isothermal anneal(c) Hold for, h	HB (approx)
1340	1525 (830)	1350 (730)	1130 (610)	20 (11)	1150 (620)	4.5	183
2340	1475 (800)	1210 (655)	1030 (555)	15 (8)	1100 (590)	6.0	201
2345	1475 (800)	1210 (655)	1020 (550)	15 (8)	1100 (590)	6.0	201
3120(d)	1625 (885)	. . .	. . .	. . .	1200 (650)	4.0	179
3140	1525 (830)	1350 (730)	1200 (650)	20 (11)	1225 (660)	6.0	187
3150	1525 (830)	1300 (705)	1190 (645)	20 (11)	1225 (660)	6.0	201
3310(e)	1600 (870)	. . .	. . .	. . .	1100 (590)	14.0	187
4042	1525 (830)	1370 (745)	1180 (640)	20 (11)	1225 (660)	4.5	197
4047	1525 (830)	1350 (730)	1170 (630)	20 (11)	1225 (660)	5.0	207
4062	1525 (830)	1280 (695)	1170 (630)	15 (8)	1225 (660)	6.0	223
4130	1575 (855)	1410 (765)	1230 (665)	35 (19)	1250 (675)	4.0	174
4140	1550 (840)	1390 (755)	1230 (665)	25 (14)	1250 (675)	5.0	197
4150	1525 (830)	1370 (745)	1240 (670)	15 (8)	1250 (675)	6.0	212
4320(d)	1625 (885)	. . .	. . .	. . .	1225 (660)	6.0	197
4340	1525 (830)	1300 (705)	1050 (565)	15 (8)	1200 (650)	8.0	223
4620(d)	1625 (885)	. . .	. . .	. . .	1200 (650)	6.0	187
4640	1525 (830)	1320 (715)	1110 (600)	15 (8)	1150 (620)	8.0	197
4820(d)	. . .	. . .	. . .	. . .	1125 (605)	4.0	192
5045	1525 (830)	1390 (755)	1230 (665)	20 (11)	1225 (660)	4.5	192
5120(d)	1625 (885)	. . .	. . .	. . .	1275 (690)	4.0	179
5132	1550 (840)	1390 (755)	1240 (670)	20 (11)	1250 (675)	6.0	183
5140	1525 (830)	1360 (740)	1240 (670)	20 (11)	1250 (675)	6.0	187
5150	1525 (830)	1300 (705)	1200 (650)	20 (11)	1250 (675)	6.0	201
52100(f)	. . .	. . .	. . .	. . .	. . .	. . .	. . .
6150	1525 (830)	1400 (760)	1250 (675)	15 (8)	1250 (675)	6.0	201
8620(d)	1625 (885)	. . .	. . .	. . .	1225 (660)	4.0	187
8630	1550 (840)	1350 (730)	1180 (640)	20 (11)	1225 (660)	6.0	192
8640	1525 (830)	1340 (725)	1180 (640)	20 (11)	1225 (660)	6.0	197
8650	1525 (830)	1310 (710)	1200 (650)	15 (8)	1200 (650)	8.0	212
8660	1525 (830)	1290 (700)	1210 (655)	15 (8)	1200 (650)	8.0	229
8720(d)	1625 (885)	. . .	. . .	. . .	1225 (660)	4.0	187
8740	1525 (830)	1340 (725)	1190 (645)	20 (11)	1225 (660)	7.0	201
8750	1525 (830)	1330 (720)	1170 (630)	15 (8)	1225 (660)	7.0	217
9260	1575 (855)	1400 (760)	1300 (705)	15 (8)	1225 (660)	6.0	229
9310(e)	1600 (870)	. . .	. . .	. . .	1100 (590)	14.0	187
9840	1525 (830)	1280 (695)	1180 (640)	15 (8)	1200 (650)	6.0	207
9850	1525 (830)	1290 (700)	1190 (645)	15 (8)	1200 (650)	8.0	223

Note: Tables are valid guides for steels slightly modified from the AISI/SAE types listed with the same carbon content. (a) In isothermal annealing to obtain a pearlitic structure, steels may be austenitized at up to 125 °F higher than temperatures listed. (b) The steel is cooled in the furnace at the indicated rate through the temperature range shown. (c) The steel is cooled rapidly to the temperature indicated and is held at that temperature for the time specified. (d) Seldom annealed; structures with improved machinability are developed by normalizing or by transforming isothermally after rolling or forging. (e) Annealing is impractical by the conventional process of continuous slow cooling; the lower transformation temperature is markedly depressed, and excessively long cooling cycles are required to obtain transformation to pearlite. (f) Predominantly pearlitic structures are seldom desired in this steel.

consequently, specific annealing cycles have become known by names characteristic of the particular process or end result.

Full annealing consists essentially of heating to a temperature above the critical temperature range, Ac_3 (Table 4-13) and holding at that temperature for a period of time; this is followed by slow cooling below the critical temperature range, Ar_1. The austenitizing temperature is usually relatively high so that full carbide solution is obtained. Slow cooling ensures that transformation of the austenite occurs only and completely in the high-temperature end of the pearlite range of the transformation diagram. This results in the formation of soft, coarse lamellar pearlite and can remove internal or residual stresses. Typical hardness levels following full annealing are given in Table 4-13 for a variety of steels. The conversion between Brinell hardness (HB) and ultimate tensile strength is given in Table 4-14.

Full annealing is a simple heat treatment and is reliable for most steels. It is, however, rather time consuming, because it involves slow cooling over the entire temperature range from the austenitizing temperature to a temperature well below that at which transformation is complete. Because of the time spent at a relatively high temperature, scaling oxidation and decarburization may also be a problem unless steps are taken to prevent or control them. Figure 4-134 illustrates the time-temperature cycle involved in this type of annealing.

Isothermal Annealing. Annealing to produce coarse pearlite, as described previously for full annealing, can also be carried out isothermally by cooling to the proper temperature, which is dictated by the transformation curve, and by allowing the steel to isothermally transform to coarse pearlite. Figure 4-135 schematically illustrates this procedure, and recommended annealing temperatures are given in Table 4-12.

This type of annealing cycle may provide considerable time savings over the conventional full annealing treatment. The time from austenitizing temperature down to transformation (isothermal) temperature is important if reproducible results are desired. This cooling rate should generally be as rapid as possible for carbon and low-alloy steels and can be quite slow for the more highly alloyed grades. The cooling rate after completion of isothermal transformation is unimportant from the standpoint of microstructure, but rapid cooling rates may produce undesirable thermal stresses. Small parts may be cooled to and held at the transformation temperature in salt or lead baths. Continuous heat treatments can also be employed for this method of annealing, where accelerated cooling to the desired temperatures and holding at these temperatures may be secured in various zones or chambers of the furnace. Scaling oxidation and decarburization may also be minimized because of the relatively short time that the parts are held in the high-temperature range.

This method does not appear to have any particular advantage over conventional full annealing for batch annealing of large furnace loads. The rate of cooling of the center of the load may be so slow as to prevent rapid cooling of the entire load to the transformation temperature. Under these circumstances, the conventional full anneal offers better assurance of obtaining the desired microstructure and end results.

Spheroidizing Anneal. This type of anneal refers to any process that on heating and cooling of the steel will produce a globular or spheroidal type of carbide. A structure in which the carbides are spheroidized has been found to have the best machinability for high-carbon steels (over 0.50%). If the steel is to be severely deformed by cold upsetting, extruding, bending, or drawing, this structure is desirable. Three types of annealing cycles are commonly used:

(a) Heating to a temperature above the Ac_1 and slow cooling between the Ac_1 and Ar temperatures, or cooling rapidly to just below the Ar_1 and holding for a prolonged period

(b) Heating to a temperature just below the Ac_1 and holding for a prolonged period, usually followed by slow

TABLE 4-14. Hardness Versus Tensile Strength

Brinell indentation diameter, mm	Brinell hardness No. (10-mm tungsten carbide ball, 3000-kg load)	Approximate tensile strength, ksi
2.60	...	
	555	298
2.65	...	292
	534	288
2.70	...	278
	514	274
2.75	...	269
		265
	495	264
2.80	...	258
		252
	477	252
2.85	...	244
		242
	461	242
2.90	...	231
	444	230
2.95	429	219
3.00	415	212
3.05	401	202
3.10	388	193
3.15	375	184
3.20	363	177
3.25	352	171
3.30	341	164
3.35	331	159
3.40	321	154
3.45	311	149
3.50	302	146

Brinell indentation diameter, mm	Brinell hardness No. (10-mm tungsten carbide ball, 3000-kg load)	Approximate tensile strength, ksi
3.55	293	141
3.60	285	138
3.65	277	134
3.70	269	130
3.75	262	127
3.80	255	123
3.85	248	120
3.90	241	116
3.95	235	114
4.00	229	111
4.05	223	...
4.10	217	105
4.15	212	102
4.20	207	100
4.25	201	98
4.30	197	95
4.35	192	93
4.40	187	90
4.45	183	89
4.50	179	87
4.55	174	85
4.60	170	83
4.65	167	81
4.70	163	79
4.80	156	76
4.90	149	73
5.00	143	71
5.10	137	67
5.20	131	65

cooling. This is also known as sub-critical annealing.

(c) Alternate heating to temperatures within and slightly below the critical range

Figure 4-136 schematically illustrates each of these cycles; temperatures and times for various isothermal and continuous cooling cycles are summarized in Table 4-15.

The spheroidal type of carbide is most easily precipitated from a heterogeneous austenite. Low annealing temperatures are employed, which do not permit the room-temperature constituents to transform completely to austenite and which prevent appreciable diffusion of carbon. Depending on the steel and the particular time-temperature cycle employed, austenitizing temperatures of not more than 100 °F (55 °C) above the lower critical temperature should be used for cycles (a) and (c). Hardnesses after spheroidization depend on carbon and alloy content. Increasing the carbon or alloy content, or both, results in increased hardness. In general, however, as-spheroidized hardness levels in the range of 160 to 220 HB can be expected from heat treatments of this sort.

Process and stress relief annealing involves heating the material to a tempera-

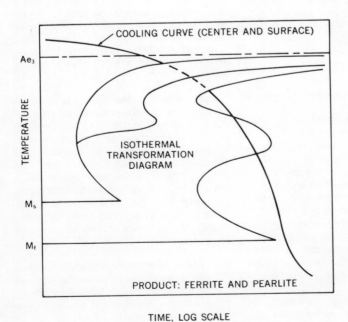

FIGURE 4-134. Schematic transformation diagram for conventional full annealing.

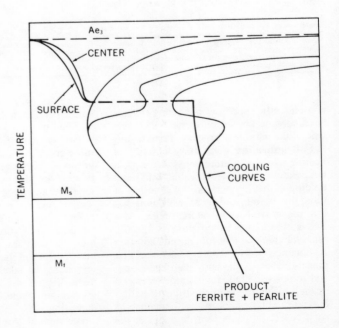

FIGURE 4-135. Schematic transformation diagram for isothermal annealing.

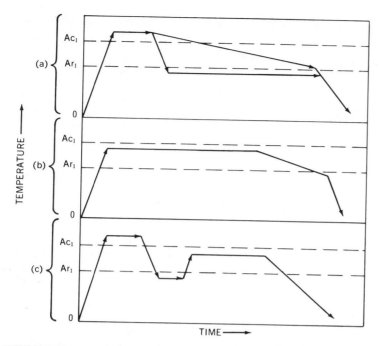

FIGURE 4-136. Schematic representation of spheroidizing annealing cycles.

TABLE 4-15. Recommended Temperatures and Time Cycles for Annealing Alloy Steels to Obtain a Spheroidized(a) Structure

Steel	Austenitizing temperature, °F (°C)	Cooling method — Conventional(b) From °F (°C)	To °F (°C)	Rate °F/h (°C/h)	Isothermal(c) Cool to °F (°C)	Hold for, h	HB (approx)
1320(d)	1480 (805)	. . .	. . .	. . .	1200 (650)	8.0	170
1340	1380 (750)	1350 (730)	1130 (610)	10 (5.5)	1180 (640)	8.0	174
1340	1320 (715)	1210 (655)	1030 (555)	10 (5.5)	1125 (605)	10.0	192
2345	1320 (715)	1210 (655)	1020 (550)	10 (5.5)	1125 (605)	10.0	192
3120(d)	1450 (790)	. . .	. . .	. . .	1200 (650)	8.0	163
3140	1370 (745)	1350 (730)	1200 (650)	10 (5.5)	1225 (660)	10.0	174
3150	1380 (750)	1300 (730)	1190 (645)	10 (5.5)	1225 (660)	10.0	187
3310(e)	. . .	. . .	. . .	. . .	. . .	. . .	. . .
4042	1400 (760)	1370 (745)	1180 (640)	10 (5.5)	1225 (660)	8.0	179
4062	1380 (750)	1280 (695)	1170 (630)	10 (5.5)	1225 (660)	10.0	197
4130	1380 (750)	1380 (750)	1230 (665)	10 (5.5)	1250 (675)	8.0	163
4140	1380 (750)	1380 (750)	1230 (665)	10 (5.5)	1250 (675)	9.0	179
4150	1380 (750)	1370 (745)	1240 (670)	10 (5.5)	1250 (675)	10.0	197
4320(d)	1425 (775)	. . .	. . .	. . .	1225 (660)	8.0	187
4340	1380 (750)	1300 (705)	1050 (565)	5 (3)	1200 (650)	12.0	197
4620(d)	1425 (775)	. . .	. . .	. . .	1200 (650)	8.0	170
4640	1370 (745)	1320 (715)	1110 (600)	10 (5.5)	1150 (620)	12.0	183
4820(d)	1370 (745)	. . .	. . .	. . .	1125 (607)	8.0	179
5045	1400 (760)	1390 (755)	1230 (665)	10 (5.5)	1225 (660)	8.0	179
5140	1380 (750)	. . .	. . .	. . .	1275 (690)	8.0	174
5120(d)	1475 (803)	1360 (740)	1240 (670)	10 (5.5)	1250 (675)	10.0	174
52100	1460 (795)	1380 (750)	1250 (675)	10 (5.5)	1275 (690)	16.0	187
6150	1400 (760)	1400 (760)	1250 (675)	10 (5.5)	1250 (675)	10.0	192
8620(d)	1450 (790)	. . .	. . .	. . .	1225 (660)	8.0	174
8640	1380 (750)	1340 (725)	1180 (640)	10 (5.5)	1225 (660)	8.0	183
8660	1380 (750)	1290 (700)	1210 (655)	10 (5.5)	1200 (650)	10.0	207
8720(d)	1450 (790)	. . .	. . .	. . .	1225 (660)	8.0	174
8750	1380 (750)	1330 (720)	1170 (630)	10 (5.5)	1225 (660)	10.0	207
9260	1400 (760)	1400 (760)	1300 (705)	10 (5.5)	1225 (660)	10.0	212
9310(e)	. . .	. . .	. . .	. . .	. . .	. . .	. . .
9840	1370 (745)	1280 (695)	1180 (640)	10 (5.5)	1200 (650)	10	192
9850	1370 (745)	1290 (700)	1190 (645)	10 (5.5)	1200 (650)	12	207

Note: Tables are valid guides for steels slightly modified from the AISI/SAE types listed with the same carbon content. (a) In isothermal annealing to obtain a pearlitic structure, steels may be austenitized at up to 125 °F higher than temperatures listed. (b) The steel is cooled in the furnace at the indicated rate through the temperature range shown. (c) The steel is cooled rapidly to the temperature indicated and is held at that temperature for the time specified. (d) Seldom annealed; structures with improved machinability are developed by normalizing or by transforming isothermally after rolling or forging. (e) Spheroidized most readily by long-time (12 to 18 h) tempering at subcritical temperature.

ture close to the lower critical temperature, Ac_1, and cooling slowly. Its function is to reduce residual or internal stresses and hardness in cold-worked materials. Recrystallization of the cold-worked structure is a consequence of the heat treatment. This treatment is frequently employed to facilitate cold shearing, cold blanking, and mild cold forming and may be used between cold forming operations to reduce hardness.

Normalizing

Normalizing is a heat treatment in which steel is heated to a temperature above its critical temperature (Ac_3) and cooled to below that range in still air. The usual practice is to normalize from 100 to 150 °F (55 to 65 °C) above the critical temperature. In some alloy steels with carbides that are only soluble with difficulty, however, considerably higher temperatures may be required to ensure solution of the carbides. This heat treatment operation is well suited to continuous furnace operations.

Because true normalizing requires uniform cooling of the forging in still air, it should not be confused with annealing (restricted cooling rates) or air hardening (a quenching procedure associated with accelerated cooling rates). Normalizing is an interesting and useful procedure and may be used to:

- Refine grain size and homogenize microstructure
- Improve machining characteristics
- Modify and refine cast dendritic structures
- Provide desired mechanical properties

Refinement of Grain Size and Homogenization of Microstructure. During forging operations, different parts of the forging will be subjected to different degrees of working or plastic deformation; finishing temperatures may vary from one forging to another, and different cooling rates will be present on cooling the forging to room temperature. A variety of as-forged grain sizes and microstructures can therefore be present in as-forged products. The normalizing operation will refine the coarse grain size resulting from a high finishing temperature and will produce a uniform, relatively fine-grained microstructure. The improvement in subsequent hardening operations is associated with the resultant finer carbide size and a distribution that is more favorable to carbide solution at the austenitizing temperatures (Table 4-16).

Improvement of machining characteristics is probably the least understood

TABLE 4-16. Effect of Heating on End Quench Hardenability of Spheroidized Steels

Expressed as Rockwell C Hardness Deficiency from the Hardness of Bars Normalized prior to Reheating and Quenching

AISI/SAE No.	Austenitizing temperature furnace(a) °F	°C	Rockwell C hardness(b) deficiency of standard end quench hardenability bars austenitized for: 15 min	30 min	60 min	120 min
1038	1550	845	14	8	2	0
1345	1500	815	16	0	0	0
2345	1450	790	11	4	0	0
2345	1550	845	2	0	0	0
3140	1500	815	11	6	1	0
3140	1525	830	2	0	0	0
3245	1550	845	20	11	8	4
4042	1500	815	28	5	2	2
4140	1550	845	14	11	7	7
4142	1550	845	18	10	3	2
4140	1650(c)	900(c)	10	0	0	0
4145	1550	845	14	5	2	2
4340	1500	815	18	10	0	0
4640	1550	845	9	0	0	0
5140	1550	845	6	1	0	0
6145	1550	845	28	18	8	5
6140	1575	855	9	3	0	0
8740	1550	845	18	1	0	0

(a) 100 °F (55 °C) above the Ac₃ temperature determined for each steel, unless otherwise noted. (b) Compared at the end quench position equivalent to the hardness of 90% martensite on the normalized and quenched bar. (c) 200 °F (95 °C) above the Ac₃ temperature.

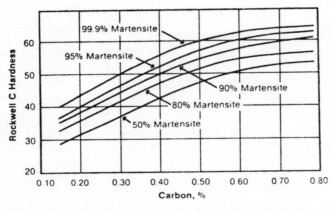

FIGURE 4-137. Approximate average relationship between carbon content, hardness, and percentage of as-quenched martensite. Source: Ref. 66.

aspect of normalizing. This is because, depending on the steel composition and prior condition and/or section size, normalizing may soften, harden, or relieve stresses. As stated earlier, the effect should not be confused in the strictest sense with annealing or hardening by quenching. For example, a steel that has been hardened by austenitizing and quenching or cold working can be softened by normalizing and also be improved in machining response. On the other hand, inherently soft and gummy steels such as 5120, 5130, 8615, and 8620, either as forged or annealed, usually can be improved in machining by an increase in hardness by normalizing.

Modification and Refinement of Cast Dendritic Structures. Unusual nonuniformity of as-cast or as-forged parts resulting in dendritic-type coarse structures will benefit from this treatment much in the same manner as refinement and homogenization of grain size and microstructure. Improved machining response is obtained as well.

Desired Mechanical Properties. To meet mechanical properties, normalizing followed by a tempering operation is often used and indeed required by specification for certain carbon and alloy steels, particularly those containing molybdenum or vanadium, or both, when intended for elevated-temperature service. Although tensile strengths may not differ much from those for annealed microstructures, the presence of bainite and coarse pearlite favors sustained stability at elevated temperatures.

Hardenability

An essential feature in the heat treatment of steel to develop the best combination of toughness and strength, or to provide maximum flexibility in the final choice of mechanical properties, is the steel's ability to be hardened to produce martensite, which is subsequently tempered to produce a spectrum of mechanical properties.

To attain the superior properties of low-temperature transformation products, it is necessary to avoid prior transformation at higher temperatures to softer products. This means, therefore, that the steel must be cooled through the high-temperature transformation ranges at a rate rapid enough to ensure that transformation does not occur at these temperatures.

If the forging is very small, so that cooling following austenitizing is uniform through the section, the required cooling rates to produce martensite (or pearlite or bainite) can be determined directly from isothermal transformation diagrams, or their continuous cooling counterparts. Steels hardened by the production of 100% martensite upon quenching manifest a certain maximum hardness that is largely dependent on carbon content (Figure 4-137).

For many steel forgings, however, the cooling rate through the entire section is not sufficient to produce a totally martensitic structure. The cooling rate is determined by thermal properties (which do not vary much from one steel to another) and the type of quenchant employed. Water quenches are more effective than oil quenches. A forging of one steel may harden completely in an oil quench, while a similar one of another steel will require a more drastic water quench, and even then only the surface layers may escape transformation to the softer product, pearlite. Mixtures of martensite and higher temperature transformation products also tend to develop as-quenched hardness dependent only on the carbon content (Figure 4-137).

The concept of "hardenability" is the means by which the ability to develop martensite to various depths in a steel forging is described. It is a function of alloy content (through its effect on the transformation diagram) and the type of quench employed (which determines the cooling history throughout the forged part). For equivalent cooling rates, highly alloyed steels develop 100% martensite to greater depths than lean alloys. For example, an

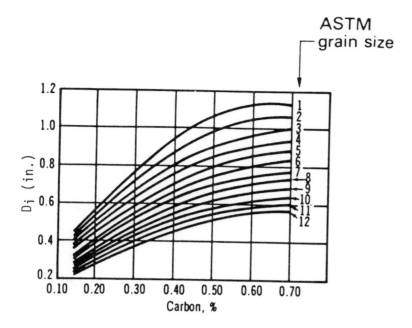

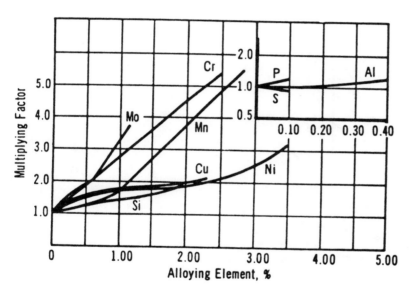

FIGURE 4-138. Multiplying factors for determination of the ideal critical diameter. Source: Ref. 67.

TABLE 4-17. Quench Severity Coefficients (H Value) For Various Quenchants

	Cooling medium		
Agitation	Oil	Water	Brine
None	0.25–0.30	0.9–1.0	2.0
Mild	0.30–0.35	1.0–1.1	2.0–2.2
Moderate	0.35–0.40	1.2–1.3	...
Good	0.4–0.5	1.4–1.5	...
Strong	0.5–0.8	1.6–2.0	...
Violent	0.8–1.1	4.0	5.0

medium of infinite severity. The D_I is estimated by multiplying a series of factors that account for the carbon content of the alloy, the austenite grain size, and the levels of each particular alloying element (Figure 4-138). D_c is obtained in turn from D_I by a correlation between the two, which is dependent on the quench medium and the degree of agitation expected (Figure 4-139, Table 4-17).

Austenitizing and Quenching. Conventional hardening of steel is accomplished, as discussed earlier, by heating the steel to 1335 to 1600 °F (725 to 870 °C), depending on composition, and cooling rapidly enough to form martensite. Recommended austenitizing temperatures are given in Table 4-18.

The effectiveness of quenching depends on the cooling characteristics of the quenching medium as related to the ability of the steel to harden. The results achieved depend on design of the system and its maintenance, quenchant used, degree of bath agitation, consideration of part configuration and size, fixturing, and, of course, steel composition and its hardenability. As far as forgings are concerned, quenching is usually direct (the most applicable), interrupted (timed quenching), or selective (isolating some area of the part for hardening).

The heat-extracting capability of the commonly used quenchants are generally ranked from fast to slow as follows: brine or caustic solutions, water, and oil. H-values or quenching power data have been developed that can be used in calculations related to hardenability, as a function of coolant circulation or agitation (Table 4-17).

Polymer solutions are modern alternatives to water and oil and require special consideration in their selection. Generally, they can provide H-factors similar to water without the undesirable vapor blanketing of water. Also, wide variations in cooling power can be obtained with many of the commercial products. Air cooling in still to forced air cannot be ignored as a quenching medium, although in a different sense. High-alloy air-hardening tool

AISI 1080 steel, when quenched, may have high hardness but low hardenability, because it does not maintain that hardness to any considerable depth in a given section. On the other hand, an AISI 4340 steel has a quenched hardness somewhat less than the AISI 1080 steel, but much greater hardenability, because this hardness is maintained at a much greater depth.

Measurable hardenability values for various steels are required, and such values are obtained by applying the Jominy end-quench hardenability test. This test has

made it possible to predict the depth to which a given grade of steel can be hardened by applying a specific quenching rate.

Another typical method of reporting hardenability is through the calculation of the critical diameter, D_c, to which a particular alloy can be expected to form 50% martensite at the center following austenitizing and quenching. The value of D_c is heavily dependent on the magnitude of D_I, or the ideal critical diameter, which is the equivalent diameter at which 50% martensite is formed when using a quenching

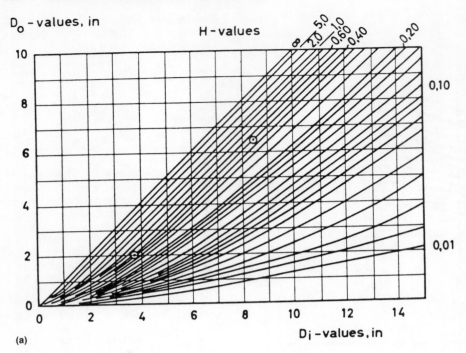

(a)

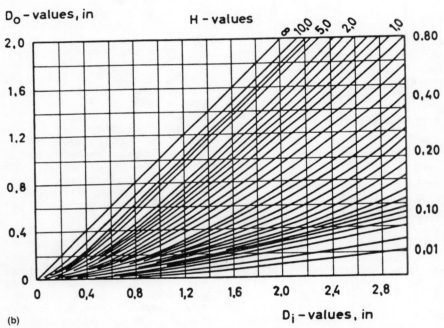

(b)

FIGURE 4-139. Correlation between critical diameter D_o, ideal diameter D_i, and H value. Diagram (b) is an enlargement of the lower left portion of diagram (a).

tion takes place, the stresses may build to a high value, causing quench cracking.

To minimize these stresses, the quenching rate should not be much in excess of that dictated by the size and hardenability of the piece. Parts should be designed so that generous fillets are used where contour changes occur, and notches caused by tool marks and so forth should be eliminated.

Tempering

As-quenched martensite is very hard and brittle. If allowed to remain in this highly stressed condition, cracks will tend to form in all but the low-carbon analyses. To prevent cracking, martensite should be tempered immediately following hardening if at all possible.

Often referred to in the shop as drawing, it is metallurgically essential to develop the final optimum combination of strength and toughness, as discussed earlier. Equally important is the prevention of quench cracking due to the enormous stresses present in the forging. An excellent rule is to temper as soon as possible while still warm. In fact, parts have been known to fracture, often violently, when tempering has been delayed or overlooked. Tempering can be a single or multiple operation, and temperature is selected on the basis of the steel grade and properties required. Certain alloy steels require avoidance of the temper brittle range. Ordinarily, after holding at temperature for a suitable time, parts may be discharged and allowed to cool safely to room temperature. Many alloy steels are subject to loss of toughness if tempering involves heating in, or substantial time in heating or cooling through, certain temperature regimes. The two most important embrittling ranges occur at approximately 500 °F (260 °C) and between 750 and 1025 °F (340 and 550 °C). Rapid cooling from an appropriate tempering temperature is required to obtain good impact property values in these cases.

Properties of Quenched and Tempered Steels. Carbon and alloy steels are used for engineering applications because of the combination of strength and toughness that can be developed. The optimum combination of strength and toughness is developed in a steel when the microstructure is tempered martensite. In addition, for applications in which wear resistance is of importance, the part may be required to have the maximum hardness. Then, the structure could be martensite that is only lightly tempered to remove quenching stresses.

Hardness is probably the property most

steels and martensitic stainless steels such as 410, 420, and 440 types will harden throughout sections as large as 3 to 7 in. in air.

Internal stresses introduced by quenching are the result of dimensional changes that occur in cooling, large temperature gradients between surface and center, and the volume change accompanying the change from gamma to alpha iron. If the steel is ductile, these changes can cause slight plastic flow of the grains, which will relieve the stresses. In higher carbon and alloy steels in which no plastic deforma-

TABLE 4-18. Austenitizing Temperatures for Direct Hardening of Carbon and Alloy Steels (SAE)

Carbon steels

Steel	Temperature °F	°C
1025	1575–1650	855–900
1030	1550–1600	845–870
1033	1525–1575	830–855
1035	1525–1575	830–855
1036	1525–1575	830–855
1037	1525–1575	830–855
1038	1525–1575	830–855
1039	1525–1575	830–855
1040	1525–1575	830–855
1041	1475–1550	800–845
1042	1475–1550	800–845
1043	1475–1550	800–845
1045	1475–1550	800–845
1046	1475–1550	800–845
1048	1475–1550	800–845
1050	1475–1550	800–845
1052	1475–1550	800–845
1055	1475–1550	800–845
1060	1475–1550	800–845
1064	1475–1550	800–845
1065	1475–1550	800–845
1070	1475–1550	800–845
1074	1475–1550	800–845
1078	1450–1500	790–815
1080	1450–1500	790–815
1084	1450–1500	790–815
1085	1450–1500	790–815
1086	1450–1500	790–815
1090	1450–1500	790–815
1095	1450–1500(a)	790–815(a)

Free-cutting carbon steels

Steel	Temperature °F	°C
1132	1525–1575	830–855
1137	1525–1575	830–855
1138	1500–1550	815–845
1140	1500–1550	815–845
1141	1475–1550	800–845
1144	1475–1550	800–845
1145	1475–1550	800–845
1146	1475–1550	800–845
1151	1475–1550	800–845

Alloy steels

Steel	Temperature °F	°C
1330	1500–1550	815–845
1335	1500–1550	815–845
1340	1500–1550	815–845
1345	1500–1550	815–845
3140	1500–1550	815–845
4037	1525–1575	830–855
4042	1525–1575	830–855

Alloy steels (continued)

Steel	Temperature °F	°C
4047	1500–1575	815–855
4063	1475–1550	800–845
4130	1500–1600	815–870
4135	1550–1600	845–870
4137	1550–1600	845–870
4140	1550–1600	845–870
4142	1550–1600	845–870
4145	1500–1550	815–845
4147	1500–1550	815–845
4150	1500–1550	815–845
4161	1500–1550	815–845
4337	1500–1550	815–845
4340	1500–1550	815–845
50B40	1500–1550	815–845
50B44	1500–1550	815–845
5046	1500–1550	815–845
40B46	1500–1550	815–845
50B50	1475–1550	800–845
50B60	1475–1550	800–845
5130	1525–1575	830–855
5132	1525–1575	830–855
5135	1500–1550	815–845
5140	1500–1550	815–845
5145	1500–1550	815–845
5147	1475–1550	800–845
5150	1475–1550	800–845
5155	1475–1550	800–845
5160	1475–1550	800–845
51B60	1475–1550	800–845
50100	1425–1475(b)	775–800(b)
51100	1425–1475(b)	775–800(b)
52100	1425–1475(b)	775–800(b)
6150	1550–1625	845–885
81B45	1550–1575	845–855
8630	1525–1600	830–870
8637	1525–1575	830–855
8640	1525–1575	830–855
8642	1500–1575	815–855
8645	1500–1575	815–855
86B45	1500–1575	815–855
8650	1500–1575	815–855
8655	1475–1550	800–845
8660	1475–1550	800–845
8740	1525–1575	830–855
8742	1525–1575	830–855
9254	1500–1650	815–900
9255	1500–1650	815–900
9260	1500–1650	815–900
94B30	1550–1625	845–885
94B40	1550–1625	845–885
9840	1525–1575	830–855

Alloy H steels

Steel	Temperature °F	°C
1330H	1600	870
1345H	1550	845
1340H	1550	845
1345H	1550	845
3140H	1550	845
4037H	1550	845
4042H	1550	845
4047H	1550	845
4063H	1550	845
4130H	1600	870
4135H	1550	845
4137H	1550	845
4140H	1550	845
4142H	1550	845
4145H	1550	845
4147H	1550	845
4150H	1550	845
4161H	1550	845
4337H	1550	845
4340H	1550	845
E4340H	1550	845
4520H	1700	925
50B40H	1550	845
50B44H	1550	845
5046H	1550	845
50B46H	1550	845
50B50H	1550	845
50B60H	1550	845
5130H	1600	870
5132H	1600	870
5135H	1600	870
5140H	1550	845
5145H	1550	845
5147H	1550	845
5150H	1550	845
5155H	1550	845
5160H	1550	845
51B60H	1550	845
6150H	1600	870
81B45H	1550	845
8630H	1600	870
8637H	1550	845
8640H	1550	845
8642H	1550	845
8645H	1550	845
86B45H	1550	845
8650H	1550	845
8655H	1550	845
8660H	1550	845
8740H	1550	845
8742H	1550	845
9260H	1600	870
94B30H	1600	870
94B40H	1550	845
9840H	1550	845

(a) This temperature range may be used for 1095 steel that is to be quenched in water, brine, or oil. For oil quenching, 1095 steel may be alternatively austenitized at 1500 to 1600 °F (815 to 870 °C). (b) This range is recommended for steel that is to be water quenched. For oil quenching, steel should be austenitized at 1500 to 1600 °F (815 to 870 °C).

often reported for quenched and tempered steels. It is a measure of the ability to support loads in service without suffering permanent shape change, or plastic deformation. Typically, its magnitude is measured in a standardized test in which an indenter is pressed into the steel using a prescribed load. The depth of penetration or the width or area of the hardness impression is used as the measure of hardness. For steels, the most common hardness tests are the Brinell, diamond pyramid, Rockwell C, Rockwell B, and Knoop hardness tests.

As mentioned previously, the maximum or as-quenched hardness of martensite is dependent primarily on the carbon content (Figure 4-137). Following tempering, the hardness of a steel may change significantly, depending on tempering temperature. Figure 4-140 shows this effect for a variety of plain-carbon steels in terms of both Rockwell C (HRC) and Vickers (VH) numbers. In this case, the change of hardness with tempering temperature (for a tempering time of 1 h) is greatest for the higher carbon alloys. Such a trend might have been expected, because the tempering process consists of the precipitation and growth of carbides. Steels with higher carbon contents thus lose more strength (in terms of Rockwell C points) because of carbide coarsening.

The addition of alloying elements, though not helpful from the standpoint of martensite hardness, does tend to enable higher tempered hardnesses to be achieved than are possible in plain-carbon steels. This effect, which may be verified by comparing the results in Figures 4-140 and 4-141, is a consequence of the retarding action that the alloying elements have on precipitation rates. The difference in hardness between the plain-carbon and alloy steels can be as much as 10 to 15 HRC points.

Tensile properties also give an idea of the ability of a steel to withstand loads in service and are derived from tests in which bars of the test material are pulled uniaxially to failure. From these experiments, measurements of yield strength, ultimate tensile strength, elongation, and reduction in area are obtained. The tensile properties of a common quenched and tempered steel, 4340, are shown in Figure 4-142. The decrease in hardness with tempering temperature, also shown in this figure, is accompanied by a general decrease in yield and tensile strength. By contrast, the ductility, both in terms of elongation as well as reduction in area, increases with the decrease in strength as the tempering temperature is increased. The observed trend is followed for all quenched and tempered steels (Table 4-19) and poses a question of tradeoffs for the designer.

Figure 4-142 also gives data on the toughness or impact energy of the 4340 steel at room temperature as a function of tempering temperature. Toughness is a measure of the ability to resist brittle fracture under load, and the simplest means of estimating its magnitude for various steels is the Charpy test. Named for the French engineer who designed it, the test consists of fracturing a notched test bar using a swinging pendulum. The "impact" energy to accomplish this, readily determined from the height to which the pendulum swings following fracture, is used as the toughness measure.

The impact energy for standard 4340 Charpy specimens reveals an interesting trend. This energy increases at first with tempering temperature, passes through a minimum, and then increases again. The minimum is due to 500 °F (260 °C) embrittlement, a name given to it because of the tempering temperature that produces the effect. At this tempering temperature,

"tramp" elements such as lead, antimony, and tin, whose content is small yet unavoidable in conventionally refined steels, segregate to the grain boundaries during the long-time tempering treatments given to quench-hardened martensitic microstructures. The precipitation of Fe_3C in platelet form at these tempering temperatures is also partly responsible for the embrittlement. Tensile ductility is not affected by this phenomenon; there is no minimum in the elongation or reduction in area versus tempering temperature plots. Only toughness is affected, and this occurs only when it is measured at ambient or cryogenic temperatures. Toughness data for other quenched and tempered steels, showing the same trend for Charpy tests conducted at −20 °F (−30 °C), are given in Figure 4-143.

A related embrittlement problem, often called temper brittleness, results from tempering martensite in, or slow cooling it through, the temperature range of 750 to 1020 °F (400 to 550 °C). As with 500 °F (260 °C) embrittlement, temper brittleness is manifested by room- or low-temperature brittle fracture, which is determined from Charpy testing, but not from tensile tests. The best way to determine the degree of property loss through such tempering treatments is by performing Charpy tests at various temperatures. In an embrittled steel, the ductile-to-brittle transition temperature (DBTT), or the temperature at which the impact energy drops sharply, will be much higher than in the unembrittled condition. This is easily detected in plots of the impact energy as a function of test temperature, such as those in Figure 4-144. Here, the embrittling effect raised the DBTT, or simply transition temperature, from −140 to −5 °F (−95 to −20 °C). This may seem unimportant, until it is realized that many steels in the unembrittled condition have transition temperatures around room temperature.

The above temperature range in which temper brittleness is induced appears to be quite large. Fortunately, the effects are great only near the midpoint of this range and when long-term tempering is used. This fact is illustrated in Figure 4-145 for 3140 steel, which shows that tempering times in excess of an hour are normally required for large increases in the transition temperature.

Fatigue is probably the most common failure mechanism of steels during service. It occurs by slow growth of a crack that usually initiates at a weak point on the surface of the structure or at a region where the loads are higher (a stress concentration). Fatigue cracks can be initiated and

grow even when the applied loads result in stresses below the yield strength. However, many metals are characterized by a threshold, or endurance limit, below which fatigue cracking can be avoided. For quenched and tempered steels, this limit is about one half the yield strength (Figure 4-146).

Resistance to fatigue can be increased by inducing compressive residual stresses on parts subject to this failure mechanism. This behavior is a result of the fact that fatigue is initiated under the action of tensile stresses. Thus, superimposing a compressive residual stress pattern may partially or totally eliminate the surface tensile stress component and allow higher tensile loads to be applied without reaching the endurance limit.

Other Heat Treatment Processes for Hardenable Steels

Martempering. The transformation into martensite occurring during the rapid cooling through the martensite temperature transformation range and the accompanying sharp temperature gradient can create high internal stresses that may result in quench cracking or distortion. A modified quenching procedure has been developed that is helpful in lowering these stresses after quenching. It involves quenching to a temperature slightly above the M_s temperature (usually into a molten salt bath), holding to permit temperature equalization, and then air cooling to room temperature. Transformation to martensite then occurs during relatively slow cooling. In the absence of the temperature gradients induced by conventional quenching, the stresses set up by the transformation are much lower than in conventional quenching operations. Standard tempering operations are then used as in conventional hardening processes. Martempering has been applied with great success to the heat treatment of tools, bearings, and dies in which cracking and especially distortion are encountered when conventional quenching techniques are employed.

If possible, the holding time for temperature equalization at just above the M_s temperature should not be so long as to allow the formation of the lower bainite structure. However, in marginal situations, bainite may be formed in the center of the section of the part. Whether or not this is permissible will depend on the service for which the part is intended.

Austempering is an isothermal heat treatment that produces lower bainite

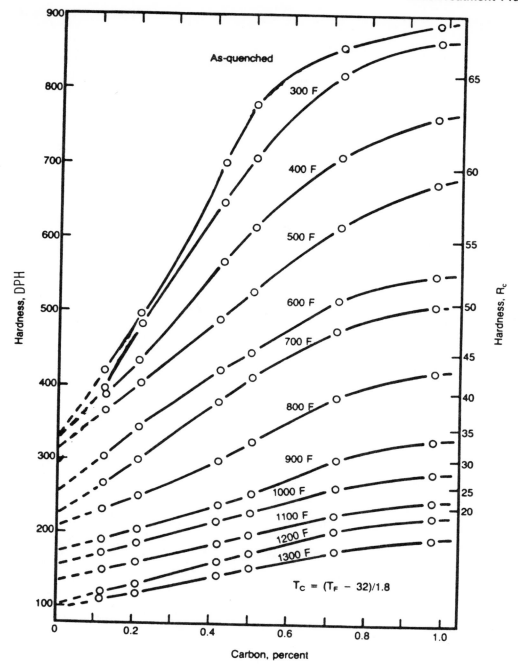

FIGURE 4-140. Hardness as a function of carbon content of martensite in iron-carbon alloys tempered at various temperatures. Source: Ref. 68.

structures and is an alternative to conventional hardening to martensite and tempering. The treatment involves quenching to the desired temperature in the lower bainite region, usually in molten salt, and holding at this temperature until transformation is complete.

It is common to hold for a time twice as long as that indicated by the isothermal

diagram to ensure complete transformation in any segregated area. After transformation has occurred, the piece may be subsequently tempered to a lower hardness level.

Austempering has an advantage over conventional quenching and tempering in that the bainite transformation takes place isothermally at a relatively high tempera-

ture. Thus, the transformation stresses are very low, distortion is held to an absolute minimum, and quench cracking almost never occurs. However, as with martempering, because of the slower cooling rates of the molten salt baths used in this technique, steel with high hardenability must be employed to prevent transformation to undesirable intermediate products during

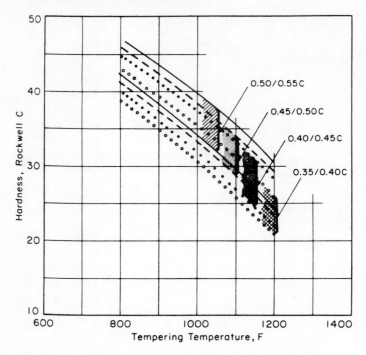

FIGURE 4-141. Hardnesses of quench hardened and tempered alloy steels. Source: Ref. 69.

cooling to the bainite temperature. A further consequence of the use of these high-hardenability steels is the longer time required for complete transformation to bainite to take place. This may make the austempering operation much more time consuming than martempering or conventional quenching and tempering. Largely because of the hardenability and time limitation discussed above, austempering has found its widest application in the heat treatment of plain high-carbon steels in small sections such as sheet, strip, and wire.

Surface Hardening or Differential Hardening. Through hardening often is not practical or even desirable from a properties standpoint. In such instances, hardening of the surface, leaving a soft core, may be beneficial. Ordinarily, forgers do not perform surface hardening operations, because they are applicable primarily to finished or nearly finished parts. Nonetheless, a brief description of these treatments is appropriate.

For design, economic, or physical dimension reasons, it may be desirable to harden only the surface of a part. In such objects, the core or inner portion of the part may be heat treated to produce a desired structure for machinability or a strength level for service, and the surface may be subsequently hardened for high strength and wear resistance. For example, the surface of the teeth of a gear must be hard to resist wear, whereas the remainder should be tough to resist shocks; a crankshaft should be hard on its bearing surfaces, whereas the web must be strong, tough, and ductile; the nose of an armor-piercing projectile must be hard, but the body must be tough.

Three principal ways in which this differential or surface hardening can be accomplished include:

- *Flame or induction hardening:* By heating to the austenite region only that part of the steel component that is to be hardened by rapid cooling
- *Case carburizing:* By making the part from a low-carbon steel that will not be appreciably hardened by quenching and by enriching the surface with carbon to yield a high carbon content in that region so that subsequent heat treatment will produce a hardened skin
- *Nitriding:* By enriching the surface layers of a suitable steel with nitrogen to form complex, but extremely hard, nitrides

Flame hardening is a process of heating the surface layers of an iron-base alloy above the transformation temperature range, Ac_3, by means of a high-temperature flame and then quenching. An essential part of this process is that the fuel gas flames impinge directly on the surface of the work. The usual depth of fully hardened material for most purposes is approximately $1/8$ in., although thicker or thinner layers of hardened material can be obtained.

Any type of hardenable steel can be flame hardened if the proper procedures are followed. With steels, the carbon content should be 0.35% or more for appreciable hardening. The best range for most flame hardening methods is between 0.40 and 0.50%. As with conventional hardening, the alloy content of the steel determines the type of quench that should be used. Consequently, the quench may vary from water or brine solution to air.

Flame-hardened steel will respond to tempering as if it were hardened by any other method. However, because flame hardening can be applied to parts of considerable size that are hardened by the progressive method, they usually can be tempered at the same time by the simul-

TABLE 4-19. Hardness Correlation for Quenched and Tempered Steels

Hardness, HRC	Hardness, HB	Tensile strength, ksi	Yield point, ksi	Elongation, %	Reduction in area, %
14	108	93–103	69–78	22.0–28.0	60.0–68.0
16	207	98–108	73–84	21.5–27.5	59.0–67.0
18	217	103–114	76–90	21.0–27.5	58.0–66.0
20	223	106–117	79–93	20.5–26.3	57.5–65.5
22	235	112–124	85–99	20.0–25.5	56.5–64.5
24	248	118–131	92–107	19.5–24.5	55.0–63.0
26	262	128–138	99–114	18.5–24.0	54.0–61.5
28	277	131–146	107–122	18.0–22.5	52.0–60.0
30	293	138–154	116–131	17.0–22.0	51.0–59.0
32	311	146–164	125–141	16.0–20.5	49.0–57.0
34	321	151–170	131–146	15.0–20.0	48.0–56.0
36	341	160–180	141–157	14.5–18.5	46.0–54.0
38	363	171–193	153–170	13.5–17.0	43.5–51.5
40	379	178–201	163–179	12.5–16.0	42.0–50.0
42	401	188–222	176–185	11.0–15.0	40.0–49.0

Note: This table is more accurate for steels with 0.30% carbon or higher; steels with less than 0.30% carbon usually have yield point values lower than those shown.

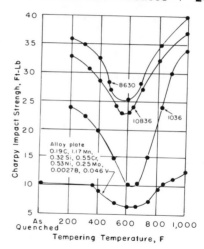

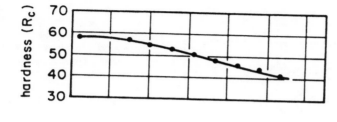

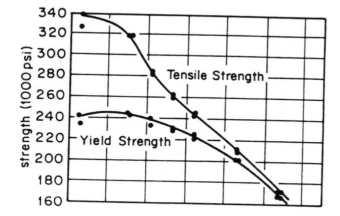

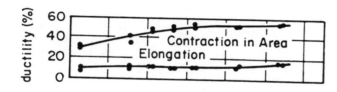

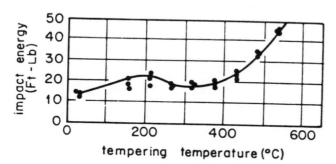

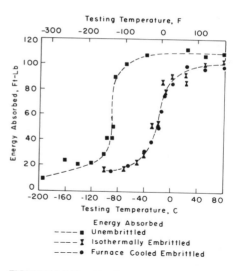

FIGURE 4-143. Tempering within the embrittlement zone (approximately 450 to 700 °F) results in a significantly lower impact strength. Values are longitudinal strengths at room temperature except for alloy plate, which was notched parallel to rolling and tested at −20 °F (−29 °C). All steels were water quenched.

FIGURE 4-142. Effect of tempering on mechanical properties of an originally martensitic structure. The steel was 4340, oil quenched, and tempered 1 h. Source: Ref. 70.

FIGURE 4-144. Notch toughness at a series of testing temperatures of a temper-brittle steel as-embrittled and as-unembrittled. Source: Ref. 71.

taneous flame hardening and tempering process.

Induction Hardening. The results obtained in induction hardening are similar to those obtained in flame hardening; the surface of the forging is heated to above the Ac$_3$ and cooled rapidly to form a hardened outer layer. However, the method of heating the surface of the part is totally different.

Any material that conducts electricity can be successfully heated by electromagnetic induction. The principle involved is similar to that of a tansformer, wherein electric current passing through the primary induces flow of current in the sec-

ondary. In this instance, the primary, called the load coil or inductor, is made from water-cooled copper tubing that conforms to the area of the part being heated. The secondary is the part itself.

High-frequency alternating current, 1000 to 500,000 cycles/s, developed by motor generators, solid-state power supplies, or radio-frequency converters, passes through the load coil and induces a flow of eddy

TEMPERING TEMPERATURE

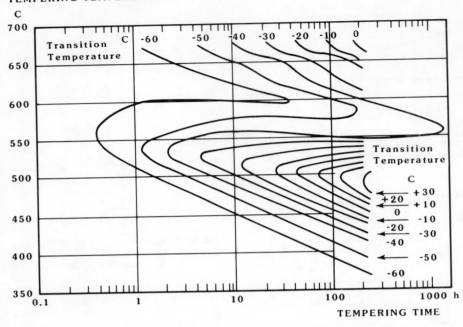

FIGURE 4-145. Dependence of transition temperature on tempering temperature and time for SAE 3140 steel. Composition is 0.40% C, 0.80% Mn, 0.60% Cr, and 1.25% Ni. Water quenched from 1650 °F (900 °C). Double tempering: first tempering for 1 h at 1250 °F (675 °C), water quenching. Second tempering: time and temperature as in diagram. Source: Ref. 71.

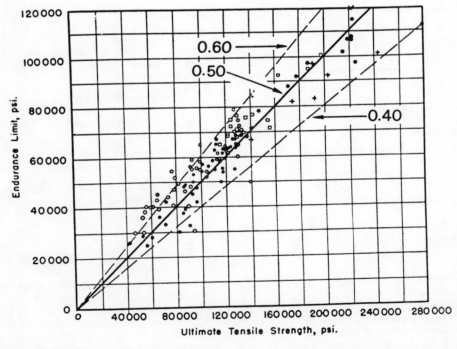

FIGURE 4-146. Relationship between fatigue strength and tensile strength for several steels. The straight lines have the slopes shown. Source: Ref. 72.

currents in the surface of the part to be heated. The resistance of the workpiece to the flow of these currents heats the metal. If the material is magnetic, heating occurs by hysteresis losses as well.

The penetration of induced current in the workpiece is inversely proportional to the square root of the frequency. Thus, a 10-kilocycle current heats approximately seven times deeper than a 500-kilocycle current. It is therefore possible to control the depth of heating by the frequency used, by the energy input, and by the time allowed for conduction of heat from the surface to the center of the part.

Medium plain-carbon steels of 0.35 to 0.60% carbon are commonly employed, although alloy steels containing approximately 0.30 to 0.50% carbon and higher can be treated successfully. Water or oil can be used as the quenching medium, depending on the hardness required and the hardenability of the steel.

The stress distribution in the hardened part is important in certain applications. A surface put in tension by hardening may fail quickly in service under fatigue conditions or bending loads, whereas if compressive stresses (typical for surface hardening via induction methods) can be induced in the outer layer, the part will probably give excellent service.

Conventional Case Carburizing or Case Hardening With Carbon. Often simply referred to as carburizing, this process is probably the most widely used when a hard, wear-resistant surface and tough core are required. Gear teeth, splines, and load-bearing areas of shafts are a few such applications.

In this surface treatment, a high-carbon layer is imparted to low-carbon steel by heating in contact with suitable carbon-bearing material. The major objective of carburizing or case hardening is to obtain a component with a tough core and a hard surface that will exhibit a high resistance to both wear and shock. The process also has the advantage that parts may be forged and machined from a soft steel and only need be ground after carburizing and hardening. This may be preferable to forging and machining from a steel that would give the necessary wear resistance without carburizing.

Brief consideration of the mechanism of carburizing shows that, in the absence of a gas phase, solid carbon does not readily diffuse into steel. For effective carburizing, the carbon must be present in the form of carbon monoxide or hydrocarbon gases. When these come into contact with the heated steel, they are dissociated at its surface and the atomic carbon is absorbed.

From this absorbed layer, the carbon diffuses into the steel.

There are three commonly employed methods of carburizing: pack carburizing, gas carburizing, and liquid carburizing. In the pack method, carburizing is performed by packing the parts in boxes with carbonaceous solids, sealing to exclude the atmosphere, and heating to a temperature above the Ac_3 of the steel. Numerous carburizing mixtures that are available contain charcoal (animal or hardwood), coke, and a variety of energizers.

Generally, the depth of case is controlled by the time and temperature of carburizing. Typical relations between these factors in ordinary practice are shown in Table 4-20. Sometimes, components are heated after carburizing to enable diffusion of carbon into the core to produce a shallower gradient in the case. This can provide a tougher and somewhat softer case more easily than attempts to limit the maximum carbon in the outer layer of the case during carburization. An illustration of the effect of diffusion is contained in Table 4-21.

In gas carburizing, the carbon of the furnace atmosphere is in the form of gaseous hydrocarbon compounds or carbon monoxide. Two common sources of carbon are commercial gases or easily vaporized hydrocarabon liquids. The preferred gas sources, when available in high purity, are natural gas and propane. Commercial practice is to use an endothermic gas as a carrier and to enrich it with one of the hydrocarbon gases to produce the necessary carburizing potential in the gas.

Carbon potential is evaluated by the measurement of the dew point, CO_2 content, or CO_2/CO ratio of the furnace atmosphere. The ratio of carrier gas to hydrocarbon gas depends on circulation, furnace size, desired carbon, work surface area, and so on. Forced circulation is used to promote maximum uniformity of carburizing.

Gas carburizing has the further advantage that the carburizing gas need not be admitted until the work is at the carburizing temperature. Also, after the carburizing cycle is complete, the work may be held at temperature in the furnace without admitting any further carburizing gas while carbon diffusion occurs.

In the liquid carburizing process, steel is carburized in a molten salt bath with the necessary carburizing agents. The case produced has a composition and structure similar to those obtained by the other two methods described. This treatment should not be confused with cyaniding. The cyanide case is high in nitrogen and low in carbon, and the reverse is true of liquid carburized cases.

The operating temperature of the bath is usually between 1550 and 1750 °F (840 and 950 °C). The low temperatures are used for light cases, up to 0.030 in. They usually contain an appreciable amount of nitrogen. At the higher temperatures, the carbon content and metallographic character of the case are similar to those observed in pack and gas carburizing and can be quite substantial.

Other Carburizing and Carbonitriding Treatments. A hard, shallow case, up to 0.020 in., can be obtained on low-carbon steels by treating the steel with a suitable cyanide compound. Although either potassium or sodium cyanide can be used, the latter is usually preferred, because it is lower in cost and has greater efficiency. The process can be likened to liquid carburizing. Carburizing is usually carried out between 1550 and 1600 °F (840 and 870 °C). Case depths rarely exceed 0.020 in. and are usually in the range of 0.005 to 0.010 in. Salt pot furnaces similar to those used for liquid carburizing are normally used.

Carbonitriding, also referred to as dry cyaniding, gas cyaniding, ni-carbing, and nitrocarburizing, is a process for case hardening a steel part in a gas carburizing atmosphere that contains ammonia gas in controlled percentages. By use of the mixed gases, both carbon and nitrogen are added to the surface layers of the steel.

Carbonitriding is usually carried out between 1425 and 1625 °F (775 and 885 °C) for parts that are quenched to develop a

hard case, and may be as low as 1300 °F (700 °C) for parts that do not require a liquid quench. Case depths usually range between 0.003 and 0.025 in.

Hardening Treatments for Carburized Parts. Insofar as hardening is concerned, a carburized piece must be considered a duplex material, having a high-carbon case and a low-carbon core, with a gradual transition from one to the other. Two critical temperatures are thus involved in hardening, and these may be 200 °F (95 °C) apart.

Thus, a double treatment would appear to be desirable to obtain optimum properties for both the case and core. First, the piece is heated to above the critical temperature of the core and suitably cooled to refine its structure. Then it is reheated to above the critical temperature of the case and quenched to harden the case.

However, with fine-grained steels and low carburizing temperatures, a single reheating and quench from above the critical temperature of the case may be sufficient. In some instances, quenching directly from the carburizing bath is employed (Figure 4-147). Table 4-22 indicates the hardening heat treatments that can be used for a variety of typical case-hardening steels.

Selective Carburizing. If certain portions of a part are required to be soft after case hardening to permit machining, straightening, and so forth, special precautions must be taken during the carburizing operation. One of two principal methods is employed, depending on circumstances.

The first method involves leaving excess stock at the portions to be left soft. After being carburized, the piece is cooled slowly, and the excess stock is machined off below the case so that this area will have a soft surface after the hardening treatment. This is a reliable method and ideal when only a few pieces are being treated.

The second method involves the imposition of a physical barrier to the carburizing agent on the areas to be left soft. This barrier could be sand, fine clay paste, sodium silicate paste, copper plate, or copper-containing pastes. Selective electrolytic copper plating of the areas required to be soft is the method most widely employed. Other methods can be used for pack and gas carburizing, but not for any liquid salt method.

Nitriding consists of subjecting steels of suitable composition to the action of a nitrogen-containing medium, commonly ammonia gas (altough treatments involving salt baths or plasma processing are also used) and thereby obtaining high surface

TABLE 4-20. Typical Relationship Between Time, Temperature, and Case Depth for Pack Carburizing Steel

| Case depth, in. | Time required, h, at: | | | |
	1600 °F (870 °C)	1650 °F (900 °C)	1700 °F (925 °C)	1750 °F (955 °C)
1/64	3.5	3	2.5	2
1/32	7	6	5	4
1/16	13	10	8	6
1/8	25	18	14	11.5

TABLE 4-21. Effect of Diffusion Treatment on Carbon Distribution

| Depth below surface, in. | Carbon content, % | |
	As-carburized	After 7 h diffusion
0.00	1.0	0.60
0.02	0.75	0.50
0.04	0.35	0.44
0.06	0.20	0.33
0.08	0.20	0.25
0.10	0.20	0.20

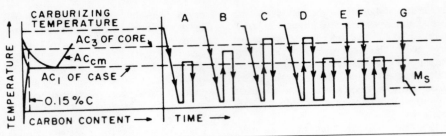

Treatment	Case	Core
A (best adapted to fine-grained steels)	Refined, excess carbide not dissolved	Unrefined; soft and machinable
B (best adapted to fine-grained steels)	Slightly coarsened; some solution of excess carbide	Partially refined; stronger than A; erratic
C (best adapted to fine-grained steels)	Somewhat coarsened; solution of excess carbide favored; austenite retention promoted in highly alloyed steels	Refined; maximum core strength and hardness. Better combination of strength and ductility than B
D (best treatment for coarse-grained steels)	Refined; solution of excess carbide favored; austenite retention minimized	Refined; soft and machinable; maximum toughness and resistance to impact
E (adapted to fine-grained steels only)	Unrefined with excess carbide dissolved; austenite retained; distortion minimized; file-proof when carbon content is high	Fully hardened
F (adapted to fine-grained steels only)	Refined; solution of excess carbide favored; austenite retention minimized	Low hardness, high toughness
G (interrupted quench; martempering)	Unrefined with excess carbide dissolved; austenite retained; distortion minimized; file-hard when carbide content is high	Fully hardened

FIGURE 4-147. Diagrammatic representation of hardening treatments following carburizing of steel and summary of the properties of case and core, provided adequate hardenability is assumed.

TABLE 4-22. Heat Treatments for Carburized Steels(a)

AISI No.	Alloy content, %	Approximate transformation temperatures, °F (°C) Case Ac_1	Core Ac_3(b)	Preferred treatments (see Figure 4-147)	Quenching medium	Retention of austenite
1015	None (0.4 Mn)	1355 (735)	1585 (868)	A, D, F	Water or 3% NaOH solution	Very slight
1019	None (0.85 Mn)	1350 (730)	1550 (843)	A, D, F	Water or 3% NaOH solution	Very slight
1117	1.5 Mn	1345 (729)	1520 (825)	A, D, F	Water or 3% NaOH solution	Slight
1320	1.75 Mn	1325 (718)	1500 (815)	E, C	Oil	Slight
3115	1.25 Ni, 0.50 Cr	1355 (735)	1500 (815)	E, C	Oil	Moderate
3310	3.50 Ni, 1.50 Cr	1330 (720)	1435 (780)	C, D, F	Oil	Strong(c)
4119	0.50 Cr, 0.25 Mo	1395 (758)	1500 (815)	E, C	Oil	Moderate
4320	1.75 Ni, 0.50 Cr 0.25 Mo	1350 (730)	1475 (800)	C, E	Oil	Strong(c)
4615	1.75 Ni, 0.25 Mo	1335 (723)	1485 (808)	E, C	Oil	Moderate
4815	3.50 Ni, 0.25 Mo	1300 (705)	1440 (780)	C, E	Oil	Strong(c)
8620	0.25 Ni, 0.50 Cr	1350 (730)	1540 (838)	E, C	Oil	Moderate
8720	0.25 Mo					

(a) All steel preferably of grain size 5 or finer. (b) Ac$_3$ of core will decrease as carbon content is increased. (c) Spheroidization of excess carbide by subcritical treatment before hardening may be used to decrease the amount of austenite retained.

hardness on the material or parts. The ammonia gas dissociates at the surface of the steel to produce atomic nitrogen, which combines with the various suitable elements in the steel to form nitrides.

The usual conditions for the nitriding process consist of subjecting the machined and heat treated parts to the action of ammonia gas at temperatures ranging from 930 to 1220 °F (500 to 660 °C). Nitriding cases are ordinarily light, typically 0.020 in. maximum. The surface hardness is, however, very high—900 to 1200 VH (Vickers diamond pyramid hardness). This hardness may be retained in heating to 900 °F (480 °C). No quenching after treatment is required.

Nitriding steels must be quenched and tempered before nitriding to develop optimum case and core properties. Tempering temperatures must exceed the nitriding temperature to eliminate the possibility of distortion of the part during the nitriding cycle. A typical procedure for annealing, heat treating, and machining of a forged shape to be nitrided is as follows:

- Normalize the forging by heating to 1800 °F (980 °C) and cool in still air
- Harden by heating to 1700 to 1750 °F (925 to 950 °C) and quench in water or oil
- Temper at 1100 to 1300 °F (590 to 700 °C), depending on desired core hardness
- Rough machine to $^1/_{32}$ in.
- Stress relieve by reheating to slightly below the tempering temperature
- Finish machine
- Nitride

Article size increases slightly during nitriding due to the increase in case volume. The amount of growth is constant under identical conditions. Once it has been determined for a given article, allowance for growth can be made in the final machining or grinding, or it can be removed by careful lapping, honing, or grinding of the nitrided part. A 0.002-in. increase in diameter in a specimen that has a case depth of 0.030 in. is considered typical.

It is frequently stated that a case-hardened part should have a hard case to withstand wear and a soft, tough core to withstand shock. This statement may be correct in certain instances, but it should be qualified. Insofar as the case is concerned, it would probably be more correct to state that the hardness over the entire cross section must be high enough that the strength at each point on the cross section will be higher than the maximum stress that will be encountered at that point in service.

Toughness in the center of a core need not be of any particular merit. Maximum toughness is required where bending and crushing stresses are the highest—that is, at or near the surface. It is frequently stated that the core of a case hardened part must be tough to withstand shock. In actual fact, all cores behave alike below the yield point, in accordance with Hooke's law. If the core is so weak that the stresses exceed the yield strength, the core will suffer plastic flow and will then fail to support the case, which may crumble.

Next to adequate cross-sectional hardness, the most important factor in parts that are subject to heavy-duty applications is that the heat treatment be controlled so as to avoid residual tensile stresses at the surface. This will not be a problem where nitrided cases are concerned; no quenching is involved, and as has been shown, nitrided heat treatment is conducted above the stress-relieving temperature of the steel.

High tensile or compressive stresses on the surface can be produced when the core hardens to form martensite after the case is cooled to room temperature. Consequently, the sequence in which the case and core form martensite is important. When any steel forms martensite, it must expand. If the case and core expand and thermally contract at the same time, they will not set up residual stresses.

Laser Surface Hardening. In recent years, industrial lasers have become available for metalworking, including surface hardening of highly stressed parts such as gears and bearings. The laser beam is a beam of light, greatly amplified and concentrated, that generates intense heat at the workpiece surface. It is easily controlled, requires no vacuum, and generates no combustion products. The process and metallurgical reactions are not fundamentally different from conventional surface hardening. However, case depths are rarely more than 0.01 in. for alloy or plain-carbon steels. Also, because the austenitizing temperature is reached in fractions of seconds, times are necessarily quite short. Thus, time may not be sufficient to obtain complete carbon in solution. Steel grades and prior microstructures are important variables. In any case, the pros and cons as well as the costs should be studied carefully to achieve optimum results.

Electron beam heat treating is another specialized selective hardening process. The surface of the part is heated rapidly above the transformation temperature of the alloy by direct bombardment or impingement of an accelerated stream of electrons. Steels with sufficient carbon and preferably sufficient hardenability for self-quenching are suitable candidates for this process. Parts must be heat treated under vacuum or inert atmosphere and can be machined or ground to final dimension with a small allowance of about 0.002 to 0.010 in. for final removal after heat treatment if necessary. Parts should be demagnetized so as not to deflect the electron beam. Advantages of this process include depth variations as a function of time, minimal distortion, small heat-affected zones, short heat treating times, maximum hardness, and minimum oxidation.

TABLE 4-24. Recommended Annealing Treatments for Stainless Steels

Designation	Treatment temperature	
	°F	°C

Conventional ferritic grades

405	1200–1500	650–815
409	1600–1650	870–900
430	1300–1450	705–790
430F	1300–1450	705–790
434	1300–1450	705–790
446	1400–1525	760–830

Low-interstitial ferritic grades

4339	1600–1700	870–925
444	1750–1850	955–1010
E-BRITE	1400–1750	760–955
SEA-CURE, SC-1	1850–1950	1010–1065
AL 29-4C	1850–1950	1010–1065
AL 29-4-2	1850–1950	1010–1065
MONIT	1850–1950	1010–1065

Note: Postweld heat treating of the low-interstitial ferritic stainless steels is generally unnecessary and frequently undesirable. Any annealing of these grades should be followed by water quenching or very rapid cooling.

Heat Treatment of Stainless Steels

Stainless steels fall into three principal categories: martensitic, austenitic, and ferritic. The first of these, the martensitic grades, is heat treated using techniques similar to those discussed above for the quenched and tempered steels; recommended austenitizing temperatures and tempering treatments are given in Table 4-23.

The other two grades of stainless steel cannot be hardened using the heat treating methods discussed above. When high strength is required in these materials, cold working (e.g., cold forging, rolling, extrusion) usually is used. However, many times these stainless steels must be put into a relatively soft and corrosion-resistant condition, and/or they must be stress relieved following welding. Annealing accomplishes this. Annealing temperatures for a variety of ferritic and austenitic stainless steels are summarized in Tables 4-24 and 4-25.

Heating Methods for Ferrous Forgings

A number of methods can be used to bring the work up to the austenitizing temperature. At high temperatures, radiation is important in heat transfer. At low temperatures (about 1000 °F, or 540 °C), convection becomes more important; consequently, positive circulation of the furnace atmosphere is recommended to promote rapid and uniform heating. Heat may also

TABLE 4-23. Procedures for Hardening and Tempering Wrought Martensitic Stainless Steels to Specific Strength and Hardness Levels

Type	Austenitizing(a) Temperature(b) °F	°C	Quenching medium(c)	Tempering temperature(d) °F min	max	°C min	max	Tensile strength ksi	MPa	Hardness, HRC, maximum
403, 410	1700–1850	925–1010	Air or oil	1050	1125	565	605	110–140	760–965	25–31
				400	700	205	370	160–220	1105–1515	38–47
414	1700–1925	925–1050	Air or oil	1100	1200	595	650	110–140	760–965	25–31
				450	700	230	370	160–220	1105–1515	38–49
416, 416(Se)	1700–1850	925–1010	Oil	1050	1125	565	605	110–140	760–965	25–31
				450	700	230	370	160–220	1105–1515	35–45
420	1800–1950	985–1065	Air or oil(e)	400	700	205	370	225–280	1550–1930	48–56
431	1800–1950	985–1065	Air or oil(e)	1050	1125	565	605	125–150	860–1035	26–34
				450	700	230	370	175–220	1210–1515	40–47
440A	1850–1950	1010–1065	Air or oil(e)	300	700	150	370	· · ·	· · ·	49–57
440B	1805–1950	1010–1065	Air or oil(e)	300	700	150	370	· · ·	· · ·	53–59
440C, 440F . .	1850–1950	1010–1065	Air or oil(e)	· · ·	325	· · ·	160	· · ·	· · ·	60 min
				· · ·	375	· · ·	160	· · ·	· · ·	58 min
				· · ·	450	· · ·	230	· · ·	· · ·	57 min
				· · ·	675	· · ·	355	· · ·	· · ·	52–56

(a) Preheating to a temperature within the process annealing range (see Table 5, *Metals Handbook*, Vol. 4, 9th ed., p. 635) is recommended for thin-gage parts, heavy sections, previously hardened parts, parts with extreme variations in section or with sharp re-entrant angles, and parts that have been straightened or heavily ground or machined, to prevent cracking and minimize distortion, particularly for type 420, 431 and 440A, B, C, and F. (b) Usual time at temperature ranges from 30 to 90 min. The low side of the austenitizing range is recommended for all types subsequently tempered to 25 to 31 HRC; generally, however, corrosion resistance is enhanced by quenching from the upper limit of the austenitizing range. (c) Where air or oil is indicated, oil quenching should be used for parts more than 1/4 in. thick; martempering baths at 300 to 750 °F (150 to 400 °C) may be substituted for an oil quench. (d) Generally, the low end of the tempering range of the 300 to 700 °F (150 to 370 °C) is recommended for maximum hardness, the middle for maximum toughness, and the high end for maximum yield strength. Tempering in the range of 700 to 1050 °F (370 to 565 °C) is not recommended, because it results in low and erratic impact properties and poor resistance to corrosion and stress corrosion. (e) For minimum retained austenite and maximum dimensional stability, a subzero treatment −100 °F ± 20 °F (−75 °C ± 10 °C) is recommended; this should incorporate continuous cooling from the austenitizing temperature to the cold transformation temperature.

TABLE 4-25. Recommended Annealing Temperatures for Austenitic Stainless Steels

Designation	Temperature(a) °F	°C
Conventional grades		
301, 302, 302B	1850–2050	1010–1120
303, 303Se	1850–2050	1010–1120
304, 305, 308	1850–2050	1010–1120
309, 309S	1900–2050	1040–1120
310, 310S	1900–1950	1040–1065
316	1900–2050	1040–1120
317	1950–2050	1065–1120
Stabilized grades		
321	1750–1950	955–1065
347, 348	1800–1950	980–1065
Carpenter 20Cb-3	1700–1750	925–955
Low-carbon grades		
304L, 304LN	1850–2050	1010–1120
316L, 316LN,		
317L	1900–2025	1040–1110
High-nitrogen grades		
201, 202	1850–2050	1010–1120
304N	1850–2050	1010–1120
316N	1850–2050	1010–1120
Nitronic 32,		
Carpenter,		
18Cr-2Ni-12Mn	1850–1950	1010–1065
Nitronic 33	1900–2000	1040–1095
Nitronic 40,		
Carpenter		
21Cr-6Ni-9Mn	1800–2150	980–1175
Nitronic 50,		
Carpenter		
22Cr-13Ni-5Mn	1950–2050	1065–1120
Nitronic 60	1900–2000	1040–1095
Carpenter 18-18	1900–2000	1040–1095
Highly alloyed grades		
317LM, 317LX,		
317L, 317L		
PLUS, 317LMO,		
7L4	2050–2100	1120–1150
JS700, JS777	1950–2100	1065–1150
904L, AL-4X,		
2RK65	1965–2055	1075–1125
Sanicro 28	. . .	. . .
AL-6X	2200–2250	1205–1230
254 SMO	2100–2200	1150–1205

(a) Temperatures given are for annealing a composite structure. Time at temperature and method of cooling depend on thickness. Light sections may be held at temperature for 3 to 5 min per 0.10 in. of thickness, followed by rapid air cooling. Thicker sections are water quenched. For many of these grades, a postweld heat treatment is not necessary. For proprietary alloys, alloy producers may be consulted for details. Although cooling from the annealing temperature must be rapid, it must also be consistent with limitations of distortion.

be developed within the metal by the passage of electric current into the metal or by induction. Further means employed for heating work include furnaces using oil, gas, coke, or coal; furnaces heated by metallic or nonmetallic electric resistors; liquid baths of lead, salt, or oil; and fluidized beds of refractory materials in which the objects are immersed.

The choice of heating method depends on a number of factors, each with its particular advantages and disadvantages. These usually revolve around the temperature required, the work to be treated, its size, economics of basic fuel or heating medium, degree of temperature control required, and tolerance of work to oxidation and decarburization.

Furnaces commonly used in heat treatment are classified in two broad categories: batch furnaces and continuous furnaces. In batch furnaces, workpieces normally are loaded and unloaded manually. A continuous furnace has an automatic conveyor system that provides a constant workload, in pounds per hour, to the unit. Figures 4-148 and 4-149 are two examples of such equipment.

Rate of heating of the work to heat treatment temperature must be controlled. A heating rate that is too rapid may set up high stresses, particularly in parts with irregular sections. A heating time of 1 h per inch of section is commonly used. However, in many cases, depending on the geometry of the part and previous heat treatment, much more rapid heating rates can be employed. Special care should be taken in reheating forgings that contain thick and

FIGURE 4-148. Two batch-type pit furnaces with cooling tanks located between furnaces. This type of equipment is commonly used to harden, normalize, or anneal various types of forgings.

FIGURE 4-149. Pusher-type continuous-cycle annealing furnace using trays that carry 500-lb loads of forgings.

thin sections of some complexity and that are made of steels with high hardenability. These may have large amounts of martensite in their structures upon cooling after forging.

The available heating rate is determined by the mass of the material being heated, the rate at which it absorbs heat, the desired temperature, and the temperature and heat transfer characteristics of the heating medium. In addition, as stated earlier, steels of high carbon and high alloy content, such as tool steels, may suffer from thermal shock. Preheating hold schedules may be required to avoid cracking. Also, any forged part with considerable machining surface stresses may be damaged if heated rapidly. Such damage is often incorrectly attributed to quenching.

Furnace Atmosphere for Conventional Heat Treatment. Heat treatment of forgings by the forging producer is normally done in an air atmosphere furnace. If freedom from both scaling (oxidation) and decarburization is required, special protective measures must be taken. One of the most popular approaches is to heat the work in a muffle-type furnace, in which the atmosphere and the heat can be controlled independently. The furnace atmosphere can be produced by burning natural gas wih a controlled amount of air in a catalytic retort to obtain a reducing or slightly oxidizing atmosphere. The gas generator is known as an endothermic (heat-absorbing) type (even though initially exothermic reactions are involved in generating the atmosphere); consequently, the atmosphere is known as endothermic. Other types of gas-generating equipment such as exothermic generators and ammonia cracking units also are used.

Endothermic gas is used in heat treating furnaces as a protective atmosphere for hardening, stress relieving, carbon restoration, and carburizing. The most recognized guides to endothermic generator operation are carbon dioxide content and dew points.* Any deviation from the relationship between the two indicates a problem (Figure 4-150).

Exothermic-type atmospheres are used when inert atmospheres are required. Applications are common in the metal-treating industry for bright annealing of stainless steel (where decarburization is not objectionable), copper, silicon steel sheets for transformer cores, aluminum, and an-

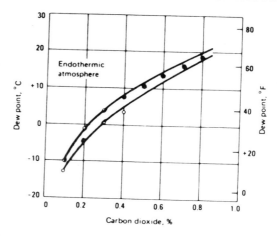

FIGURE 4-150. Variation in the relationship between dew point and carbon dioxide in generation of an endothermic atmosphere, as obtained from four plants. The generator in each plant was operated at a different temperature, all within the range of 1840 to 2000 °F (1005 to 1095 °C).

nealing of coils of wire and steel sheet. Atmospheres can be very rich (20% or higher total combustibles) or very lean (less than 1% total combustibles). Exothermic atmospheres are monitored most accurately and/or are controlled by an infrared analyzer sensitized to an appropriate range of carbon monoxide (CO), carbon dioxide (CO_2), and hydrogen, depending on atmosphere richness. Some traces of methane (CH_4) also may be present.

Nitrogen (N_2) gas also is used in large percentages for its abundant availability, low cost, and inert characteristics. However, for some metals such as stainless steels and nickel alloys, absorption at surfaces in atomic form can be deleterious to corrosion resistance, welding, and/or mechanical properties.

Basically, all methods of atmosphere control can effectively be divided into two groups: those involving control of the atmosphere-generating system and those involving control of the atmosphere once it is inside the furnace. Both are important in maintaining a controlled condition throughout the heating processes. Methods of control and analysis of the atmospheres vary, and the sophistication depends on need and economics. These include infrared CO_2 controllers, oxygen probes, dew point controls, etc. Safety is of utmost importance and, of course, must be designed for any system used.

Temperature-Control Systems. The results obtained in heat treatment are strongly dependent on the degree of control in maintaining or varying the required temperatures. In fact, many specifications dictate tolerances beyond which the process may be considered unacceptable, even

though the properties are within specification. Modern equipment is therefore equipped with sophisticated temperature-control systems. Such systems include temperature sensors, controllers, final control elements, measurement instruments, and set-point programmers.

In general, temperatures are sensed by several types of pyrometers. Their selection, placement, and monitoring circuits are designed to meet specific shop needs and involve serious planning and engineering cost considerations.

When sensors are part of a system involving digital or analog measuring instruments, autographic recorders, and set-point programmers, many forgings can be run through a heat-treat cycle from start to finish automatically and without interruption. Time, temperature, location, discharge, and any cycle function are observed simply on a control panel. Figures 4-151 and 4-152 are block diagrams of two closed-loop systems.

Digital instrumentation is replacing analog instrumentation because of its superior performance, flexibility, and simplicity. Reduced costs for microprocessing technology have made digital instrumentation a practical choice. Such controls ultimately pay off not only in the quality control aspect of heat treatment, but also in the conservation of energy.

Computer-Aided Heat Treatment

The heat treatment of high-temperature steels and titanium- and nickel-base alloys often requires strict control of temperature and time at temperature to maximize me-

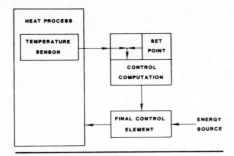

FIGURE 4-151. Basic control loop.

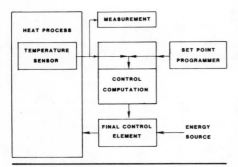

FIGURE 4-152. Basic control loop with auxiliary devices.

chanical property response. Computerized systems expand the capabilities of a heat treating facility to maintain the temperature control, cycle time, and documentation needed for vendor certification.

The benefits that can be derived from the installation of an automated setup depend to a large extent on whether it will be used on a batch or continuous process and on the type of record and control system it will be replacing. The degree of sophistication built into the system, its ease of use, and the operator's willingness to work with the system are three other important factors that ultimately determine the success of the system.

In justifying the installation of an automated system, often referred to as a heat-treat computer system (HTCS), the main areas of consideration include:

- Increased productivity
- Improved quality assurance
- Simplified record retention and recall
- Energy savings

Installing a computer system will not automatically guarantee these benefits, but it will establish the potential. Actual benefits will depend ultimately on how the system is used.

As the price of product liability increases, demands for tighter control of the

heat treatment and for more thorough and accurate records also increase. Control and recording systems that meet requirements of a few years ago may not be able to meet the exacting requirements demanded for some of today's new materials, or for new applications for existing materials. Therefore, having a flexible control system that can be easily adapted, rather than replaced, to meet new control and recordkeeping requirements is important, particularly with the high price of instrumentation.

A simple control system on a heat treat furnace can be broken down into five basic elements:

- Controller
- Recorder
- Temperature-sensing device
- Actuator for adjusting the energy source
- Overtemperature alarm

The controller must maintain the closest tolerance required for the components processed through the furnace. The resolution of the monitoring and recording equipment must meet or exceed the tolerance required of the controller.

Heat-Treat Computer Systems. The computerized system used by Ladish Co. at its Cudahy Plant is typical of the new generation of computer-controlled heat treatment facilities. It is used to control and provide heat-treat records on 21 critical heat-treat furnaces. The furnaces are gas-fired box furnaces. Each furnace has eight burners and is under- and over-fired. All furnaces run with excess air to meet temperature uniformity requirements; tolerances of ± 5 °F (± 3 °C) are not uncommon. A valve and valve drive mounted in the main gas line of the furnace provide control. Each furnace has five permanently mounted sidewall thermocouples, with up to nine additional thermocouples added to the load.

The computer system used to control these furnaces consists of two digital computers, each containing 256 K of RAM (random access memory) along with a 10M byte disk, an input/output device, and a graphics terminal and copier. The thermocouples from each furnace are connected to the computer through a cable containing 16 pairs of thermocouple lead. The lead wire has special limits of error, with each pair of wires shielded and twisted. The cable itself has an overall shield, a ground wire, and is coated with teflon for a high-temperature environment.

Prevention of a complete system failure is important in system design. The system is designed to operate using only one com-

puter, with the second computer acting as a back-up. Should the primary unit fail or halt for any reason, the back-up unit takes over control in a matter of seconds. Each computer has its own source of isolated and regulated power, along with 20 min of battery back-up on the memory. The input/output unit also has redundant controller cards.

Data on a load to be heat treated is preloaded into the computer by heat-treat operators as the loads are configured. Part information, including temperatures, times at temperature, and heating and cooling rates, is entered into the computer. A "fill-in-the-blank" dialogue makes it easy for operators to enter data. The location of parts in the load by serial number as well as the location of any load thermocouples is also entered. When a load is placed physically in a furnace, the operator recalls the preloaded data and enters a numerical code to signal which furnace has been loaded. The computer then takes over, running the prescribed heat-treat cycle for that load.

In controlling the furnace, the computer does not use any one thermocouple as the control thermocouple. Instead, it controls off any or all thermocouples, depending on the number of thermocouples on the load, which part of the cycle the furnace is in, and the temperatures within the furnace. During heat-up, the computer controls off the hottest thermocouple in the furnace to prevent temperature overshoot. The computer opts to heat or cool to setpoint as fast as possible, unless a specific heating or cooling rate has been specified. Once all thermocouples are in tolerance, the computer switches into an averaging mode of control. The computer is programmed with an adaptive control algorithm, to optimize the process as much as possible. Control variables permit each furnace to be tuned individually for optimum control response.

All thermocouples are monitored at 5-s intervals, with output to valve drives sent out as $\frac{1}{2}$-s pulses. When a furnace is loaded, the computer stores the maximum and minimum reading of each thermocouple at periodic intervals during that interval. The interval at which readings are stored depends on the length of the heat-treat cycle. These maximum/minimum readings then become part of the permanent history for that load.

The temperature error for each thermocouple being used on the load or in the furnace is preprogrammed into the computer according to set numbers. The computer then automatically corrects the reading of the thermocouples so that the operator sees the actual temperature at the thermocouple on his display. The computer also keeps track of the number of hours that the sidewall couples have been in use. The hours of use are calculated as a function of the operating temperature of the furnace, taking into account the accelerated deterioration of thermocouples at higher temperatures. The computer alerts the operators when thermocouples must be replaced. Load thermocouples are used only once and discarded. The computer also keeps track of the number of hours a furnace has been in use since its last temperature uniformity survey. An alert is activated when a furnace approaches its "due date."

All temperature-related alarms that remain in effect for 90 s or more are automatically logged as a part of the permanent history for that load. The computer records the date, time, and thermocouple number, as well as the maximum or minimum temperature of the thermocouple while it was out of tolerance. The computer also marks the load as a "referred" load, meaning that it was not heat treated according to the specifications entered into the computer for that load. All referred loads must be reviewed by either a heat treating supervisor or a metallurgical engineer with a comment added to the permanent history of the load concerning its disposition.

Operator communication with the system takes place through any one of five cathode ray tube terminals located in the operators' control room, or via six terminals mounted at various locations beside the furnaces. The operators can use 16 numeric and 15 alphabetical commands to bring up displays, enter information into, or obtain information from the computer. All commands that allow data entry into the computer require personnel to first enter a valid security number. All personnel who regularly work with the system have been assigned a security number linked to their name in the computer. Different levels of security limit some commands to individuals on a need-to-know basis.

Five minutes before a load is due to come out of a furnace, the computer signals the operator by putting out a "get geady to pull" alert. The computer will not allow a load to be pulled until it meets the required time at temperature. If a load must be removed prematurely, it must be aborted and will be marked "referred." After the operator signals the computer that the load has been pulled, the computer prints out a heat treatment history for the load. This history is also transferred to magnetic tape for storage for a minimum of 4 years.

All data are stored on tape according to a unique charge number for each load processed. The charge number is automatically assigned by the computer and is a 12-digit number consisting of the date the job was loaded, the shift, the number of loads heat treated in the furnace that shift, and the 4-digit furnace number. These data can be recalled at any time by simply mounting the appropriate tape and entering the desired charge number.

Thermocouple readings for a load can either be printed out or plotted and copied using the graphics terminal and copier. With the terminal, an operator can either obtain curves showing the maximum and minimum readings of any sidewall or load thermocouple, a curve of the average or the maximum/minimum readings, or a plot showing the curves for the maximum and minimum of all the load and sidewall thermocouples. The temperature range being plotted is variable and is specified by the individual making the plots. Figure 4-153 shows a time versus temperature plot, in which the maximum and minimum readings of the load thermocouples have been plotted.

At the end of each shift, the computer prints out a shift report detailing the loads heat treated that shift. It also puts out an operations report at the start of first shift covering the efficiency of operations by furnace during the previous three shifts. The report also contains week-to-date and month-to-date totals concerning such things a production time, idle time, and total pounds heat treated.

With the computer system, heat treating tolerances are the same as they were before the computer, but there is much less room for error within the system. Problems associated with pens that become dry or loose and are therefore unable to indicate the correct temperature are eliminated. Furthermore, temperature ranges of charts are no longer a problem, and interpretation of what temperature actually is being indicated by a mark on a chart is no longer necessary. The problem associated with the filing and retrieving of thousands of individual charts and records also has been eliminated. Customers like the system, because it assures them that their parts are being heat treated according to their specifications and that system errors are reduced as much as possible.

Controlled Cooling of Steel Forgings to Develop High Strength

In previous sections, the control of mechanical properties by heat treatment fol-

lowing forging has been discussed at length. Recently, a new class of steel has been introduced. In these steels, very desirable strength properties can be developed by controlled cooling from hot forging temperatures without any further heat treatment. They contain small (microalloying) additions of certain elements that remain in solid solution during forging, but which precipitate from solution during suitable cooling cycles. The processes involved in precipitation, very important in the heat treatment of aluminum and nickel-base alloys, are discussed in the next section.

In the new microalloyed steels, carbon contents usually are around 0.35 to 0.55%. The element most often used in the precipitation-strengthening reaction is vanadium, which forms vanadium carbides and carbonitrides. At typical forging temperatures of 2000 to 2200 °F (1100 to 1200 °C), the vanadium additions, usually on the order of 0.1 wt%, remain in solution. Following forging, the carbides and carbonitrides come from solution as the steel forging cools through the 1600 to 1700 °F (870 to 925 °C) temperature regime. The cooling rate is very important; too rapid a cooling rate prevents adequate precipitation, and too slow a rate can result in precipitate coarsening and loss of strength. In addition, the austenite transforms to ferrite and pearlite as the temperature drops below the upper and lower critical temperatures. The strengths that are achieved in this microstructure of ferrite, pearlite, and vanadium carbide/carbonitride precipitates can be relatively high (Table 4-26). Although the strengths (and toughnesses) are not as high as those in quenched and tempered steels of equal carbon content, the microalloyed forging steels find wide application in the automotive industry, where many parts require only moderate strength and negligible toughness.

If distortion occurs during the cooling of forged components, straightening must be carried out. A subsequent stress relief annealing treatment may then be necessary. The annealing temperature must not be high enough to cause coarsening of the precipitates, however.

Heat Treatment of Nonferrous Forgings

The range of heat treatments applied to forgings of other materials such as aluminum, nickel, and titanium alloys is not as varied as the range for steels discussed earlier. For these materials, the major heat treating processes are annealing, solution

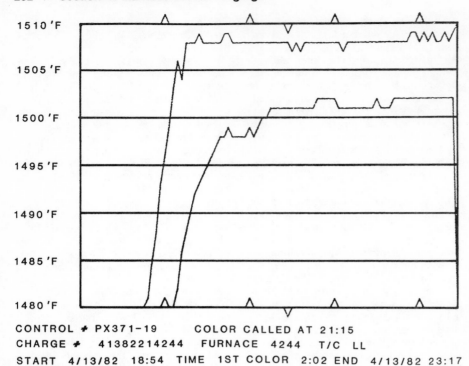

CONTROL ≠ PX371-19   COLOR CALLED AT 21:15
CHARGE ≠ 41382214244   FURNACE 4244   T/C LL
START 4/13/82   18:54   TIME 1ST COLOR 2:02   END   4/13/82 23:17
CONTROL: 1505'F +5 −5   LADISH CO.

FIGURE 4-153. Time-versus-temperature plot. Courtesy of Ladish Co.

heat treatment, and precipitation (or aging). Each of these is discussed briefly and illustrated with examples from several alloy groups.

Annealing. As with steel, annealing is a high-temperature heat treatment employed to produce a product of maximum ductility and minimum strength. These characteristics are usually achieved by one or a combination of processes, including the following:

● *Recovery and recrystallization:* The strengthening effects of cold (or hot) work may be eliminated by processes known as recovery and recrystallization. In the former, softening is accomplished by removing individual crystalline defects, known as dislocations, that

increase hardness as their number increases during metalworking operations. Recrystallization is a softening process in which totally strain-free crystalline grains are produced.

● *Overaging:* Many nonferrous metals develop high strength by heat treatments in which a fine dispersion of a second phase is produced by a process known as precipitation (discussed below). By making this dispersion as coarse and widely spaced as possible (i.e., by overaging it), strength is reduced and ductility is increased.

Examples of annealing treatments for several alloy groups are given in a subsequent section.

Solution heat treatment and precipitation hardening are processes or reactions that make possible the strengthening and hardening of certain metals. These metals consist of alloy systems that exhibit a decrease in solid solubility of one constituent in another, as the heat-treat temperature is decreased from its maximum point. Precipitation hardening is a principal means of hardening nonferrous alloys, particularly those of aluminum, copper, magnesium, and nickel. It has also become important in certain high-strength or high-temperature ferrous alloys, where it supplements the normal steel hardening process (described above), as in precipitation-hardenable stainless and maraging steels, or serves as the means of hardening in iron-base superalloys that cannot rely on the martensitic reaction.

Precipitation hardening depends on the slow precipitation of a constituent (or "phase") from a supersaturated solid solution produced by a solution heat treatment operation (Figure 4-154). When the precipitate forms "coherently" (atomic lattice of the precipitate is continuous with the matrix), the forced fit of the precipitate on the matrix lattice creates large strains over moderate distances and, in effect, results in hardening. The particles of precipitate grow with time and reach a critical size at which they can no longer sustain the increasing strains. They break away from the matrix and become "noncoherent," causing the hardness and strength of the metal to decrease.

A typical precipitation-hardening system is illustrated in Figure 4-155, in which the decreasing solubility of B in A is indicated by line 1–2. An alloy of composition X, heated to a temperature t within the A field, and held until solid-solution equilibrium conditions have been reached, will consist of the single phase A. If this single phase were cooled rapidly or quenched to room temperature so that equilibrium conditions could no longer be maintained, then B would be retained in a supersaturated solid solution of A. This solid solution is unstable and tends to reach

TABLE 4-26. Chemical Composition and Mechanical Properties of Microalloyed Steels Controlled Cooled Following Forging

| Chemical composition specified, % | | | | | | | Yield strength | | Mechanical properties | | |
C	Mn	Si	S	P	Total N	V	0.2%, MPa (ksi)	UTS, MPa (ksi)	Elongation, %	Reduction in area, %	Hardness, HB
0.42–0.47	0.6–1.0	0.6 max	0.045–0.065	0.035 max	· · ·	0.08–0.13	500 (73) min	800–900 (116–131)	8 min	20 min	· · ·
0.40–0.46	1.0–1.2	0.6 max	0.04–0.06	0.035 max	0.03	0.08–0.13	500 (73) min	800 (116) min	15 min	· · ·	245–290
0.44–0.54	0.6–1.0	0.15–0.6	0.045–0.065	0.035 max	· · ·	0.08–0.13	450 (65) min	780–900 (113–131)	· · ·	· · ·	· · ·

Source: Ref. 73.

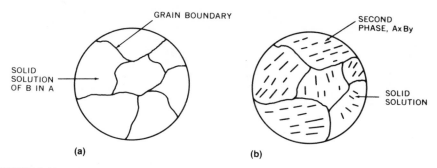

FIGURE 4-154. Representation of the change in structure upon exceeding the solid solubility of a solute. (a) Microstructure of a solid-solution alloy. (b) Structure of alloy after solubility limit has been exceeded.

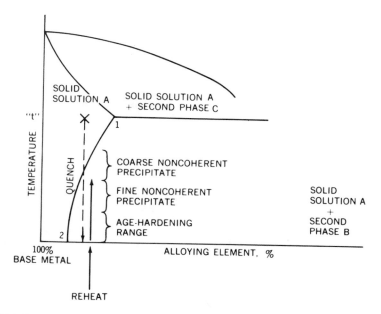

FIGURE 4-155. Diagram of steps in the precipitation hardening of an alloy: typical precipitation-hardening system.

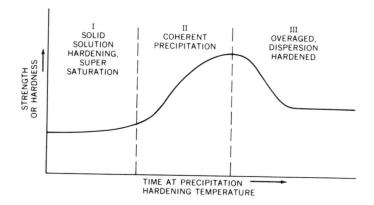

FIGURE 4-156. Typical precipitation-hardening curve.

equilibrium by precipitation of B. Precipitation in some alloy systems may occur naturally with time at room temperature. Other systems may require reheating to elevated temperature to cause the reaction to occur or be accelerated to occur over reasonable periods of time.

When the solid solution A is first quenched, there are no precipitated particles, and the alloy is in a soft and ductile condition. As coherent precipitation begins, hardness and strength increase, as illustrated by a typical "aging" curve shown in Figure 4-156. This increases with continuing precipitation and growth of the particles until the optimum particle size and strength are reached at some established aging time. At that time, heat treatment is complete, and the alloy is cooled to room temperature. Further heating beyond this point causes "overaging," or agglomeration of the precipitate, and both hardness and strength decrease, as shown in Figure 4-156.

Depending on the alloy system, the precipitation phase may be a practically pure metal, another phase, or an intermetallic compound. A highly alloyed system may involve the precipitation and hardening reaction of more than one phase. The precipitation of the particles by reheating the quenched alloy is called precipitation or aging heat treatment, and the resulting hardening is usually termed precipitation hardening or age hardening.

Heat Treatment of Aluminum Alloys

Wrought aluminum alloys have a wide range of compositions and are classified as either heat-treatable or non-heat-treatable. The non-heat-treatable alloys, which have the designations 1xxx, 3xxx, and 5xxx, do not undergo the precipitation processes described above to any great extent. They are hardened solely by strain hardening and may be softened by annealing at an appropriate temperature (Table 4-27). These two conditions, or tempers, are indicated by the letters "H" (strain-hardened) and "O" (annealed). For the H condition, two or more digits often are used in the temper designation to indicate the degree of strain hardening and any heat treatment that may or may not have been applied to the product. For example, for non-heat-treatable alloys, H1xx is used to signify a strain-hardened condition and H2xx a strain-hardened product that has been partially annealed.

The heat-treatable aluminum alloys are designated by four-digit numbers in the series 2xxx, 4xxx, 6xxx, and 7xxx. These

TABLE 4-27. Annealing Temperatures for Some Common Non-Heat-Treatable Aluminum Alloys

Alloy designation	Composition(a), wt%	Annealing temperature(b) °F	°C
1100	Commercial-purity aluminum	650	345
3003	1.2 Mn	775	413
5052	2.5 Mg, 0.25 Cr	650	345
5083	4.5 Mg, 0.7 Mn	775	413
5154	3.5 Mg, 0.25 Cr	650	345
5357	1.0 Mg, 0.25 Mn	650	345

(a)Nominal balance of alloys is aluminum. (b)Hold time at temperature not required.

alloys may be supplied in the annealed (O) or strain-hardened (H) tempers. Annealing in these cases may involve simultaneous relief of cold work as well as overaging. The alloy is heated to a high temperature and slowly cooled through a temperature range in which precipitates coarsen rapidly to produce overaging. Annealing treatments for some common heat-treat-able aluminum alloys and their resulting properties are given in Table 4-28.

Heat-treatable aluminum alloys that are solution treated and aged are designated with T and W tempers. The two most common are T4 (solution heat treated and quenched) and T6 (solution heat treated and then "artificially" aged at a temperature above room temperature). The W temper applies to aluminum alloys, such as 7075, that spontaneously age at room temperature following solution annealing.

During aging, a variety of precipitates form from the supersaturated solid solution. In order of occurrence, these are phases known as GP zones (disk-like clusters of atoms), θ'', θ', and θ in the prototype of the aluminum alloy systems, the Al-Cu system. Maximum strength is provided by the formation of θ'' and θ' particles. In most commercial alloys, multistage precipitation and strength changes analogous to those that occur in the Al-Cu system are found as well.

Selection of time-temperature cycles for

precipitation heat treatment should receive careful consideration. Unfortunately, the cycle required to optimize one property, such as ultimate tensile strength, usually differs from that required to maximize others, such as yield strength and corrosion resistance. Thus, the cycles that are typically used (Table 4-29) represent compromises that provide the best combinations of properties.

Heat Treatment of Superalloys

The heat treatment of wrought superalloy products such as forgings consists largely of solution annealing and precipitation-hardening treatments. The superalloy class includes alloys rich in nickel, nickel and iron, and iron. All of these alloys consist of an FCC matrix at room and elevated temperatures. This phase is typically referred to as γ, or austenite, analogous to the high-temperature FCC phase formed during heat treatment of steels.

Alloying additions lead to the precipitation of various phases, including γ' [Ni$_3$(Al,Ti)], γ'', and various carbides such as MC (M = Ti, Cb, etc.), M$_6$C (M = Mo, W), or M$_{23}$C$_6$ (M = Cr). By and large, the primary strength of superalloys is derived from the γ' and γ'' dispersion developed through heat treatment. In nickel-base superalloys (e.g., Waspaloy and Astroloy), aluminum and to some degree titanium combine with nickel to form γ'. In nickel-iron-base alloys (for example, Inconel 718, Incoloy 901) and iron-base superalloys (for example, A-286), titanium, columbium, and, to a lesser extent, aluminum combine with nickel to form γ' or γ''. Furthermore, the nickel-iron and iron-base superalloys are all prone to the formation of yet other phases, such as those referred to as η (Ni$_3$Ti) and δ(Ni$_3$Cb).

The solution annealing and precipitation temperature regimes for several of the important superalloys are shown in the "pseudobinary" phase diagrams in Figure 4-157. For both Waspaloy and Incoloy 901, the solvus temperatures depend primarily on the aluminum and titanium contents, and not on other alloying elements such as molybdenum and chromium, which provide solid-solution strength to the γ matrix.

Likewise, the solution and precipitation temperatures in Inconel 718 are strongly dependent on the columbium content. It may also be noticed from Figure 4-157 that the heat treatment of the superalloys must be carried out at very high temperatures. These temperatures are usually only several hundred degrees Fahrenheit below

TABLE 4-28. Annealing Treatment (O Temper) and Resulting Tensile Properties for Some Common Heat-Treatable Aluminum Alloys

Alloy designation	Nominal composition, wt%	Annealing temperature(a), °F (°C)/time (h)	Yield strength, ksi	Ultimate tensile strength, ksi	Elongation, %
2014	4.4 Cu, 0.8 Si, 0.8 Mn, 0.4 Mg	775 (413)/2–3	14	27	18
2024	4.5 Cu, 1.5 Mg, 0.6 Mn	775 (413)/2–3	11	27	22
4032	12.5 Si, 1 Mg, 0.9 Cu, 0.9 Ni	775 (413)/2–3	...	...	...
6061	1 Mg, 0.6 Si, 0.25 Cu, 0.25 Cr	775 (413)/2–3	8	18	30
7075	5.5 Fn, 2.5 Mg, 1.5 Cu, 0.3 Cr	775 (413)/2–3	15	33	16

(a) Followed by furnace cooling at a maximum rate of 50 °F/h (28 °C/h) to 500 °F (260 °C), except for 7075 alloy, which may be air cooled. (b) Properties are only typical; they may vary with product form.

TABLE 4-29. Solution Heat Treatment and Precipitation Schedules and Resulting Tensile Properties for Some Common Heat-Treatable Aluminum Alloys

Alloy designation	Solution annealing temperature, °F (°C)/time, min	Precipitation (T6) temperature, °F (°C)/time, h	Solution annealed condition Yield strength, ksi	Ultimate tensile strength, ksi	Elongation, %	T6 condition Yield strength, ksi	Ultimate tensile strength, ksi	Elongation, %
2014	935 (500)/ up to 60 in salt bath	340 (645)/ 8–12	42	62	20	60	70	13
2024	920 (495)/ 10–60 in salt bath	375 (190)/ 11–13	47	68	19	57	69	10
4032	950 (510)/ 4	340 (170)/ 8–12	...	...	...	46	55	9
6061	985 (530)/ 10–60 in salt bath	345 (175)/ 6–10	21	35	25	40	45	17
7075	895 (480)/ 10–60 in salt bath	250 (120)/ 24–28	...	...	...	73	83	11

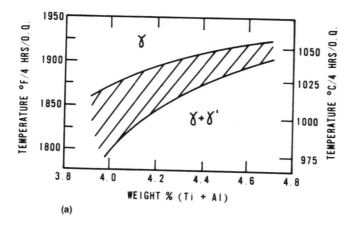

(a)

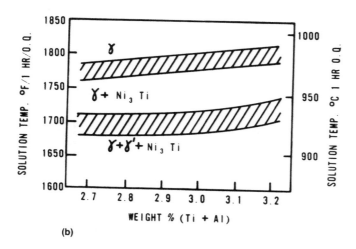

(b)

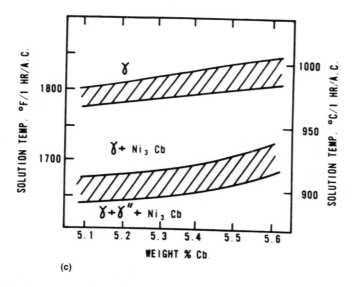

(c)

FIGURE 4-157. Portions of pseudobinary phase diagrams for (a) Waspaloy, (b) Incoloy 901, and (c) Inconel 718. Source: Ref. 74.

those at which incipient melting occurs. For this reason, forging of these alloys is quite difficult. However, these same characteristics enable superalloy forgings to be used at very high temperatures that are often substantially above those at which high-strength quenched and tempered steels are appropriate. The heat treatment and mechanical properties of several superalloys are summarized in Table 4-30.

Heat Treatment of Titanium Alloys

Like iron, titanium exists in more than one crystalline structure (allotrope), depending on the temperature. The transformation from one form to another serves as the basis for many of the heat treatments used to control the properties of titanium-base alloys. Below 1625 °F (885 °C), the crystal structure of pure titanium is hexagonal-closed packed, referred to as alpha (α). Above this temperature, the stable form is a body-centered cubic phase called beta (β). The temperature above which titanium as well as its alloys becomes totally beta phase is known as the beta transus temperature.

Various alloying elements are added to pure titanium to improve various properties. These alloying elements may raise or lower the beta transus temperature, or may not affect it. Elements that raise it, thus stabilizing the alpha phase to higher temperatures, are known as alpha stabilizers and include aluminum and oxygen. On the other hand, elements that stabilize the beta phase to lower temperatures, the so-called beta-stabilizing elements, include molybdenum, vanadium, iron, chromium, niobium, and manganese. Alloying additions such as zirconium and tin have essentially no effect on the beta transus temperature.

The addition of beta-stabilizing elements, in addition to lowering the transus temperature, is often effective in enabling some beta phase to be retained in a metastable state below the transus temperature down to room temperature. The amount of beta phase retained serves as a convenient means of classifying titanium into groups such as alpha, near-alpha, alpha-beta, near-beta, and beta as the amount of beta phase increases from none to 100%. Some of the more common titanium alloys in each group and their beta transus temperatures are shown in Table 4–31.

The most common heat treatments for titanium forgings are stress relieving, annealing, and solution treatment and aging. Each of these is discussed in the paragraphs that follow.

TABLE 4-30. Heat Treatment and Tensile Properties of Several Wrought Superalloys

Alloy	Nominal composition, wt%	Solution treatment	Aging treatment	Room temperature					1400 °F (760 °C)				
				Yield strength		Ultimate tensile strength		Elongation, %	Yield strength		Ultimate tensile strength		Elongation, %
				ksi	MPa	ksi	MPa		ksi	MPa	ksi	MPa	
Waspaloy	0.5 C, 19.5 Cr, 13.5 Co, 3.00 Ti 1.30 Al, 0.005 B, 4.3 Mo, 0.05 Zr balance Ni	1975 °F (1080 °C)/ 4 h + air cool	1550 °F (840 °C)/ 24 h + air cool + 1400 °F (760 °C)/ 16 h + air cool	115	795	185	1275	25	98	675	115	795	28
Astroloy	0.05 C, 15.0 Cr, 15.0 Co, 3.50 Ti, 4.40 Al, 0.030 B, 5.25 Mo, balance Ni	2150 °F (1175 °C)/ 4 h + air cool 1975 °F (1080 °C)/ 4 h + air cool	1550 °F (840 °C)/ 24 h + air cool 1400 °F (760 °C)/ 16 h + air cool	152	1050	205	1415	16	132	910	168	1160	21
Incoloy 901	0.05 C, 13.5 Cr, 42.7 Ni, 34.0 Fe, 2.5 Ti, 0.25 Al, 0.015 B, 6.1 Mo	2000 °F (1095 °C)/ 2 h + water quench	1450 °F (790 °C)/ 2 h + air cool + 1325 °F (720 °C)/ 24 h + air cool	130	895	175	1205	14	92	635	105	725	19
Inconel 718	0.04 C, 19.0 Cr, 52.5 Ni, 0.90 Ti, 0.50 Al, 0.005 B, 3.05 Mo, 5.30 Cb, balance Fe	1800 °F (980 °C)/ 1 h + air cool	1325 °F (720 °C)/ 8 h + furnace cool + 1150 °F (620 °C)/ 8 h + air cool	163	1125	198	1365	21	116	800	124	855	30
A-286	0.05 C, 150.0 Cr, 26.0 Ni, 2.00 Ti, 0.20 Al, 0.005 B, 1.25 Mo, 0.30 V, balance Fe	1800 °F (980 °C)/ 1 h + oil quench	1525 °F (720 °C)/ 16 h + air cool	105	725	146	1005	25	62	430	64	440	15

Source: Ref. 38.

Stress relieving treatments are used to decrease or eliminate undesirable residual stresses that arise from nonuniform hot forging deformation or cold forging and sizing and from machining operations that may result in redistribution of such stresses. The treatment normally lasts 15 min to 4 h and is performed at temperatures of 900 to 1300 °F (480 to 700 °C) for alpha, near-alpha, and alpha-beta alloys, or 1250 to 1500 °F (675 to 815 °C) for near-beta and beta titanium alloys. Sometimes, stress relieving can be avoided if some other heat treating operation such as annealing or solution treatment and aging is to be carried out, provided that the cooling rate is uniform in these operations, as it must be in the former one.

Annealing is used in titanium alloys to soften as well as to control properties by adjusting the amounts, distributions, and shapes of the various phases. Because alpha and alpha-beta alloys have low hardenability (i.e., the ability to retain beta phase during cooling is low), they are used frequently in the annealed condition in service. A variety of annealing treatments include those referred to as mill, duplex, triplex, recrystallization, and beta. Mill annealing is a general-purpose treatment given to most mill products. Performed at relatively low temperatures, it is not a full anneal and may leave traces of a worked structure. However, it does impart a degree of uniformity in the material processed. As with other annealing treatments, parts are usually air or furnace cooled following treatment.

Duplex and triplex annealing treatments, as their names imply, are processes performed at two or three temperatures, respectively. The different temperatures are employed to alter the shapes, sizes, and distributions of the alpha and beta phases to enhance tensile, fatigue, creep, fracture, or other properties. The final step of such treatments is usually a "stabilization" anneal at a temperature only slightly above the maximum intended service temperature that is intended to produce a thermally stable microstructure.

Recrystallization and beta annealing also are used to impart certain properties needed for service. In recrystallization annealing, a totally recrystallized structure is obtained by heat treating at high temperatures. If working and annealing are done below the transus temperature, a structure of equiaxed alpha phase in a matrix of acicular alpha and beta (which possesses a good blend of tensile and low-cycle fatigue properties) is obtained in near-alpha and alpha-beta alloys. Working and recrystallization of beta and near-beta alloys below the transus temperature lead to the development of a microstructure of alpha particles in a beta matrix whose strength and toughness may be varied over a wide range, depending on the working and heat treating temperatures. When working and/ or annealing is conducted above the beta transus (as in beta annealing), an acicular alpha microstructure, which has good creep and toughness properties, is obtained in near-alpha and alpha-beta alloys. An all-beta microstructure is obtained from such treatments in beta alloys.

Solution Treating and Aging. A wide range of properties can be obtained in alpha-beta and beta titanium alloys by solution treating and aging. These treatments are employed to control the strength and fracture-related properties (ducility, creep, fatigue, and toughness) in both alpha-beta alloys and beta alloys. For titanium, solution treating refers to solutioning of alpha phase in the beta phase and aging consists of subsequent beta decomposition.

Solution treating and aging are similar to recrystallization annealing and stabilization in that the former in each case is a high-temperature process and the latter is a low-temperature process. Solution treatment and aging, however, are the usual terms when the primary objective is strengthening, as in aluminum- and nickel-base alloy systems.

For alpha-beta alloys, solution treating is normally done at a temperature 50 to 150 °F (30 to 85 °C) below the transus temperature (to retain an equiaxed alpha structure), if ductility is important, or above the transus temperature (to produce an acicular alpha structure), if high fracture

TABLE 4-31. Beta Transus Temperatures of Titanium Alloys

Alloy	Beta transus °F ± 25	°C ± 15
Commercially pure Ti, 0.25 max O₂	1675	910
Commercially pure Ti, 0.40 max O₂	1735	945
Alpha and near-alpha alloys		
Ti-5Al-2.5Sn	1925	1050
Ti-8Al-1Mo-1V	1900	1040
Ti-6Al-2Sn-4Zr-2Mo	1820	995
Ti-6Al-2Cb-1Ta-0.8Mo	1860	1015
Ti-0.3Mo-0.8Ni-(Ti code 12)	1615	880
Alpha-beta alloys		
Ti-6Al-4V	1830(a)	1000(b)
Ti-6Al-6V-2Sn (Cu + Fe)	1735	945
Ti-3Al-2.5V	1715	935
Ti-6Al-2Sn-4Zr-6Mo	1720	940
Ti-5Al-2Sn-2Zr-4Mo-4Cr (Ti-17)	1650	900
Ti-7Al-4Mo	1840	1000
Ti-6Al-2Sn-2Zr-2Mo-2Cr-0.25Si	1780	970
Ti-8Mn	1475(c)	800(d)
Beta or near-beta alloys		
Ti-13V-11Cr-3Al	1330	720
Ti-11.5Mo-6Zr-4.5Sn (beta III)	1400	760
Ti-3Al-8V-6Cr-4Zr-4Mo (beta C)	1460	795
Ti-10V-2Fe-3Al	1480	805
Ti-15V-3Al-3Cr-3Sn	1400	760

(a) ±30. (b) ±20. (c) ±50. (d) ±35.

TABLE 4-32. Recommended Solution Treating and Aging (Stabilizing) Treatments for Titanium Alloys

Alloy	Solution temperature °F	°C	Solution time, h	Cooling rate	Aging temperature °F	°C	Aging time, h
Alpha or near-alpha alloys							
Ti-8Al-1Mo-1V	1800–1850(a)	980–1010(a)	1	Oil or water	1050–1100	565–595	· · ·
Ti-6Al-2Sn-4Zr-2Mo	1750–1800	955–980	1	Air	1100	595	8
Alpha-beta alloys							
Ti-6Al-4V	1750–1775(b,c)	955–970(b,c)	1	Water	900–1100	480–595	4–8
	1750–1775	955–970	1	Water	1300–1400	705–760	2–4
Ti-6Al-6V-2Sn (Cu + Fe)	1625–1675	885–910	1	Water	900–1100	480–595	4–8
Ti-6Al-2Sn-4Zr-6Mo	1550–1650	845–870	1	Air	1075–1125	580–605	4–8
Ti-5Al-2Sn-2Zr-4Mo-4Cr	1550–1600	845–870	1	Air	1075–1125	580–605	4–8
Ti-6Al-2Sn-2Zr-2Mo-2Cr-0.25Si	1600–1700	870–925	1	Water	900–1100	480–595	4–8
Beta or near-beta alloys							
Ti-13V-11Cr-3Al	1425–1475	775–800	¹⁄₄–1	Air or water	800–900	425–480	4–100
Ti-11.5Mo-6Zr-4.5Sn (beta III)	1275–1450	690–790	¹⁄₅–1	Air or water	900–1100	480–595	8–32
Ti-3Al-8V-6Cr-4Mo-4Zr(beta C)	1500–1700	815–925	1	Water	850–1000	455–540	8–24
Ti-10V-2Fe-3Al	1400–1435	760–780	1	Water	925–975	495–525	8
Ti-15V-3Al-3Cr-3Sn	1450–1500	790–815	¹⁄₄		950–1100	510–595	8–24

(a) For certain products, use solution temperature of 1650 °F (890 °C) for 1 h then air cool or faster. (b) For thin plate or sheet, solution temperature down to 890 °C (1650 °F) can be used for 6 to 30 min, then water quench. (c) This treatment is used to develop maximum tensile properties in this alloy.

toughness or resistance to stress corrosion is required. When solutioning is done below the transus temperature in alpha-beta alloys, proper control of temperature is essential. If the beta transus is exceeded, tensile properties (especially ductility) are reduced and cannot be fully restored by subsequent thermal treatment.

Solution treating temperatures for beta alloys are usually above the transus. Because the alloy is single phase above the transus (as in beta annealing or beta solution treatment of alpha-beta alloys), care must be exercised to prevent beta grain growth at high temperatures or long times. For some of the newer beta alloys, such as Ti-10V-2Fe-3Al, which are finish forged below the beta transus (to develop a microstructure with some alpha phase) sub-transus solution treatment is common.

Following solution treatment, the cooling rate must be controlled to avoid the decomposition of beta phase. If the rate is too low, appreciable diffusion may occur, and decomposition of the altered beta phase during aging may not provide the desired strength properties. For most applications, air cooling is used for beta and near-beta alloys, and water or brine quenching is used for alpha-beta alloys that are not as highly beta-stabilized and thus are said to have low hardenability.

Following solution treatment and quenching, titanium parts are reheated to aging temperatures typically between 800 and 1200 °F (425 and 650 °C). Aging causes decomposition of the supersaturated beta phase retained on quenching and an increase in strength. A summary of solution treating and aging times and temperatures for titanium alloys is presented in Table 4-32. The particular time/temperature combination selected depends on the required tensile and fracture properties. In general, the strength of alpha-beta alloys can be varied over only a rather limited range in contrast to the rather wide range for beta alloys (Figure 4-158).

During aging of highly beta-stabilized alpha-beta alloys, a metastable transition phase, known as the omega (ω) phase may be produced, which produces unacceptable brittleness. It can be avoided by severe quenching following solution treating and rapid reheating to aging temperature above 800 °F (425 °C).

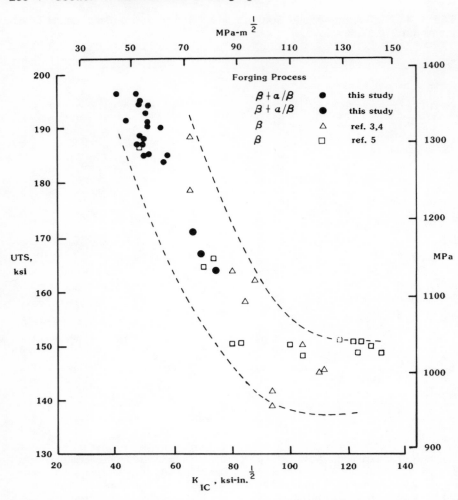

FIGURE 4-158. Fracture toughness as a function of tensile strength for Ti-10V-2Fe-3Al. Source: Ref. 75.

Controlled Quality Forging

*Contributing Authors: Ray Huffaker, Assistant Vice President,
Ladish-Pacific Company, Los Angeles, California;
and A. J. DeRidder, Assistant Vice President, Quality Department,
Ladish Company, Cudahy, Wisconsin*

Because so many forgings are used in critical applications in which strength, reliability, and safety are important considerations, effective systems of quality control are maintained in plants throughout the forging industry. In forging, quality control pertains particularly to metallurgical soundness, assurance of specified property levels and dimensional accuracy, and proper identification and certification. It begins with metallurgical review of material specifications to ensure quality—both internal soundness and external surface preparation—for acceptable performance and property levels. Quality control continues through virtually every step in the production and inspection of forgings.

Control, Inspection, and Testing of Incoming Stock

Methods and systems vary, but each forge plant takes positive measures to maintain inventory control and prevent mixing of materials. When forging stock is received at the forge plant, it is coded, marked, and assigned to the specific job. When required, incoming stock is held in quarantine until it is cleared to be sent to a bonded store by the laboratory. Pertinent information is entered into a computerized inventory and release system to maintain control over each job lot through processing and operations.

Visual inspection of bars and billets to be forged is commonly performed during handling and stock cutting operations (Figure 4-159). Where surface irregularities exist that could carry through the forging process and result in defective final forgings, the material is rejected, or defects are removed. Material that is programmed for critical applications requiring high strength and performance is ultrasonically inspected to ensure internal soundness. In particular applications, stock is also subjected to magnetic or fluorescent particle inspection or pancaking to detect small cracks or seams that may be present (Figure 4-160).

Metallurgical tests of various kinds to verify composition, structure, and properties of raw materials may also be performed, depending of customer specifications. These include standard tests for hardenability, tensile, impact, and stress-rupture properties. When exceptional accuracy is required, vacuum fusion equipment, such as that shown in Figure 4-161, may be used. Ultra-precise determinations of material compositions can be provided quickly by automatic x-ray spectrographs and electronic recording spectrometers.

After a thorough understanding of end use requirements has been obtained, instructions are prepared by the forge plant chief metallurgist or chief engineer. These instructions specify the grade of material to be used for a particular job and detail the preparation, heating, forging, testing, and inspection procedure to be followed in processing the order. Furnace temperatures and cooling practices are specified to ensure proper treatment of the material during processing.

Process Control

As soon as the dies for a forging are completed, they are carefully checked for dimensional accuracy. Die proofs of lead or plaster are cast to provide an exact sample of the configuration the die impressions will produce at forging temperature. After the forge plant operator has verified the accuracy of the die impressions (Figure 4-162), the die proof is often sent to the customer for approval prior to forging production.

FIGURE 4-159. Bars being checked visually for signs of surface irregularities during handling and shearing.

FIGURE 4-160. Inspection of surfaces of forging stock using automatic seam depth indicator at the mill.

During die-tryout, after customer approval has been obtained and dies are inserted in forging equipment, match surfaces are checked to ensure proper alignment of both die halves. Sample forgings are made and inspected immediately for dimensional accuracy. Depending on customer specifications, samples may be sent to the laboratory for verification of metallurgical structure and properties. Production begins only after required metallurgical and dimensional approvals are received. For critical applications, a trial quantity consisting of three to five pieces is processed through the production sequence for dimensional, metallurgical, and nondestructive test evaluation prior to proceeding with production quantities.

During production, furnace temperature and the temperature of the forging stock are monitored to provide the optimum thermal range and ensure conformance to planned forging procedures. At regular intervals, "hot inspectors" verify dimensional accuracy by taking forgings from the production unit for checking, often before the forgings have cooled. Using equipment such as templates, scales, and calipers, inspectors can rapidly detect un-

acceptable variations, stop production, and see that necessary adjustments are made. Samples are also checked by layout during and at the end of a production run to monitor die wear (Figure 4-163).

Control of Heat Treatment

Control of heat treatment conditions such as furnace temperature and atmosphere and quenching conditions are monitored to ensure that optimum properties are developed. A hardness survey is performed after completion of all required heat treating cycles to ensure conformance to customer specifications.

An appropriate number of samples (varying from a few pieces to 100% of the run, according to a preselected sampling plan) are taken for quality control tests, which may include dye penetrant, ultrasonic, magnetic, or fluorescent particle tests and various tests for metallurgical properties. Upon completion of finishing operations (where the forging is cleaned to remove scale and lubricant, permitting reliable inspection of surfaces and dimensions), special fixtures and gages are often used to provide final confirmation of dimensional accuracy (Figure 4-164).

Methods of Validating Forging Integrity

Various test methods are available to confirm metallurgical soundness of forgings and may be classified according to their effect on the workpiece—destructive or nondestructive. The extent of testing and use of various test methods is ultimately determined by the customer's specifications. Thus, the following test methods are commonly used selectively and generally only when specified.

Nondestructive Testing

The procedures developed to evaluate the complete integrity of forgings should be able to uncover any randomly occurring flaws without destroying or altering the usefulness of the part. This has led to several highly important commercial techniques generally described as nondestructive testing (NDT), which includes the simple and obvious procedures of critical visual examination of the forging surface and extends to the sophisticated devices used in ultrasonic and magnetic particle methods. Some NDT procedures have the potential capability of determining physical or mechanical properties and detecting microsized imperfections.

Etching. Acid or caustic etching is used to prepare the surfaces of forgings for ef-

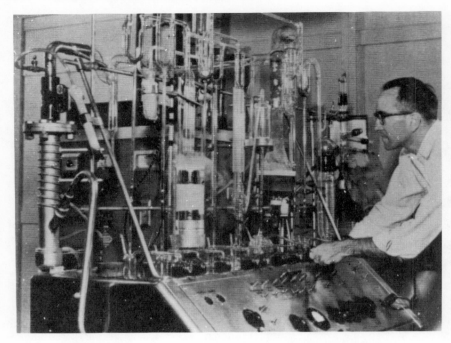

FIGURE 4-161. Vacuum fusion apparatus in a forge plant laboratory.

fective visual inspection. Surface replication techniques are utilized to review grain size and microstructural conformance.

Visual Penetrant Techniques. When irregularities occur in the surface of a forging, they are readily observed by applying a liquid such as penetrating oil to the forging surface. If the liquid has low viscosity and low surface tension and if the surface is clean of materials that could prevent wetting, then the liquid is drawn into the flaw by capillary action. After penetration is completed, the forging surface is cleaned by wiping or by using an emulsifying agent, or water, and a developing material is applied. The developer acts like a blotter to draw out the liquid, accentuating the indication size and defining shape and location.

The penetrants contain either a dye, usually bright red to contrast with the white developer, or a dissolved fluorescent filler, which makes them easily visible under ultraviolet light (Figure 4-165).

Magnetic particle inspection (Figure 4-166) can detect most discontinuities at the surface of metal and, to some extent, those that lie completely under the surface. It depends on the magnetic properties of the forging and is suitable only for metallic materials that can be intensely magnetized. The interpretation of each test indication, however, requires great skill and judgment.

The principle used in magnetic particle inspection is distortion of a magnetic field whenever a flaw is encountered. If the flaw is at or near the metal surface, the field distortion causes leakage flux at the surface nearest to the defect. The simplest means for indicating this leakage flux is to observe the concentration of magnetic particles in such areas. The visibility of the particles can be enhanced by dyeing the particles or coating them with a fluorescent material.

Variations in the basic technique include different means of applying the magnetic field, such as using both alternating single phase and poly-phase current as well as rectified or direct current, applied continuously or in bursts. The power can be applied directly to the forging, or through a coil wrapped around the part. The difference between the two methods is that the direction of magnetic flux is turned 90°. By applying the power leads at different locations on the forging or by using an external coil, the field direction can be varied to search for flaws regardless of their orientation. A third type of magnetization, using two currents applied simultaneously, can produce a swinging field for the same purpose.

The magnetic particles can be applied either during or after magnetization. The latter procedure, termed the residual method, is not as sensitive as the former. The particles can be dry or applied in a suspension of water or light petroleum distillate. The fluorescent particle method uses the wet application technique, and when the indications are read in a darkened space, visibility of even small defects is excellent.

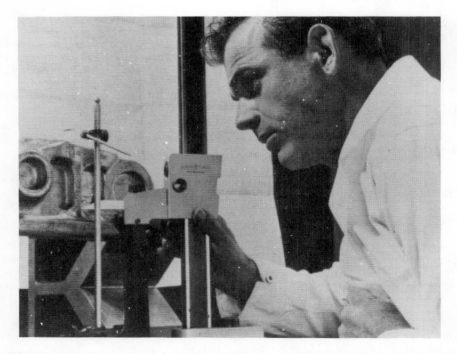

FIGURE 4-162. Final verification of dimensional accuracy of die impressions is obtained by checking the die proof.

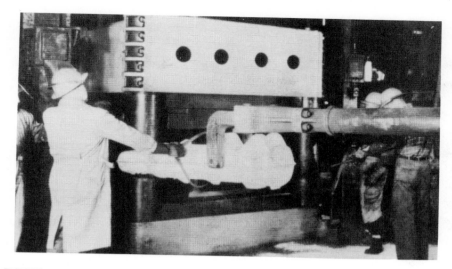

FIGURE 4-163. Hot inspector using calipers to check dimensions on forging as manipulator removes it from the trim press.

FIGURE 4-164. Dimensional check of a forging using special fixtures.

Permanent records of the magnetic particle indication may be made by lacquer coating of the indication on the part, transparent tape recording of the indication, photographic recording of the indication, and fluorescent indication photography. For heavy-duty requirements such as aircraft landing gear parts, standards of acceptance have been established for rating the forgings and the billets. In general, however, the standards for most forgings are established by agreement between the forging producer and purchaser.

Ultrasonic Inspection. When acoustic energy in the ultrasonic range passes through a metal, sound waves tend to travel in straight lines rather than diffuse in all directions as they do in the audible range (Figure 4-167). A defect in the path of the "beam" will cause a reflection of some of the energy, thus depleting the energy transmitted. This casts an acoustic shadow, which can be monitored by a detector placed opposite the transducer (energy source). If the acoustic energy is introduced as a very short burst, then the reflected energy coming back to the originating transducer can also be used to show the size and location of the defect. The transducer must be coupled to the material under test by some liquid medium, because air is too compressible. These pro-

cesses are shown schematically in Figure 4-168 and are referred to as the pulse-echo and the transmission methods.

Most current forging inspection methods utilize waves (called longitudinal waves) propagated into the test piece perpendicular to the surface. Other wave forms can be introduced into the metal at an angle other than 90° and can be most useful. At oblique angles, the beam ricochets off both faces of the metal. If a defect is encountered, the reflected wave will follow several paths, one of which will retrace the path of the original wave back to the transducer, where it can be imaged and observed on a cathode ray tube.

Eddy Current Testing. When a coil carrying a high-frequency current is brought close to an electrically conductive material, eddy or induced currents are generated in the conductor. These in turn create a magnetic field. As in magnetic particle inspection, a flaw can be detected by measuring distortions in the field. The process is exceptionally sensitive to minor variations in the material under test, even detecting changes in electrical conductivity caused by variations in heat treatment. Electrical conductivity testing has become an accepted standard for verification of heat treatment response in many aluminum alloys.

Electrical conductivity, as used for quality control of forged and heat-treated aluminum alloys, has been in commercial forging practice for a number of years. Its purpose is to gage the general corrosion resistance of aluminum alloys in various heat treatment (temper) conditions and within prescribed limitations to reveal satisfactory or suspect material conditions.

Electrolytic Polishing. In the early 1950s, electropolishing was established as a commercial method of microstructural evaluation. The test equipment utilizes an electrical power source unit with an electrolyte pump unit, which in effect polishes a finely ground surface by anodic disposition of metallic ions. This equipment provides a means of quality control of forged titanium components, but can also be useful on other nonferrous and ferrous materials. Electropolishing eliminates the need to destroy a forging to evaluate the microstructure. As an adjunct to electropolishing, replicating of the electropolished area after etching affords a means of removing from the work area a replica of the electropolished area after visual inspection for microphotography and subsequent documentation. The replicas are placed on a glass slide and then in a vacuum evaporator. A high-energy current is

FIGURE 4-165. Inspection of forgings with fluorescent dye penetrants under fluorescent light.

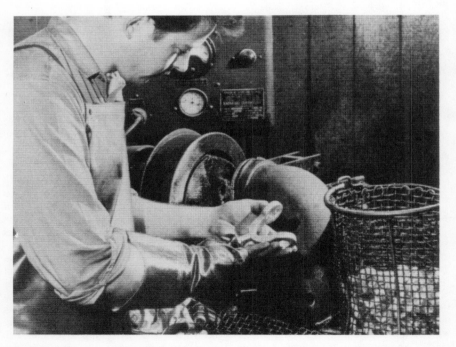

FIGURE 4-166. Forging being checked for indications of surface irregularities after magnetic particle inspection.

applied to a tungsten filament wrapped in aluminum, which vaporizes the aluminum and deposits it uniformly on the glass slide, creating, in effect, a duplication of the microstructure that can be examined and photographed.

X-Ray Testing. Testing of forgings by x-ray techniques is possible, but seldom employed, because defects for which x-ray tests would be used are not often present in forgings. Other test methods such as ultrasonic techniques are extremely practical and effective, whereas x-ray tests are particularly costly in relation to their effectiveness.

Destructive Evaluation

Macroetch. The hot acid deep-etch is one of the most revealing procedures for assessing forging quality and appropriateness of the forging sequence. However, it must be used on a spot-check basis, because it is a destructive test.

The forging is sawed into sections, usually split longitudinally with regard to the forging fiber, and immersed for approximately 30 min in an acid solution. The fiber structure and any discontinuities at the plane of the saw cut are readily apparent. The forger can, by experience, alter the die configuration, forging sequence, or temperature of deformation to optimize the directionality of properties associated with the fibered structure.

Photomicrography. Structural characteristics not visible to the unaided eye can be observed by microscopic study (Figure 4-169) of a polished and etched section of the forging. Metallographic techniques are many and varied, and the structures revealed are directly related to the mechanical performance of the material. The several phases of steels such as ferrite, austenite, carbide, martensite, and bainite have identifying features that permit recognition. In addition, the grain size and shape, the amount of carbide present, the existence of decarburization or oxidation, the presence of carburization, nitriding, or graphitization, and the degree of cold work can be identified by a skilled metallographer. The test is important to the control of forging quality, but like macroetching techniques, it cannot be used to ensure that random defects are not present.

Fracture Face Information. The fracture face produced in test bars and prototype forgings tested to destruction contains valuable information. For tensile test bars, the ductility of the steel is indicated by the presence or absence of a cup-and-cone-shaped surface. Brittle grain-boundary phases or intermetallics are revealed by the tendency for the forging to form intergranular rather than transgranular fractures. The former is revealed by flat, shiny facets of the grains or dendrites, while the latter has a silky gray or fibrous appearance. Sharp, minute fractures, such as those associated with high hydrogen con-

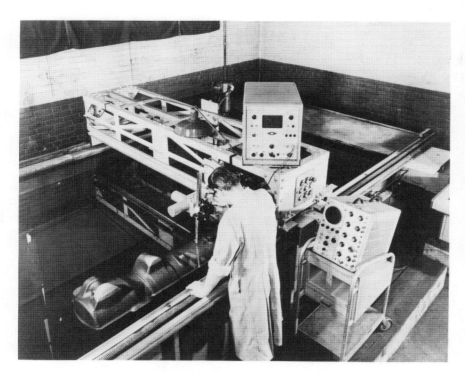

FIGURE 4-167. Large landing gear cylinder forging undergoing sonic inspection.

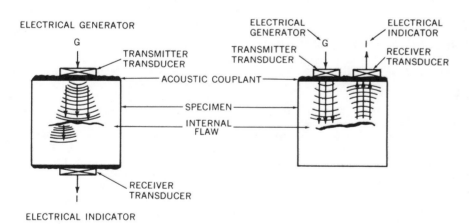

FIGURE 4-168. Schematic of transmission method (left) and pulse-echo method (right) used in sonic inspection of forgings.

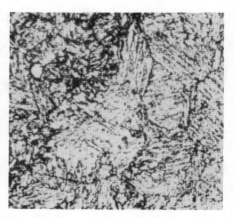

FIGURE 4-169. Photomicrograph of 18% Ni, 9% Co maraging steel solution treated and aged. Magnification is 500×; etchant, 5% nital.

Test Material From Forgings

Most basic material properties of interest in forgings are acquired through tests of a destructive nature. Test bars of a specific size and configuration are tested to define these properties. Due to scrap and cost considerations, test material is necessarily minimal. However, quality control specifications require rigorous testing programs to ensure a high level of product reliability. These conditions have resulted in alternate test material locations such as prolongs and test rings.

Prolongs and test rings are usually forged integrally with the part, although there are situations requiring a separately forged test bar. Their placement and size are critical from a metallurgical standpoint. Properties determined from alternate test material may not be totally representative of that from cross-sectional test material. However, some proportionality can be established and maintained to ensure meaningful comparison.

Because properties identified in a test bar are related to and affected by their location in the forging and their orientation relative to the grain flow of the forging, it is extremely important that these factors be carefully standardized and controlled. In determining location and orientation of the test bar in a forging, it is essential that (1) the test bar be readily removed as a block, (2) the forging reduction in the bar be representative of that in the part, (3) the directionality be proper for the property sought, and (4) the test ring or prolongation not alter a minimum dimension of the forging. Test bar location, metal section, method of attachment, and location of prolong (if required) are commonly resolved between purchaser and forgings

tent, frequently are apparent as bright facets within a normal structure.

Mechanical Property Evaluation

The critical application of highly reliable forgings has made the qualification of various material properties a necessity. A number of techniques are used to catalog these properties, but their goal is to estab-lish some measure of structural integrity or basic material response. The types of tests commonly performed include hardness tests, tension tests, impact tests, fatigue tests, creep and stress-rupture tests, and fracture toughness tests. While each measures a different aspect of material behavior, together they are an accurate representation of the physical and mechanical behavior of a forging under complex loading conditions.

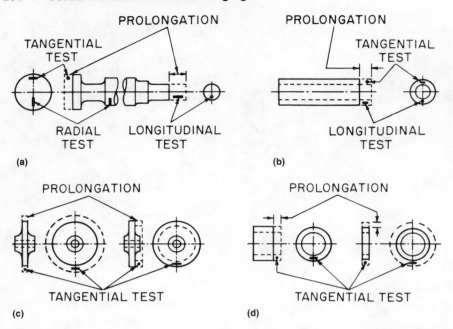

FIGURE 4-170. Typical locations of test bar prolongations and integral rings. (a) Shafts and rotors. (b) Hollow forgings. (c) Disk forgings. (d) Ring forgings.

producer and are usually specified (Figure 4-170).

Test Specimen Preparation

Numerous test specimen preparation techniques presently exist, including lathe turning and grinding for the production of circular cross section specimens. New developments in process automation and control increase this number. Each technique can have a different effect on the surface integrity of the finished specimen. These differences are important, because recent data suggest good correlation between material properties and material removal methods in materials such as titanium and nickel-base alloys and some ferrous materials.

For example, titanium alloys are known to show an increase in fatigue strength when gentle or low-stress grinding techniques are employed compared to abrasive or even conventional grinding techniques. A similar correlation exists for other aerospace forging materials in properties such as tensile strength and creep and stress-rupture life.

Because the postmachining residual stress profile in a test specimen has a definite effect on material properties, concern over its control and measurement is increasing. Low-stress grinding, longitudinal polishing, and x-ray diffraction verification of residual surface stress intensity are commonly used to comply with the residual surface stress levels of test specimens specified by purchasers of critical application forgings.

Statistical Testing Methods

Purchasers involved in the procurement of large numbers of standardized forgings are becoming more cognizant of the powerful use of statistical methods. Cost reduction, improvements in quality levels, and time savings are but a few of the benefits. Statistical methods can be utilized in many areas of production, but are especially useful in the area of mechanical property testing.

Through the use of statistical methods, mechanical testing of critical parts, many of which require testing on a one-to-one basis, can be reduced significantly. This reduction is accomplished by first analyzing the target data base to determine its mean and variability and then comparing these to specified requirements. If sufficient margin exists between the statistically predicted maximum and/or minimum values and the specification requirements, then reduced testing may be initiated. In some cases, changes in the manufacturing process may be required to establish the desired margin.

The reduced testing frequency can be continued as long as the test results fall within the statistically developed limits and the manufacturing process is unchanged. Should any value fall outside the statistical limits, the entire group represented by the randomly selected sample would be 100% tested. Investigation of the variation would determine its cause and establish appropriate corrective action.

Principal Mechanical Testing Procedures

Of the many tests that can measure one forging property or another, a few tests have found widest acceptance for quality control and greatest value for predicting the engineering performance of components. The principal testing methods are hardness testing, tension testing, impact testing, creep and stress-rupture testing, fatigue testing, and fracture mechanics testing.

Hardness testing is a means of determining resistance to deformation—penetration, abrasion, machining, etc. In metals, the most commonly used methods depend on the material's resistance to penetration by a much harder body. Tests of this nature supply a relatively nondestructive means of evaluating hardness. Penetration hardness tests are also empirically related in some materials to properties such as tensile, fatigue, and impact strength and can be used to indicate possible deficiencies in these properties.

The most commonly used penetration tests are the Brinell, Rockwell, and Vickers hardness tests. All require that the material be of sufficient thickness, have a good surface finish that is free from scale and other foreign material, and have a firm support relative to the indentation mechanism.

The Brinell hardness test involves forcing a hardened spherical indenter into a surface under a known load. By measuring the diameter of the resulting cavity, a Brinell hardness number (HB) can be calculated. The standard indenter is 10 mm in diameter and is used with standard loads of 3000, 1500, and 500 kg. The 10-mm indenter and 3000-kg load are typically used on steels (Figure 4-171).

The Rockwell hardness test employs either a hardened spherical indenter or a spherical-tipped conical diamond called a Brale. A minor load, usually 10 kg, is used to establish initial penetration. Then a major load of 150, 100, or 60 kg is applied. The difference in penetration depth between the two loads is a measure of the hardness, which can be read directly from the attached scale that measures penetration (Figures 4-172 and 4-173). Various combinations of load and indenters are possible. The two most commonly used are the Rockwell B scale (HRB) using a

FIGURE 4-171. Hydraulic Brinell hardness tester. Courtesy of Page-Wilson Corp.

FIGURE 4-172. Rockwell hardness testing machine. Courtesy of Wilson Mechanical Instrument Division, American Chain & Cable Co., Inc.

$1/_{16}$-in.-diameter hardened spherical indenter and a major load of 100 kg and the Rockwell C scale (HRC) using a Brale indenter and a major load of 150 kg.

The Vickers hardness test is similar to the Brinell test, except that the indenter tip is a diamond in the shape of a square pyramid and uses a load between 1 and 120 kg. The resulting impression is quite small and requires the use of medium-power magnification for measurement of the impression diagonal length.

Although the Vickers test is considered the most versatile and accurate of the three methods, its increased equipment cost and lower production test rate make the Brinell and Rockwell tests more appealing. Standard test methods for Brinell, Rockwell, and Vickers hardness testing can be found in ASTM E 10, E 18, and E 92, respectively.

Tension testing employs the application of a constantly increasing uniaxial load to a specimen until fracture occurs. The testing machine typically has a hydraulic, screw, or lever-type drive to apply load and an accurate transducer to measure this load. A sensitive extensometer, attached to the specimen at two specific gage points, measures the amount of elongation that occurs during the test. Specimens are of a specific configuration and commonly have a circular cross section. The two curves shown in Figure 4-174 represent the two types of behavior typical of metals subjected to the tension test. Curves can be drawn either by plotting load versus elongation or stress versus strain.

The engineering stress versus strain curve is the most useful for quality control and involves division of load and elongation by the original cross-sectional area and the original gage length, respectively. Of specific importance on the stress-strain curve are the modulus of elasticity, yield strength, and ultimate strength.

The modulus of elasticity is defined as the ratio of stress to strain in the linear portion of the stress-strain curve. This property is not very sensitive to metallurgical structure, heat treatment, or processing imposed on the material. It is, however, sensitive to temperature.

The yield strength is determined by constructing a line parallel to, but at some offset from, the modulus line. Though the offset value is arbitrary, an industry-wide value is 0.2%.

The ultimate strength indicates the point at which plastic deformation of the test specimen switches from uniform to localized "necking." It indicates the maximum stress that the material is capable of supporting.

FIGURE 4-173. Indicating dial on Rockwell hardness testing machine. Courtesy of Wilson Mechanical Instrument Division, American Chain & Cable Co., Inc.

Also important from the tension test are two measures of material ductility: elongation and reduction in area. Elongation is defined as the ratio of the difference in length between the original and final gage lengths and the original gage length expressed in percent. Reduction in area is similarly the ratio of the difference between the original and final cross-sectional area and the original cross-sectional area expressed in percent.

Testing rates used in tension tests can influence the magnitude of the properties obtained, especially in strain-rate-sensitive materials. This is typical of most titanium and nickel-base alloys. Generally, increasing the testing rate will increase strength and decrease ductility. Decreasing the testing rate decreases strength and increases ductility. Purchase specifications usually dictate what rates are to be used, but in their absence, the latest version of ASTM E 8 should be consulted.

Test specimen configuration has been subjected to a large degree of standardization. The size of the component being tested dictates the test specimen size. Typically, the 0.252-in.-diameter specimen provides uniform reproducible results where test material is sufficient. Various diameters and methods of finishing the specimen ends exist (Figure 4-175). Again, whenever purchase specifications are nonspecific, ASTM E 8 should be consulted.

Because of the sensitive nature of the tension test, automation is becoming increasingly popular. Computerized control of testing rates and data acquisition allows for more accurate and reproducible results. This is especially helpful when production-type statistical testing is per-

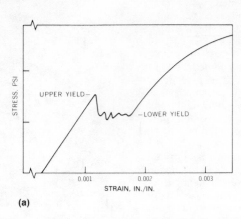

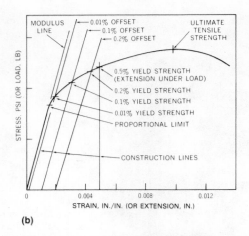

FIGURE 4-174. Stress-strain curves derived from tensile tests illustrating nomenclature of engineering properties. (a) An enlarged section of the curve with the sharp knee typical of ferritic steels. (b) Construction lines for tensile test nomenclature.

formed. Specialized systems are capable of automatically performing calculations of modulus of elasticity, yield strength, ultimate strength, and elongation.

Impact testing is a dynamic test in which a carefully prepared specimen is struck by a single blow from a hammer, and the energy absorbed to completely fracture the specimen is recorded. The energy absorption, measured in foot-pounds, is determined by measuring the follow-through of the impacting pendulum. The test provides useful information about the metallurgical condition of material subjected to impact (Figure 4-176 a–c).

The specimen configurations shown in Figures 4-177 and 4-178 employ a notch of carefully controlled dimensions. The notch acts as a stress concentration point; if the material distributes the stress uniformly, the impact value will be high. The Charpy-type test specimen is supported horizontally on each end and is struck in the center in simple beam loading, with the impact occurring on the face opposite and directly behind the notch. The Izod-type test specimen is held as a vertical cantilever beam and is broken by a blow delivered at a fixed distance above the notch.

In addition to recording the energy absorbed by the specimen, measurement of the relative deformation and fracture appearance are made. The deformation of the specimen from its untested geometry, lateral expansion, and an estimate of the relative area of fibrous surface (shear) as compared to the brittle (cleavage) surface give an indication of the ductility of the material. Because the notched-beam impact test shows a tendency toward embrittlement as temperature decreases, it is well suited for determining the transition temperature from ductile to brittle fracture in steels.

Typical industry standards require certification of impact testing machines when they are used to qualify components for nuclear and other high-liability applications. ASTM E 23 presents standard testing and verification procedures for Charpy impact specimens. Verification involves the annual testing of standardized specimens and comparison of the average value obtained from the machine with the nominal value of the standardized specimens.

Creep and Stress-Rupture Testing. Creep is defined as time-dependent deformation that occurs with application of a load. The rate of creep is directly related to the applied load and temperature. Creep-measuring equipment is relatively simple. The test involves holding a specimen at a constant temperature, while simultaneously applying a static tensile load. Elongation of the specimen is measured periodically and plotted as a function of time. Test duration may run into thousands of hours, making automated temperature control and data acquisition systems appealing.

The creep curve has the typical appearance shown in Figure 4-179. Upon initial loading, elongation A occurs. The primary creep rate B is exemplified by a gradually decreasing rate. The secondary creep rate C is relatively constant and usually the minimum creep rate. The third-stage creep rate D is constantly increasing in nature and with sufficient time will lead to specimen failure. If the specimen is unloaded by the distance E prior to fracture, the elastic deflection upon loading will be recovered. The amount of permanent deformation is represented by F. Test method

ASTM E 139 provides a standard method of performing the creep test.

The stress-rupture (or creep-rupture) test is identical in nature to the creep test, except that the loads are typically higher and the tests are continued to failure. Instead of measuring creep, the hours to failure are monitored. The stress-rupture test can be conducted utilizing a variety of specimens, including smooth, notched, and combination specimens. ASTM E 292 covers the standard test method for stress-rupture testing.

In industry today, neither creep nor stress-rupture tests are carried out the full duration of the expected service life of the component. Long-time tests are based on extrapolation of short-time tests. Extrapolations based on log-log plots of stress versus time to failure and stress versus minimum creep rate and other parameters such as the Larson-Miller parameter have been developed for meaningful extrapolations of creep and stress-rupture data.

Fatigue Testing. In many types of service applications, forgings are required to withstand repeated or cyclic loading in various detrimental environments. Fatigue refers to the mechanical damage materials encounter under these conditions. Fatigue fractures are typified by a crack initiating at some surface or internal irregularity. The crack propagates with each load application until the affected material cross section is small enough to fail in a catastrophic manner. With this and other material properties in mind, parts are designed, where necessary, with a minimum expected fatigue life when subjected to a specific set of stress, temperature, corrosion, or other parameters.

Fatigue testing involves subjecting a specimen of specific configuration to cyclic stresses until failure. The stresses are generally less than those that would cause failure under static loading conditions. By conducting a number of tests at successively lower levels of stress, it is possible to find a value that will not produce failure, regardless of the number of applications. This stress value is called the fatigue limit of the material.

There are three types of fatigue tests commonly employed in the collection of fatigue data. The first is the rotating beam fatigue test, in which the specimen is subjected to a uniform bending moment over its entire length while being rotated. Any given fiber of the specimen is subjected alternately to tension and compression stresses of equal magnitude. A machine widely used to perform this type of test is the R.R. Moore machine. There are variations of this type of test, but all incorporate rotating bending to produce fatigue. Fatigue life results obtained with this machine are compared to service data in Figure 4-180.

A second type of test is the repeated bending or direct flexure test, in which the specimen is repeatedly flexed but not rotated. This test has the advantage of not requiring a specimen with a special surface preparation and thus is useful in testing such items as flat rolled stock.

The third, an increasingly popular method, is the axial fatigue test, in which the load is applied directly along the longitudinal axis of a nonrotating specimen (Figure 4-181). Although machines used to perform this type of test are particularly susceptible to misalignment, which induces unwanted bending moments in the specimen, their versatility makes them very appealing.

Axial loading-type machines are designed to perform tests where the mean stress is not zero (the stress is not completely reversed). They also have the ability to perform tests in various modes of control: load, strain, or displacement control. Because of easy specimen access at low testing frequencies, axial fatigue machines are easily automated. Computerization and automation of control and data acquisition have important implications in high-technology applications.

In most types of fatigue tests, specimen preparation is probably the most critical aspect. Because fatigue behavior is very sensitive to surface conditions, much data scatter and unreliability can be expected if special precautions are not taken. Typically, for axial fatigue specimens, low-stress grinding techniques followed by alternate polishing in the circumferential and longitudinal directions with successively

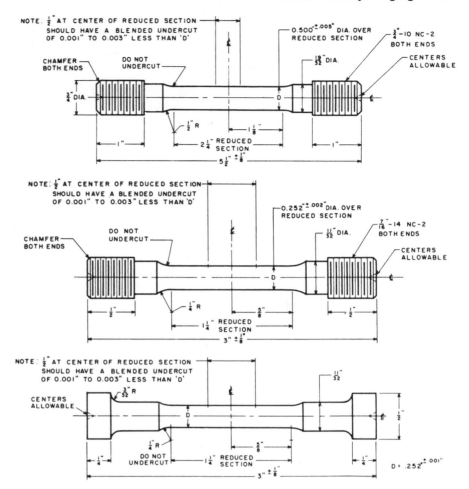

FIGURE 4-175. Various diameters and methods of finishing specimen ends.

finer grits produce the most satisfactory specimens. Testing parameters are usually prescribed in customer specifications, but test method ASTM E 606 is commonly followed.

Fracture Toughness Testing. The considerable use of high-strength alloys in forging and interest in the concept of notch toughness have led to the development of methods of quantitatively predicting the unstable extension of a crack or surface flaw. The selection of steels with a suitable notch toughness was previously a matter of finding one with a suitably low transition temperature. Very high-strength steel alloys and other superalloys generally do not exhibit significant increases or transitions in toughness with increasing temperature. As a matter of fact, the energy levels necessary for unstable crack propagation may be significantly less in these materials than those of the more ductile low-strength steels. The basic con-

cepts of linear elastic fracture mechanics and stress field analysis have helped in the evolution of such standard test methods as ASTM E 399.

Testing of fracture toughness specimens consists of two parts. The first is fatigue precracking the specimen to a predetermined crack length in a series of stages. In the second, the specimen is pulled apart in tension at some specific rate. Through data obtained during precracking and testing and various calculations and validity checks, a value representative of the resistance to crack propagation is determined (Figure 4-182). Because of efforts expended in the development of linear elastic fracture mechanics, accompanied by the determination of fatigue crack propagation rates and critical flaw sizes, estimates of service life are now possible. High-strength high-toughness materials can now be used with much greater assurance for applications requiring resistance to brittle failure.

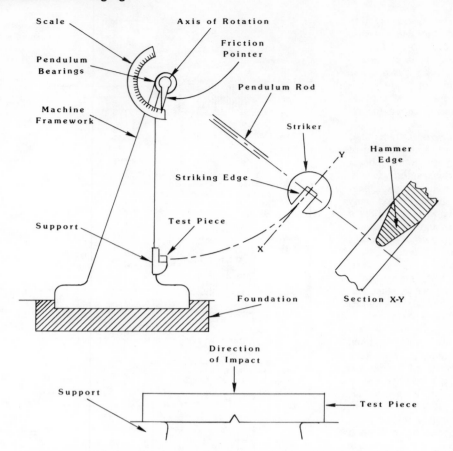

FIGURE 4-176a. Schematic of Charpy impact test machine. Source: Ref. 76.

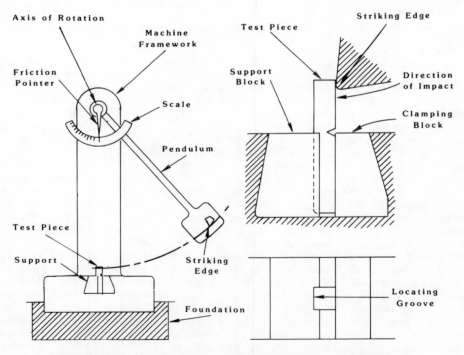

FIGURE 4-176b. Schematic of Izod impact test machine. Source: Ref. 76.

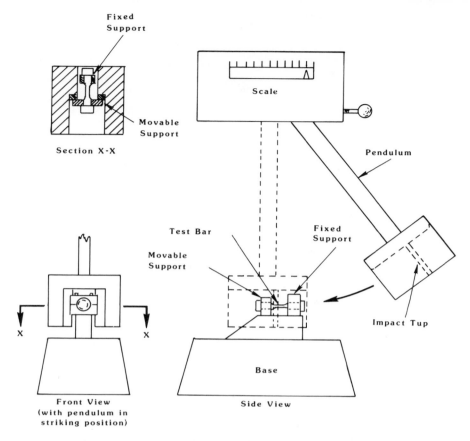

FIGURE 4-176c. Schematic of impact test machine adapted for determination of forgeability. Sequence: (1) Heated tension test bar is placed in fixed support and movable support. (2) Pendulum is released and impact tup strikes movable support. (3) Tension test bar breaks, and energy (in ft-lb) is read from scale. Source: Ref. 77.

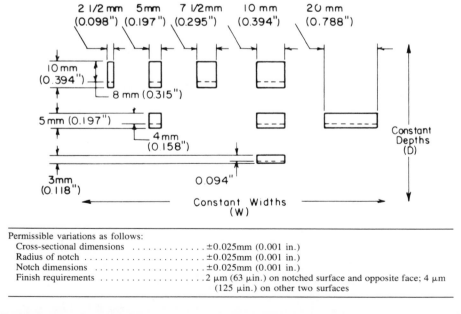

Permissible variations as follows:
Cross-sectional dimensions . . . . ±0.025mm (0.001 in.)
Radius of notch . . . . ±0.025mm (0.001 in.)
Notch dimensions . . . . ±0.025mm (0.001 in.)
Finish requirements . . . . 2 μm (63 μin.) on notched surface and opposite face; 4 μm
(125 μin.) on other two surfaces

FIGURE 4-177. Charpy (simple-beam) subsize (Type A) impact test specimens. Source: Ref. 78.

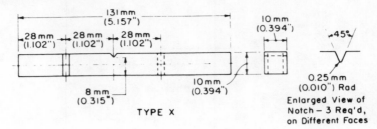

TYPE X

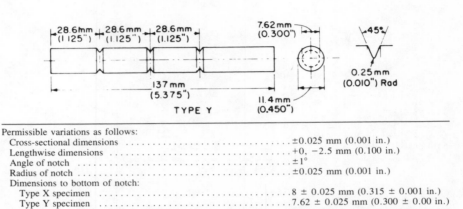

TYPE Y

Permissible variations as follows:
Cross-sectional dimensions ....±0.025 mm (0.001 in.)
Lengthwise dimensions ....+0, −2.5 mm (0.100 in.)
Angle of notch ....±1°
Radius of notch ....±0.025 mm (0.001 in.)
Dimensions to bottom of notch:
  Type X specimen ....8 ± 0.025 mm (0.315 ± 0.001 in.)
  Type Y specimen ....7.62 ± 0.025 mm (0.300 ± 0.00 in.)

FIGURE 4-178. Izod (cantilever-beam) impact test specimens (Types X and Y). Source: Ref. 78.

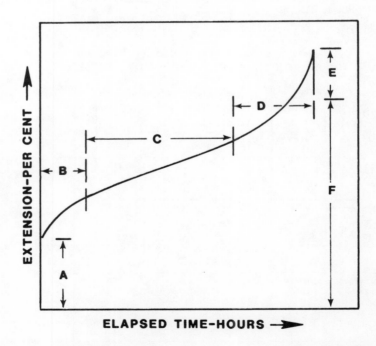

FIGURE 4-179. Schematic creep curve. Extension plotted against elapsed time. A, elastic extension; B, creep at decreasing rate; C, creep at approximately constant rate; D, creep at increasing rate; E, elastic contraction; F, permanent change of length.

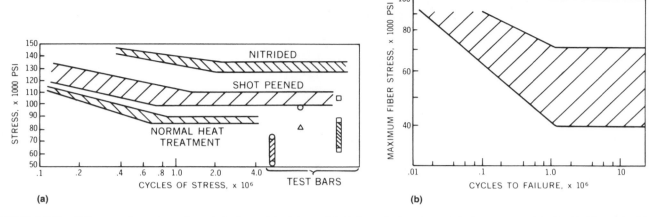

FIGURE 4-180. Fatigue performance of crankshafts and test bars. (a) Fatigue limits of crankshafts compared to R. R. Moore test bars. (b) Fatigue resistance of bars trepanned from crankshafts.

FIGURE 4-181. Computerized low-cycle fatigue machine. Courtesy of Battelle's Columbus Laboratories.

FIGURE 4-182. Fracture toughness equipment. Courtesy of Battelle's Columbus Laboratories.

References

32. Schey, J.A., *Principles of Forging Design,* American Iron and Steel Institute, Washington, DC, 1965.

33. Lange, K., Ed., *Handbook of Metal Forming,* McGraw-Hill, New York, 1985.

34. Billigmann, J. and Feldmann, H., *Upsetting and Forming,* Carl Hanser Verlag, Munich, 1973 (in German).

35. Geiger, R., Possibilities and Limits in Process Combinations, in *New Developments in Bulk Forming,* Metal Forming Institute, University of Stuttgart, June 1977 (in German).

36. Mayrhofer, K., *Cold Extrusion of Steel and Nonferrous Metals,* Springer-Verlag, Berlin, 1983 (in German).

37. Turno, A., Closed-Die Forging and Pressing of Components with Teeth, *Industrie-Anzeiger,* Vol. 94, 1972, p. 1997 (in German).

38. *High-Temperature, High-Strength Nickel Base Alloys,* International Nickel Co., Inc., New York, 1977.

39. Simmons, W.F., "Description and Engineering Characteristics of Eleven New High-Temperature Alloys," DMIC Memorandum 255, Defense Metals Information Center, Battelle's Columbus Laboratories, Columbus, OH, June 1971.

40. Fix, D.K., Titanium Precision Forgings, in *Titanium Science and Technology,* Vol. 1, Plenum Press, New York, 1972, p. 441–451.

41. Watmough, T., Kulkarni, K.M., and Parikh, N.M., "Isothermal Forging of Titanium Alloys Using Large Precision-Cast Dies," Technical Report AFML-TR-70-161, IIT Research Institute, Chicago, July 1970.

42. Kulkarni, K.M., Isothermal Forg-

ing—From Research to a Promising New Manufacturing Technology, in *Proceedings of the 6th North American Metalworking Research Conference,* Society of Manufacturing Engineers, Dearborn, MI, 1978, p. 24–32.

43. Chen, C.C. and Coyne, J.E., Recent Developments in Hot Forging of Titanium Alloys, in *Titanium '80,* Metallurgical Society of AIME, Warrendale, PA, 1980, p. 2513.

44. Moskowitz, L.N., Pelloux, R.N., and Grant, N., Properties of IN-100 Processed by Powder Metallurgy, in *Superalloys—Processing,* Proceedings of the 2nd International Conference, MCIC Report 72-10, Metals and Ceramics Information Center, Battelle's Columbus Laboratories, Columbus, OH, 1972.

45. Radcliffe, S.V. and Kula, E.B., Deformation, Transformation, and Strength, in *Deformation Processing,* W.A. Backofen *et al.,* Ed., Syracuse University Press, 1964, p. 321.

46. "Precision Forging Machines," technical literature, GFM Corp., Steyr, Austria.

47. Altan, T. and Sabroff, A.M., "Comparison of Mechanical Presses and Screw Presses for Closed-Die Forging," Technical Paper MF70-125, Society of Manufacturing Engineers, Dearborn, MI, 1970.

48. Foucher, J., "Influence of Dynamic Forces Upon Open Back Presses," Doctoral Dissertation, Technical University, Hannover, 1959 (in German).

49. Altan, T., Important Factors in Selection and Use of Equipment for Metalworking, in *Proceedings of the 2nd Inter-American Conference on Materials Technology,* Mexico City, Aug. 24, 1970.

50. Reimenschneider, F. and Nickrawietz, K., Drives for Forging Presses, *Stahl und Eisen,* Vol. 79, 1959, p. 494 (in German).

51. Altan, T., Boulger, F.W., Becker, J.W., Akgerman, N., and Henning, H. J., *Forging Equipment, Materials, and Practices,* MCIC-HB-03, Metals and Ceramics Information Center, Battelle's Columbus Laboratories, Columbus, OH, 1973.

52. Kienzle, O., Development Trends in Forming Equipment, *Werkstattstechnik,* Vol. 49, 1959, p. 479 (in German).

53. Rau, G., A Die Forging Press With a New Drive, Metal Forming, July 1967, p. 194–198; see also *Industrie-*

Anzeiger, Vol. 88, 1966, p. 1841–1844 (in German).

54. *Engineers Handbook*, Vol. 1–2, VEB Fachbuchverlag, Leipzig, 1965 (in German).

55. Spachner, S.A., "Use of a Four-Bar Linkage as a Slide Drive for Mechanical Presses," SME Paper MF70-216, Society of Manufacturing Engineers, Dearborn, MI, 1970.

56. Maekelt, H., *Mechanical Presses*, Carl Hanser Verlag, Munich, 1965 (in German).

57. Bohringer, H. and Klip, K.H., Development of the Direct-Drive Percussion Press, *Sheet Metal Industries*, Vol. 43, Nov. 1966, p. 857.

58. Altan, T. and Sabroff, A.M., Important Factors in the Selection and Use of Equipment for Forging, Part I, II, III, and IV, *Precision Metals*, June, July, Aug., and Sept. 1970.

59. Klaprodt, Th., Comparison of Some Characteristics of Mechanical and Screw Presses for Die Forging, *Industrie-Anzeiger*, Vol. 90, 1968, p. 1423 (in German).

60. Lange, K., Machines for Warmforming, in *Hutte, Handbook for Plant Engineers*, Vol. 1, Wilhelm Ernst & John Verlag, Hannover, 1957, p. 657 (in German).

61. Kortesoja, V.A., *Properties and Selection of Tool Materials*, American Society for Metals, 1975.

62. Kannappan, A., Wear in Forging Dies—A Review of World Experience, *Metal Forming*, Vol. 36 (No. 12), Dec. 1969, p. 335; Vol. 37 (No. 1), Jan. 1970, p. 6.

63. Naujoks, W. and Fabel, D.C., *Forging Handbook*, American Society for Metals, 1953.

64. Woldman, N.E., *Metal Process Engineering*.

65. *Atlas of Isothermal Transformation and Cooling Transformation Diagrams*, American Society for Metals, 1977.

66. Hodge, J.M. and Orehoski, M.A., *Trans. AIME*, Vol. 167, 1946, p. 627.

67. Jatczak, C.F., *Metal Progress*, Vol. 100 (No. 3), 1971, p. 60.

68. Grange, R.A., Hribal, C.R., and Porter, L.F., *Met. Trans. A*, Vol. 8, 1977, p. 1775.

69. Grossman, M.A. and Bain, E.C., *Principles of Heat Treatment*, American Society for Metals, 1964.

70. Klingler, L.J., Barnett, W.J., Fromberg, R.P., and Troiano, A.R., *Trans. ASM*, Vol. 46, 1954, p. 1557.

71. Buffum, D.C. and Jaffe, L.D., *Trans. ASM*, Vol. 43, 1951, p. 644.

72. Dolan, T.J. and Yen, C.S., *Proc. ASTM*, Vol. 48, 1948, p. 664.

73. Whittaker, D., *Metallurgia*, Vol. 46 (No. 4), April 1979, p. 275.

74. Muzyka, D.R., MiCon 78: Optimization of Processing, Properties, and Service Performance Through Microstructural Control, in *ASTM STP 672*, M. Abrams *et al.*, Ed., ASTM, Philadelphia, 1979, p. 526.

75. Boyer, R.R., *Journal of Metals*, Vol. 32 (No. 3), 1980, p. 61.

76. Fenner, A.J., *Mechanical Testing of Materials*, National Engineering Laboratory, East Kilbride, Glasgow, Philosophical Library, 1965.

77. Sabroff, A.M., Boulger, F.W., Henning, H.J., and Spretnak, J.W., "Fundamentals of Forging Practice," Supplement to Technical Documentary Report No. ML-TDR-64-95, Contract No. AF 33(600)-42963, Battelle Memorial Institute, Columbus, OH, March 1965.

78. *Annual Book of ASTM Standards*, ASTM, Philadelphia, 1984.

GLOSSARY

Definitions of Selected Technical Terms and Nomenclature Relating to Forging

Age. An operation in which forgings are subjected to low-temperature treatment for specific periods of time to effect the complete or partial precipitation of the solutes in the alloy, resulting in controlled hardening of the metal.

Age hardening (aging). The latter part of a two-step heat-treating operation applied to certain alloys for strengthening and hardening (see also *Solution heat treatment*). Aging involves heating to a relatively low temperature for a specified period of time, and results in controlled precipitation of the constituent dissolved during the solution heating treatment.

Aging. The change in the properties of a metal that occurs at relatively low temperature following a final heat treatment or a final cold working operation; aging tends to restore equilibrium in the metal and eliminate any unstable condition induced by a prior operation.

Aircraft quality. Denotes stock of sufficient quality to be forged into highly stressed parts for aircraft or other critical applications. Such materials are of extremely high quality, requiring closely controlled, restrictive practices in their manufacture in order that they may pass rigid requirements, such as magnetic particle inspection (Ref: Aerospace Material Specification 2301).

Air-lift hammer. A type of gravity drop hammer where the ram is raised for each stroke by an air cylinder. Because length of stroke can be controlled, ram velocity and thus energy delivered to the workpiece can be varied.

Alloy. A material having metallic properties and composed of two or more chemical elements of which at least one is a metal. In practice, the word is commonly used to denote relatively high-alloy grades of material—for example, "alloy" steels as differentiated from "carbon" steels. Materials are alloyed to enhance physical and mechanical properties such as strength, ductility, and hardenability.

Angularity. The conformity to, or deviation from, specified angular dimensions in the cross section of a shape or bar.

Annealing, full. A heat-treating operation wherein metal is heated to a temperature above its critical range, held at that temperature long enough to allow full recrystallization, then slowly cooled through the critical range. Annealing removes working strains, reduces hardness, and increases ductility.

Anvil (base). Extremely large, heavy block of metal that supports the entire structure of conventional gravity- or steam-driven forging hammers. Also, the block of metal on which hand (or smith) forgings are made.

Anvil cap (sow block). A block of hardened, heat-treated steel placed between the anvil of the hammer and the forging die to prevent undue wear to the anvil.

Auxiliary operations. Additional processing steps performed on forgings to obtain properties, such as surface conditions or shapes, not obtained in the regular processing operation.

Axial rolls. In ring rolling, vertically displaceable, tapered rolls, mounted in a horizontally displaceable frame opposite from but on the same centerline as the main roll and rolling mandrel. The axial rolls control the ring height during the rolling process.

Backward extrusion. Forcing metal to flow in a direction opposite to the motion of a punch or die.

Backing arm. A device for supporting the ring rolling mill mandrel from above during the roll process.

Bar. A section hot rolled from a billet to a form, such as round, hexagonal, octagonal, square, or rectangular, with sharp or rounded corners or edges, with a cross-sectional area of less than 16 in.^{2}; a solid section that is long in relation to its cross-sectional dimensions, having a completely symmetrical cross section and whose width or greatest distance between parallel faces is $^3/_8$ in. or more.

Bar end. See *End loss.*

Base. See *Anvil.*

Batch-type furnace. A furnace for heating of materials in which the loading and unloading is done through a single door or slot.

Bed. Stationary platen of a press to which the lower die assembly is attached.

Bend. Operation to preform (bend) stock to approximate shape of die impression for subsequent forging; also includes final forming.

Bend or twist (defect). Distortion similar to warpage, but resulting from different causes; generally caused in the forging or trimming operations. When the distortion is along the length of the part, it is called "bend"; when across the width, it is called "twist." Low-draft and no-draft forgings are more susceptible to bending, as they must be removed from the dies by some form of mechanical ejection. Dull trimming tools and improper nesting will cause bending in the trimming operation. When bend or twist exceeds tolerances, it is considered a defect. Corrective action entails either hand straightening, machine straightening, or cold restriking.

Bender. (1) Bends stock in the required directions for preliminary forging to approximate the ultimate shape; the die portion forming the longitudinal axis in one or more planes. (2) A die impression, tool, or mechanical device designed to bend forging stock to conform to the general configuration of die impressions subsequently to be used.

Bending. A preliminary forging operation to give the piece approximately the correct shape for subsequent forming.

Billet. (1) A semifinished section hot rolled from a metal ingot, with a rectangular cross section usually ranging from 16 to 36 in.^{2}, the width being less than twice the thickness. Where the cross section exceeds 36 in.^{2}, the term "bloom" is properly but not universally used. Sizes smaller than 16 in.^{2} are usually termed "bars"; a solid semifinished round or square product which has been hot worked by forging, rolling, or extrusion. (2) A semifinished, cogged, hot-rolled, or continuous-cast metal product of uniform section, usually rectangular with radiused corners. Billets are relatively larger than bars.

Blank. A piece of stock (also called a "slug" or "multiple") from which a forging is to be made.

Blast cleaning (blasting). A process for cleaning or finishing metal objects by use of an air jet or centrifugal wheel that propels abrasive particles (grit, sand, or shot) against the surfaces of the workpiece at high velocity.

Blister. A raised spot on the surface of the metal caused by expansion of gas in a subsurface zone during thermal treatment.

Block. The forging operation in which metal is progressively formed to general desired shape and contour by means of an impression die (used when only one block operation is scheduled).

Block and finish. The forging operation in which the part to be forged is blocked and finished in one heat through the use of a die having both a block impression and a finish impression in the same die. This also covers the case where two tools mounted in the same machine are used, as in the case of aircraft pistons. Only one heat is involved for both operations.

Block, first and second. Blocking operation performed in a die having two blocking cavities in the same die; the part being forged is successively blocked in each impression all in one heat. As many as three blocker dies are sometimes needed for some forgings and up to three operations are sometimes required in each die.

Block, first, second, and finish. The forging operation in which the part to be forged is passed in progressive order through three tools mounted in one forging machine; only one heat is involved for all three operations.

Blocker. Gives the forging its general shape, but omits any details that might restrict the metal flow; corners are well rounded. The primary purpose of the blocker is to enable the forming of shapes too complex to be finished after the preliminary operations; it also reduces die wear in the finishing impression.

Blocker (blocking impression). The impression in the dies (often one of a series of impressions in a single die set) that imparts to the forging an intermediate shape, preparatory to forging of the final shape.

Blocker die. A die used for preliminary forming of a die forging.

Blocker-type forging. A forging that approximates the general shape of the final part with relatively generous finish allowance and radii. Such forgings are sometimes specified to reduce die costs where only a small number of forgings are desired and the cost of machining each part to its final shape is not excessive.

Blocking. A forging operation often used to impart an intermediate shape to a forging, preparatory to forging of the final shape in the finishing impression of the dies. Blocking can ensure proper "working" of the material and contribute to greater die life.

Blow. The impact or force delivered by one workstroke of the forging equipment.

Board hammer. A type of gravity drop hammer where wood boards attached to the ram are raised vertically by action of contrarotating rolls, then released. Energy for forging is obtained by the mass and velocity of the freely falling ram and the attached upper die.

Boss. A relatively short protrusion or projection on the surface of a forging, often cylindrical in shape.

Bottom draft. Slope or taper in the bottom of a forged depression that tends to assist metal flow toward the sides of depressed areas.

Bow. Longitudinal curvature.

Box annealing. A heat-treating process whereby metal to be annealed is packed in a closed container to protect its surfaces from oxidation. Sometimes used to describe the process of placing forgings in a closed container immediately after forging operations are completed, permitting forgings to cool slowly.

Brinell hardness. The hardness of a metal or part, as represented by the number obtained from the ratio between the load applied on and the spherical area of the impression made by a steel ball forced into the surface of the material tested.

Brinell hardness testing. Method of determining the hardness of materials; involves impressing a hardened ball of specified diameter into the material surface at a known pressure (10-mm ball, 500-kg load for aluminum alloys). The Brinell hardness number results from calculations involving the load and the spherical area of the ball impression. Direct-reading testing machines designed for rapid testing are generally used for routine inspection of forgings, and as a heat treat control function.

Broken surface. Surface fracturing, generally most pronounced at sharp corners.

Buffing. A light polish produced by use of fine abrasives applied by cloth wheels running at high speed.

Burr. A thin ridge or roughness left on forgings by cutting operation such as slitting, shearing, trimming, blanking, or sawing.

Buster (preblocking impression). A type of die impression sometimes used to combine preliminary forging operations such as edging and fullering with the blocking operation to eliminate blows.

Carbonitriding. A process of case hardening a ferrous material in a gaseous atmosphere containing both carbon and nitrogen.

Carbon steel. Steel that owes its properties chiefly to various percentages of carbon without substantial amounts of other alloying elements.

Carburizing (carburization). Adding carbon to the surface of low-carbon steel by heating the metal below its melting point (usually 1600 to 1800 °F) while in contact with carbonaceous solids, liquids, or gases.

Case. The surface layer of an alloy that has been made substantially harder than the interior by some form of hardening operation.

Case hardening. A heat treatment or combination of processes in which the surface layer of a ferrous alloy is made substantially harder than the interior. Carburizing, cyaniding, nitriding, and heating and quenching techniques are commonly used. Case hardening can provide a hard, wear-resistant surface on a forging, while retaining a softer, tougher core.

Cast. See *Die proof*

Cavity, die. The machined recess in a die that gives the forging its shape.

Centering arms. In ring rolling, externally mounted rolls, adjusted to the outside diameter of the ring during rolling. The rolls maintain and guide the ring in a centerline position to achieve roundness.

Ceramic fiber. A lightweight, soft fiber available in blanket and other forms in various temperature grades up to 3000 °F for insulating furnaces, producing quick heating due to low thermal conductivity.

Chamfer. Break or remove sharp edges or corners of forging stock by means of straight angle tool or grinding wheel.

Charpy test. A pendulum-type impact test where the specimen is supported as a simple beam and is notched opposite the point of impact. The energy required to break the beam is used as an index of impact strength measurement.

Check. Crack in a die impression, generally due to forging pressure and/or excessive die temperature. Die blocks too hard for the depth of the die impression have a tendency to check or develop cracks in impression corners.

Chip-mill. An intermediate inspection and repair operation in which surface defects in forgings are located and removed by means of chipping hammers, rotor mills, and similar tools (not to be confused with final inspection, where similar operations are performed).

Chipping. A method for removing seams and other surface defects with a chisel or gouge, so that the defects will not be worked into the finished product.

Chisel. Forging tool used to cut metal by notching. Cold chisels are used to notch cold metal so that it can be broken by a hammer blow; hot chisels are often used to make a complete cut in hot metal.

Chop. A die forging defect; metal sheared from a vertical surface and spread by the die over an adjoining horizontal surface.

Chucking lug. A lug or boss to the forging so that "on center" machining and forming can be performed with one setting or chucking; this lug is machined or cut away on the finished item.

Cleaning. The process of removing scale, oxides, or lubricant—acquired during heating for forging or heat treating—from the surface of the forging. (See also *Blasting, Pickling, Tumbling*.)

Close-tolerance forging. A forging held to unusually close dimensional tolerances. Often little or no machining is required after forging.

Closed die forging. See *Impression die forging*.

Closure, die. A term frequently used to mean variations in thickness of a forging.

Cogging. The reducing operation in working the ingot into a billet by the use of a forging hammer or a forging press.

Coin sizing. A cold squeezing operation for refining face distance dimensions on forgings.

Coin straighten. a combination coining and straightening operation performed in special cavity dies so designed as to also impart a specific amount of working in specified areas of the forging to relieve stresses developed during heat treatment.

Coining. The process of applying necessary pressure to all or some portion of the surface of a forging to obtain closer tolerances or smoother surfaces or to eliminate draft. Coining can be done while forgings are hot or cold and is usually performed on surfaces parallel to the parting line of the forging.

Coining dies. Dies in which the coining or sizing operation is performed.

Cold coined forging. A forging that has been restruck cold in order to hold closer face distance tolerances, sharpen corners or outlines, reduce section thickness, flatten some particular surface, or, in non-heat-treatable alloys, increase hardness.

Cold heading. Working metal at room temperature in such a manner that the cross-sectional area of a portion or all of the stock is increased.

Cold inspection. A visual (usually final) inspection of the forgings for visible defects, dimensions, weight, and surface condition at room temperature. The term may also be used to describe certain nondestructive tests, such as magnetic particle, dye penetrant, and sonic inspection.

Cold shut. A defect characterized by a fissure or lap on the surface of a forging that has been closed without fusion during the forging operation.

Cold trimming. Removing flash or excess metal from the forging in a trimming press when the forging is at room temperature.

Cold working. Permanent plastic deformation of a metal at a temperature below its recrystallization point—low enough to produce strain hardening.

Compression strength. The maximum load per unit of cross-sectional area obtained, before plastic deformation or rupture, by compressing.

Concavity. A concave condition applicable to the width of any flat surface.

Concentricity. Adherence to a common center.

Conventional forging. A forging characterized by design complexity and tolerances that fall within the broad range of general forging practice.

Core. The softer interior portion of an alloy piece that has been surface (case) hardened; or, that portion of a forging removed by trepanning or punching.

Counterblow forging equipment. A category of forging equipment wherein two opposed rams are activated simultaneously, striking repeated blows on the workpiece at a midway point. Action may be vertical, as in the case of counterblow forging hammers, or horizontal, as with an "impacter."

Counterlock. A jog in mating surfaces of dies to prevent lateral die shift caused by side thrust during forging of irregularly shaped pieces.

Crank. Forging shape generally in the form of a "U" with projections at more or less right angles to the upper terminals. A crank shape; crank shapes are designated with the number of throws—such as one, two, or three throw cranks (1K, 2K, 3K).

Critical point. The temperature in metal at which recrystallization or other phase transformation takes place.

Critical temperature. The temperature at which allotropic transformation (changes in structure) takes place in metal.

Critical (temperature) range. Temperatures at which changes in the phase of a metal take place. Changes are determined by absorption of heat when the metal is heated, and liberation of heat when it is cooled.

Crop end. See *End loss*.

Cross. Forged shape of a general four-pointed star or cross; may have hole in center. If one arm is much longer, the shape is termed "Y." Abbreviation is "C."

Cross forging. Preliminary working of forging stock in alternate planes, usually on flat dies, to develop mechanical properties, particularly in the center portions of heavy sections.

Cross hatch. Light broken surface; see also *Broken surface*.

Cross, long. Cross-shaped forging with two opposite arms longer than the other two. Abbreviation is "LC."

Cutoff. A pair of blades positioned in dies or equipment (or a section of the die milled to produce the same effect as inserted blades) used to separate the forging from the bar after forging operations are completed. (Used only when forgings are produced from relatively long bars instead of from individual, precut multiples or blanks.)

Cutters. Cutters are used with power hammers, instead of chisels. They often have long, straight blades, but sometimes the blades are curved or in the shape of a 90° angle; blades are attached to handles of varying lengths.

Cutting. Cutting stock to specified length or weight on circular saws, band hacksaws, or shear presses.

Cyaniding. A process for surface hardening by absorption of carbon or nitrogen by an iron-base alloy brought about by heating to a suitable temperature in contact with a cyanide salt, followed by quenching.

Decarburization. The loss of carbon from the surface of steel by heating above lower critical temperature or by chemical action. Decarburization is usually present to a slight extent in steel forgings. Excessive decarburization can result in defective products.

Descaling. The process of removing oxide scale from heated stock prior to or during forging operations, using such means as extra blows, wire brushes, scraping devices, or water spray.

Die cavity. The machined recess that gives a forging its shape.

Die check. A form of die wear, die check is a crack in a die impression due to forging and thermal strains at relatively sharp corners; upon forging, these cracks become filled with metal, producing sharp ragged edges on the part. Usual die wear is the gradual enlarging of the die impression due to erosion of the die material, generally occurring in areas subject to repeated high pressures during forging operations.

Die forging. A forging that is formed to the required shape and size by working in machined impressions in specially prepared dies.

Die impression. The portion of the die surface that shapes the forging.

Die layout. The transfer of the forging drawing or sketch dimensions to templates or die surfaces for use in sinking dies.

Die life. The productive life of a die impression, usually measured in terms of the number of forgings produced before the impression has worn beyond the permitted tolerances.

Die line. A line or scratch resulting from the use of a roughened tool or the drag of a foreign particle between tool and product.

Die lock. For locked dies, a dimension expressing extreme variation in parting line level measured in a direction parallel to ram stroke.

Die lubricant. A compound sprayed, swabbed, or otherwise applied on die surfaces or forging during forging to reduce friction between the forging and the dies. Lubricants may also ease release of forgings from the dies and provide thermal insulation.

Die match. The condition of dies, after having been set up in the forging equipment, where every point in one die half is within specified alignment with every point in the mating die half.

Die number. The number assigned to a die for identification and cataloging purposes, usually the same number that is assigned for the same purpose to the product made from that die.

Die proof (cast). A casting of the die impression made to confirm the exactness of the impression.

Die set. Two (or, for a mechanical upsetter, three) machined dies to be used together during the production of a die forging.

Die shift. The condition occurring after the dies have been set up in the forging unit, and in which a portion of the impression of one die is not in perfect alignment with the corresponding portion of the other die. This results in "mismatch" in the forging, a condition that must be held within the specified tolerance.

Die sinker. A skilled toolmaker who machines die impressions.

Die sinking. Machining the die impressions for producing forgings of required shapes and dimensions.

Die straighten. A straightening operation performed in either a hammer or a press using flat or cavity dies to remove undesired deformation and bring the forging within the straightness tolerance.

Dies (die blocks). The metal blocks into which forging impressions are machined and from which forgings are produced.

Dies, forging. Forms for the making of forgings; generally consist of a top and bottom die. The simplest will form a completed forging in a single impression; the most complex, made up of several die inserts, may have a number of impressions for the progressive working of complicated shapes. Forging dies are usually in pairs, with part of the impression in one of the blocks and the balance of the impression in the other block.

Dies, gripper. Clamping or lateral dies used in a forging machine or mechanical upsetter.

Discontinuities. Includes cracks, laps, folds, cold shuts, and flow-through, as well as internal defects such as inclusion, segregation, and porosity; internal discontinuities can be detected and evaluated using ultrasonic testing equipment.

Disk. Blanks for gears, rings, or hubs are examples of this type of forging; parts may or may not have holes. Abbreviation is "D."

Double forging. A forging designed to be cut apart and used as two separate pieces.

Dowel. A metal insert placed between mating surfaces of the die shank and die holder in the forging equipment to ensure lengthwise die match.

Draft. The amount of taper on the sides of the forging necessary for removal of the workpiece from the dies. Also, the corresponding taper on the side walls of the die impression.

Draft angle. The angle of taper, expressed in degrees, given to the sides of the forging and the side walls of the die impression.

Draftless forging. A forging with zero draft on vertical walls.

Draw stock. The forging operation in which the length of a metal mass (stock) is increased at the expense of its cross section; no "upset" is involved. The operation covers converting ingot to pressed bar using "V," round, or flat dies.

Drawing. A forging operation in which the cross section of forging stock is reduced and the stock lengthened between flat or simple contour dies. See also *Fuller*.

Drawing (tempering). A heat-treating process where metal is reheated, after hardening or normalizing, to a temperature below the lower limit of the critical range, then cooled to secure desired properties—particularly toughness. Tool hardeners generally prefer the term "tempering."

Drop forging. A forging produced by hammering metal in a drop hammer between dies containing impressions designed to produce the desired shape. See also *Impression die forging*.

Drop hammer. A term generally applied to forging hammers wherein energy for forging is provided by gravity, steam, or compressed air. See also *Air-lift hammer, Board hammer, Steam hammer*.

Dye penetrant testing. Inspection procedures for detecting surface irregularities using penetrating liquids containing dyes or fluorescent substances. See also *Zyglo*.

Edger (edging impression). The portion of the die impression that distributes metal, during forging, into areas where it is most needed to facilitate filling the cavities of subsequent impressions to be used in the forging sequence. See also *Fuller*.

Edging. The forging operation of working a bar between contoured dies while turning it 90° between blows to produce a varying rectangular cross section.

Elastic limit. The maximum stress a metal will withstand without permanent deformation.

Elongation. The amount of permanent stretch in a tensile test specimen before rupture. It is usually expressed as a percentage change of the original gage length, such as 25% over a 2-in. gage length.

End loss (crop end). Bar end left over after cutting bar lengths of stock into forging multiples. See also *Multiple*.

Endurance limit. A limiting stress, below which metal will withstand without fracture an indefinitely large number of cycles of stress; above this limit, failure occurs by the generation and growth of cracks until fracture results.

Extrusion. The process of forcing metal to flow through a die orifice in the same direction in which energy is being applied (forward extrusion); or in the reverse direction (backward extrusion), in which case the metal usually follows the contour of the punch or moving forming tool. The extrusion principle is used in many impression die forging applications.

F.A.O. An abbreviation of "finish all over"; it designates that a forging must have sufficient size over the dimensions given on the drawing so that all surfaces may be machined in order to obtain the dimensions shown on the drawing. The amount of additional stock necessary for machining allowance depends on the size and shape of the part, and is agreed on by the vendor and the user.

Fatigue. The progressive fracture of a metal by means of a crack that enlarges under repeated cycles of stress.

Feather (fin). The thin projection formed on a forging by trimming or when the metal under pressure is forced into hairline cracks or die interfaces.

Fiber. A characteristic of wrought metal, including forgings, indicated by a fibrous or woody structure of a polished and etched section, and indicating directional properties. Fiber is chiefly due to the extension of the constituents of the metal synonymous with flow lines and grain flow in the direction of working.

Fillet. The concave intersection of two surfaces. In forging, the desired radius at the concave intersection of two surfaces is usually specified.

Finish. The material machined off the surface of a forging to produce the finish machine component. Also, the surface condition of the component resulting from machining.

Finish all over (F.A.O.). Specification designating that the forgings must be made sufficiently larger than dimensions shown to permit machining on all surfaces to given sizes.

Finish allowance. Amount of stock left on the surface of the forging for machining.

Finish forging. See *Conventional forging*.

Finisher or finishing impression. The die impression that imparts the final shape to the forging.

First block, second block, and finish. The forging operation in which the part to be forged is passed in progressive order through three tools mounted in one forging machine; only one heat is involved for all three operations.

Flame hardening. A process of surface hardening a ferrous alloy by heating it above the transformation range with a high-temperature flame, followed by rapid cooling.

Flange. See *Rib*.

Flash. Necessary metal in excess of that required to completely fill the finishing impression of the dies. Flash extends out from the body of the forging as a thin plate at the line where the dies meet and is subsequently removed by trimming. Cooling faster than the body of the component during forging, flash can serve to restrict metal flow at the line where dies meet, thus ensuring complete filling of the finishing impression.

Flash extension. Portion of flash remaining after trimming. Flash extension is measured from the intersection of the draft and flash at the body of the forging to the trimmed edge of the stock.

Flash land. Configuration in the finishing impression of the dies designed either to restrict or encourage growth of flash at the parting line, whichever may be required in a particular instance in order to ensure complete filling of the finishing impression.

Flash line. See *Parting line*.

Flash pan. Machined-out portion of dies to permit flow-through of excess metal.

Flat die forging (open die forging). Forging worked between flat or simple contour dies by repeated strokes and manipulation of the workpiece. Also known as "hand" or "smith" forging.

Flattening. The forging operation of flattening the forging stock prior to further working.

Flatter. Forging tool used to make a smooth, flat surface. See also *Set hammer*.

Flop forging. A forging in which the top and bottom die impressions are identi-cal, permitting the forging to be turned upside down during the forging operations.

Flow lines. Patterns in a forging resulting from the elongation of nonhomogeneous constituents and the grain structure of the material in the direction of working during forging; usually revealed by macroetching. See also *Grain flow*.

Flow-through. A forging defect caused by metal flow past the base of a rib with consequent rupture of the grain structure.

Flying rolling mandrel. A rolling mandrel not supported at the top for rolling rings with lower rolling forces. An increased production rate is achieved by omitting the use of the backing arm.

Fold. A forging defect caused by folding the metal back on its own surface during its flow in the die cavity.

Forgeability. Term used to describe the relative ability of material to deform without rupture. Also describes the resistance to flow from deformation.

Forging. The product of work on metal formed to a desired shape by impact or pressure in hammers, forging machines (upsetters), presses, rolls, and related forming equipment. Forging hammers, counterblow equipment, and high-energy-rate forging machines impart impact to the workpiece, while most other types of forging equipment impart squeeze pressure in shaping the stock. Some metals can be forged at room temperature, but the majority of metals are made more plastic for forging by heating.

Forging machine (upsetter or header). A type of forging equipment, related to the mecahnical press, in which the main forming energy is applied horizontally to the workpiece, which is gripped and held by prior action of the dies.

Forging plane. The plane that includes the principal die face and that is perpendicular to the direction of the ram travel. When parting surfaces of the dies are flat, the forging plane coincides with the parting line. See also *Parting plane*.

Forging quality. Term describing stock of sufficiently superior quality to make it suitable for commercially satisfactory forgings.

Forging rolls. Power-driven rolls with shaped contours and notches for introduction of the work, used in preforming operations.

Forging stock. A wrought rod, bar, or other section suitable for subsequent change in cross section by forging.

Forging strains. Strains that have been set up in the metal by the process of forging; they are usually relieved by subsequent annealing or normalizing.

Former. Part of a master used in machining impressions in dies. See *Master, Model, Template.*

Forming. A process whereby planes of a definite shape are changed without materially changing the cross section.

Forming dies. Dies in which a rough impression has been machined or gouged, for use between the flat dies of a steam hammer; used when the quantity of forgings required does not warrant the cost of drop forging dies and a closer shape than can be obtained with flat dies.

Foundation. The mass of structural material on which forging equipment is placed to support the weight and to absorb residual energy of the forging operation.

Fracture test. Examination of the broken surface of a test specimen or forging to determine the structure of the metal or certain of its properties.

Fuller (fullering impression). Portion of the die that is used in hammer forging primarily to reduce the cross section and lengthen a portion of the forging stock. The fullering impression is often used in conjunction with an edger (or edging impression).

Gate (sprue). A portion of the die that has been removed by machining to permit a connection between multiple impressions or between an impression and the bar of stock.

Gathering stock. Any operation whereby the cross section of a portion of the forging stock is increased above its original size.

German die. A die in which a rough impression has been machined or gouged and used between the flat dies of the steam hammer. Primarily used to obtain a relative closer shape in the forging than can be obtained with flat dies when the quantity of forgings required is not sufficient to permit investment in production forging dies.

Gouge. A gross type of scratch.

Grain. The characteristic crystalline structural unit of metals and alloys.

Grain flow. Fiber-like lines appearing on polished and etched sections of forgings that are caused by orientation of the constituents of the metal in the direction of working during forging. Grain flow produced by proper die design can improve required mechanical properties of forgings.

Grain separation. In forging aluminum, rapid metal flow sometimes causes a separation or rupture of grain. Metal flow is affected by lubricant, die and metal temperature, part shape, alloy, and hammer operator technique; consequently, any one or combination of these factors can cause grain separation. The irregular crevices are seldom more than a few thousandths of an inch deep and can be removed by grinding or polishing.

Grain size. The average size of the crystals or grains in a metal as measured against an accepted standard.

Gravity hammer. A class of forging hammer wherein energy for forging is obtained by the mass and velocity of a freely falling ram and the attached upper die. Examples are board hammers and air-lift hammers.

Grinding. Process of removing metal by abrasion from bar or billet stock to prepare stock surfaces for forging. Occasionally used to remove surface irregularities and flash from forgings.

Gripper dies. The lateral or clamping dies used in a mechanical upsetter.

Guide. The parts of a drop hammer or press that guide the up-and-down motion of the ram in a true vertical direction.

Gutter. A depression around the periphery of the die impression outside the flash pan that allows space for the excess metal; surrounds the finishing impression and provides room for the excess metal used to ensure a sound forging. A shallow impression outside the parting line.

H-shape forging. A forging in the approximate form of an "H."

Hammer forging. A forging that is made on the flat die of a steam hammer. A forged piece produced in a forging hammer, or the process of forming such a piece. See also *Board hammer, Power-drive hammer, Rope hammer.*

Hand forge (smith forge). The forging operation in which the forming is accomplished on dies that are generally flat. The piece is shaped roughly to the required contour with little or no lateral confinement; operations involving mandrels are included.

Hand forging. A forging made by hand on an anvil or under a power hammer without dies containing an exact finishing impression of the part. Such forgings approximate each other in size and shape but do not have the commercial exactness of production die forgings. Used where the quantity of forgings required does not warrant expenditure for special dies, or where the size or shape of the piece is such as to require means other than die forging. A forging worked between flat or simply shaped dies by repeated strokes and manipulation of the piece. Also known as smith forging or flat die forging.

Hand straightening. A straightening operation performed on a surface plate to bring a forging within the straightness tolerance. Frequently, a bottom die from a set of finish dies is used instead of a surface plate. Hand tools used include mallets, sledges, blocks, jacks, and oil gear presses in addition to regular inspection tools.

Handling holes. Holes drilled in opposite ends of the die block to permit handling by the use of a crane or bar.

Handling marks. Nicks and gouges formed on forgings if improperly handled; most prevalent for forgings in the as-forged condition prior to heat treatment.

Hardenability. In ferrous and age-hardenable alloys, the property that determines the depth and distribution of hardness induced by quenching.

Hardening. A heat treatment consisting of heating an alloy to a temperture within or above the critical range, maintaining that temperature for the prescribed time (usually 15 to 30 min), then quenching or otherwise rapidly cooling. For age-hardening alloys, a two-stage process consisting of solution heat treatment and aging.

Hardie. Forging tool resembling a chisel, except that it is supported in the anvil and the metal to be cut rests on its cutting edge.

Hardness. (1) General term, covering the resistance of metal to plastic deformation by force. (2) Hardness numbers obtained by use of any of the several hardness tests for metals.

Hardness testing. See *Brinell hardness testing, Rockwell hardness testing, Scleroscope hardness testing.*

Header. See *Forging machine.*

Heat (forging). Amount of forging stock placed in a batch-type furnace at one time.

Heat of metal. The quantity of material manufactured from one melt at the metal producer's facility. Metal from a single heat is extremely uniform in chemical analysis.

Heat treatment. A combination of heating, holding, and cooling operations applied to a metal or alloy in the solid state to produce desired properties.

Heat-treat stain. Discoloration of the metal surface caused by oxidation during thermal heat treatment.

Helve hammer. A power hammer in which power is delivered through a helve or handle; used in light work, tool making, and supplementary operations.

High-energy-rate forging (high-velocity or high-speed forging). The process of producing forgings on equipment capable of extremely high ram velocities resulting from the sudden release of a compressed gas against a free piston.

Hog out. A product machined from bar stock or from a hand forging rather than from an impression die forging. The process is commonly known as "hogging out" material.

Homogenizing. See preferred term, *Preheating.*

Hot inspection. An in-process visual examination of forgings, using gages, templates, or other nondestructive inspection equipment to ensure quality.

Hot isostatic pressing (HIP). The consolidation of encapsulated metal powder at elevated temperature by high-pressure inert gas.

Hot stamp. Impressing markings in a forging while the forging is in the heated, plastic condition.

Hot working. The mechanical working of a metal at a temperature above its recrystallization point—a temperature high enough to prevent strain hardening.

Hub. A boss that is in the center of the forging and forms a part of the body of the forging.

Hydraulic press. A forging press with a hydraulically operated ram.

Impact extrusion. A relatively rapid extrusion. See also *Extrusion.*

Impact testing. Tests to determine the energy absorbed in fracturing a test bar at high velocity. See also *Charpy test, Izod test.*

Impression. A cavity machined into a forging die to produce a desired configuration in the workpiece during forging.

Impression die forging. A forging that is formed to the required shape and size by machined impressions in specially prepared dies that exert three-dimensional control on the workpiece.

Inclusion. Impurities in metal, usually in the form of particles in mechanical mixture.

Induction hardening. Process of hardening the surface of a forging by heating it above the transformation range by electrical induction, followed by rapid cooling.

Insert. A piece of steel that is removable from a die. The insert may be used to fill a cavity, or to replace a portion of the die with a grade of steel better suited for service at that point.

Insert, die. A relatively small die containing part or all of the impression of a forging, and which is fastened to the master die block.

Inspection. The process of checking a forging for adherence to standards given in the specifications.

Ironing. (1) A press operation used to obtain a more exact alignment of the various parts of a forging, or to obtain a better surface condition. (2) An operation to increase the length of a tube by reduction of wall thickness and outside diameter. See also *Coining, Swaging.*

Isothermal annealing. A process of heating ferrous material above its critical temperature, then cooling to and holding a fixed temperature until transformation to a desired microstructure.

Izod test. A pendulum-type impact test in which the specimen is supported at one end as a cantilever beam and the energy required to break off the free end is used as a measure of impact strength.

Jominy. A hardenability test for steel to determine the depth of hardening obtainable by a specified heat treatment.

Key. A wedge used to secure dies into the forging equipment.

Knockout mark. A small protrusion, such as a button or ring of flash, resulting from the depression of a knock out pin from the forging pressure, or the entrance of metal between the knockout pin and the die.

Knockout pin. A power-operated plunger installed in a die to aid removal of the finished forging.

Lap. A surface defect appearing as a seam, caused by the folding over of hot metal, fins, or sharp corners and by subsequent rolling or forging (but not welding) of these into the surface. See also *Fold.*

Layout. (1) Transferring drawing or sketch dimensions to templates or dies for use in sinking dies. (2) A detailed inspection operation in which significant dimensions of a forging are checked against blueprint specifications.

Layout sample. A plaster, lead, or forged alloy sample taken from new dies to verify accuracy by layout and precise measurement. See also *Cast.*

Lead proof. A reproduction in lead, or a lead alloy, of the die impression, obtained by clamping the two dies together in alignment and pouring molten metal into the finish impression.

Liners. Thin strips of metal inserted between the dies and the units into which the dies are fastened.

Lock. One or more changes in the plane of the mating faces of the dies. In a compound lock, two or more changes are in the mating faces. A counterlock is a lock placed in the dies to offset a tendency for die shift caused by a necessary steep lock, a condition in which the parting line is not all in one plane.

Locked dies. Dies with mating faces that lie in more than one plane.

Lubricant residue. The carbonaceous residue resulting from lubricant burned on the surface of the forged part.

L shape. Right-angle pieces, or those similar to a crank arm.

L, spread. When projections of an "L" shape are not necessarily at 90° angles, when angles vary, or when a cross shape has adjacent arms that are longer than the other two, it becomes a spread L. Abbreviation is "SL."

Loose material. During forging operations, pieces of flash often break loose, necessitating cleaning of the dies between forging blows; this is usually accomplished by lubricating the die while air is blown on it. Insufficient cleaning results in pieces of flash becoming imbedded in the surface of the forging. Such forgings are often salvaged by removing the loose pieces and hot reforging to fill out the depressions.

Machinability. That property of metal which governs the ease with which metal may be removed by chip formation.

Machine forging (upsetter forging). The process of forging in a forging machine (upsetter), in which the metal is moved into the die impression by pressure applied in a horizontal direction by the moving die in the ram.

Machining allowance. See *Finish allowance.*

Macroetch. A testing procedure for conditions such as porosity, inclusions, segregations, caburization, and flow lines from hot working. After applying a suitable etching solution to the polished metal surface, the structure revealed by the action of the reagent can be observed visually.

Macrograph. A photographic reproduction of any object that has been magnified not more than 10 diameters.

Macrostructure. The structure and internal condition of metals as revealed on a polished and etched sample, examined either by the naked eye or under low magnification (up to 10 diameters).

Magnaflux® test. See *Magnetic particle testing.* Trade name of Magnaflux Corp.

Magnetic particle testing. A nondestructive test method of inspecting areas on or near the surface of ferromagnetic materials. The metal is magnetized, then iron powder is applied. The powder adheres to lines of flux leakage, revealing surface and near-surface discontinuities. Magnetic particle testing is used for both raw material acceptance testing and product inspection. Quality levels are usually agreed on in advance by the producer and purchaser.

Magnaglo®. A type of magnetic particle testing where the magnetic powder is fluorescent and the inspection is performed under black light. See also *Magnetic particle testing*. Trade name of Magnaflux Corp.

Mandrel forging. The process of rolling and forging a hollow blank over a mandrel in order to produce a weldless, seamless ring or tube.

Martempering. The process of quenching an austenitized ferrous alloy in a medium at a temperature in the upper portion of the temperature range of martensite formation, or slightly above that range, and holding in the medium until the temperature throughout the alloy is substantially uniform. The alloy is then allowed to cool in air throughout the temperature range of martensite formation.

Master. Wood, metal, or plastic reproduction of one side of a proposed forged shape, used to control cutters on tracer-controlled die sinking equipment. See also *Former, Plaster*.

Master block. A forging die block primarily used to hold insert dies.

Match. A condition in which a point in one die-half is aligned properly with the corresponding point in the opposite die-half within specified tolerance.

Matched edges (match lines). Two edges of the die face that are machined exactly at 90° to each other, and from which all dimensions are taken in laying out the die impression and aligning the dies in the forging equipment.

Matching draft. Increased draft used on the shallow side of a forging to match its surface at the parting line with a similar surface of less draft on the deeper side.

Mechanical press. A forging press with an inertia flywheel and with a crank and clutch or other mechanical device to operate the ram.

Mechanical properties. Those properties of a material that reveal the elastic and inelastic reaction when force is applied, or that involve the relationship between stress and strain; for example, the modulus of elasticity, tensile strength, and fatigue limit. Mechanical properties are dependent on chemical composition, forging, and heat treatment.

Mechanical upsetter. A three-element forging press, with two gripper dies and

a forming tool, for flanging or forming relatively deep recesses.

Mechanical working. Subjecting metal to pressure, exerted by rolls, hammers, or presses, in order to change the shape or physical properties of the metal.

Metal discontinuities. See *Discontinuities, metal*.

Microstructure. The structure and internal condition of metals as revealed on a ground and polished (and sometimes etched) surface when observed at high magnification (over 10 diameters).

Mismatch. The misalignment or error in register of a pair of forging dies; also applied to the condition of the resulting forging. The acceptable amount of this displacement is governed by blueprint or specification tolerances. Within tolerances, mismatch is a condition; in excess of tolerance, it is a serious defect. Defective forgings may be salvaged by hot reforging operations.

Model. See *Former, Master, Plaster*.

Modulus of elasticity. The ratio, within the limits of elasticity, of the stress to the corresponding strain.

Multiple. A piece of stock for forging that is cut from bar or billet lengths to provide the exact amount of material needed for a single workpiece.

Natural draft. Taper on the sides of a forging, due to its shape or position in the die, that makes added draft unnecessary.

Nitriding. Producing surface hardness in ferrous metals by adding nitrogen to the outside layer while heating the metal, in contact with ammonia gas or other suitable nitrogenous material, below the critical temperature range.

No-draft forging. A forging with extremely close tolerances and little or no draft, requiring a minimum of machining to produce the final part. Mechanical properties can be enhanced by closer control of grain flow and retention of surface material in the final component.

Nonferrous. Metals or alloys that contain no appreciable quantity of iron; applied to such metals as aluminum, copper, magnesium, and their alloys.

Nonfill (underfill). Occurs when the die impression is not completely filled with metal. Some causes are: improper distribution of metal in preforming operations such as fullering, edging, and blocking; excessive removal of material by chipping defects prior to finish forging; improper lubrication of die impression; low forging pressure; rough or uneven die finish; inadequate hammer or press capacity.

Normalizing. A heat treatment in which ferrous alloys are heated to approximately 100 °F above the critical range, holding that temperature for the required time, and cooling to room temperature in air.

Notch sensitivity. The reduction in the impact, endurance, or static strength of a metal, caused by stress concentration as a result of scratches or other stress raisers on the surface.

Off gage. Deviation of thickness or diameter of a solid product beyond the standard or specified dimensional tolerances.

Oil stain. A stain produced by the incomplete burning of lubricant on the surface of a product.

Open die forging. See *Flat die forging*.

Optical pyrometer. An optical viewing device used to measure elevated temperature.

Orange peel. A surface roughening encountered in forming products from material that has a coarse grain size.

Overheating. Can occur in preheat furnaces prior to forging or in the heat-treating operation. The condition results when metal temperature exceeds the critical temperature of the alloy involved and a change in phase occurs; this is also known as the transformation temperature. Externally, overheated material will often form blisters or a web of fine cracks; internally, overheating causes precipitation of melted constituents around grain boundaries and the formation of rounded pools of melted constituents often called "rosettes."

Overetch. In the normal processing of aluminum forgings, a caustic etch operation is employed for the dual purpose of cleaning parts and emphasizing defects to facilitate visual inspection. Immersion of parts for too long or use of a too concentrated solution will produce a rough, slightly pitted surface.

Parting. The line around the periphery of a forging at which the flash has been forced out of the impression.

Parting line. The dividing plane between the two dies used in forging metal. The line along the edge of a forging where the dies meet, or the line along the corresponding edge of the die impression.

Pickling. The process of removing oxide scale from forgings by treating in a heated acid bath.

Pick-up. Small particles of oxidized metal adhering to the surface of a product.

Pierce. In ring rolling, the process of providing a through hole in the center of an upset forging as applied to ring blank preparation.

Pipe. A cavity formed in metal (especially ingots) during the solidification process by the contracting of that part of the liquid metal which is the last to solidify.

Pit. A sharp depression or hole in the surface of metal.

Plan view area. The area of the plan view of a forging; sometimes used to indicate the relative size of the forging.

Plane, forging. The plane that includes the principal die face and that is perpendicular to the direction of the ram stroke; when the parting is flat, the forging plane coincides with the parting line. See also *Forging plane*.

Planishing. A finishing operation to remove the trim line of a forging or to obtain closer tolerances. Usually done by hot or cold rolling, pressing, or hammering.

Plaster cast. See *Lead cast*.

Plastic deformation. The ability of a material to permanently distort without fracture under the action of an applied stress.

Platter. The entire mass of metal upon which the hammer performs work, including the flash, sprue, tonghold, and as many forgings as are made at a time.

Plug. A protruding portion of a die impression for forming a corresponding recess in the forging.

Poisson's ratio. The ratio of lateral unit deformation to longitudinal unit deformation within the elastic limit during a uniaxial tension or compression test. Also called the "factor of lateral contraction"; a body that is stretched lengthwise becomes thinner crosswise. Poisson established by experiment that, within the elastic limit, the ratio of the length of stretch or squeeze to the length by which a body of given material is decreased or increased in crosswise thickness is a constant; for aluminum, Poisson's ratio is an average of approximately 0.33.

Polish (stock preparation). Grind or polish surfaces to remove scratches, scars, and marks left by cutting equipment; operation usually performed by a flexible shaft machine with an abrasive disk.

Polishing. A mechanical finishing operation to apply a gloss or luster to the surface of a product.

Power-driven hammer. A forging hammer with a steam or air cylinder for raising the ram and augmenting its downward blow.

Precision forging. A forging produced to closer tolerances than normally considered standard by the industry.

Preform. The forging operation in which stock is preformed or shaped to a predetermined size and contour prior to subsequent die forging operations; the operation may involve drawing, bending, flattening, edging, fullering, rolling, or upsetting. The preform operation is not considered to be a scheduled operation unless a separate heat is required; usually, when a preform operation is required, it will precede a forging operation and will be performed in conjunction with the forging operation and in the same heat. In ring rolling, a term generally applied to ring blanks of a specific shape to be used for profile (contour) ring rolling.

Preheating. A high-temperature soaking treatment used to change the metallurgical structure in preparation for a subsequent operation, usually applied to the ingot.

Preparation charge. A one-time charge covering the cost of sinking dies and preparing required auxiliary tooling for producing forgings to a particular design. In usual practice, this charge conveys to the customer the exclusive right to purchase forgings produced on this tooling. The dies themselves are the property of the forger, who also has the responsibility for maintaining and replacing the dies as required for satisfactory production of forgings.

Prepierce. In ring rolling, a vertically mounted piercing (punching) tool used for preparation of ring blanks on the ring blank press. A tapered tool of various diameters and lengths.

Press forging. The shaping of metal between dies by mechanical or hydraulic pressure. Usually this is accomplished with a single work stroke of the press for each die station.

Process annealing. Heating iron-base alloys to a temperature at, or close to, the lower limit of the critical range and then cooling as desired, usually for stress relief.

Profile (contour) rolling. In ring rolling a process to produce seamless rolled rings with a predesigned shape either on the O.D. or I.D., requiring less volume of material and less machining to produce finished parts.

Proof. Any reproduction of a die impression in any material, frequently a lead or plaster cast. See also *Die proof*.

Proportional limit. The greatest stress that a material is capable of sustaining without a deviation from the law of proportionality of stress to strain.

Punch. A shearing operation to remove a section of metal as outlined by the inner parting line in a blocked or finished forging; the operation is generally performed on a trim press using a punch die. A tool used in punching holes in metal. The movable die in a press or forging machine.

Punchout. A pierced hole in a forging.

Punchout rigging. The parts required to fasten the punch and plates to the press.

Pusher furnace. A continuous-type furnace where stock to be heated is charged at one end, carried through one or more heating zones, and discharged at the opposite end.

Quantity tolerance. Allowable variation of quantity to be shipped on a purchase order, agreed on by the forging producer and purchaser when the order is placed. (A schedule of suggested standard quantity tolerances for various order quantities is available from the Forging Industry Association.)

Quench. Rapid cooling of metal from above the critical range in some quenching medium, usually water or oil.

Radial-axial ring rolling mill (RAW). A type of ring forging equipment for producing seamless rolled rings by controlling the outside diameter, the inside diameter, and the ring height (axial height).

Radial ring rolling mill (RW). A type of ring forging equipment for producing seamless rolled rings by controlling only the outside and inside diameters.

Radial roll (main roll, king roll). The primary driven roll of the rolling mill for rolling rings in the radial pass. Roll supported at both ends.

Radial rolling force. The action produced by the horizontal pressing force of the rolling mandrel acting against the ring and the main roll. Usually expressed in metric tons.

Ram. The moving or falling part of a drop hammer or press to which one of the dies is attached; sometimes applied to the upper flat die of a steam hammer.

Radius. To remove sharp edges or corners of forging stock by means of a radius or form tool (radius OE: radius one end; radius BE: radius both ends).

Reducing atmosphere. Combustion in a furnace where there is no excess oxygen or a deficiency of oxygen; also termed "wet fire."

Reduction of area (contraction of area). The difference, in a tension specimen, between the size of the original sectional area and that of the area at the point of rupture. It is generally stated as the percentage of decrease of cross-sectional area of a tension specimen after rupture.

Refining temperature or heat. A temperature employed in heat treating to refine grain structure—in particular, grain size.

Reflectoscope. A nondestructive inspection instrument in which internal quality of forgings or stock is evaluated through the utilization of high-frequency sound.

Refractory. Heat-resistant material, usually nonmetallic, used for furnace linings.

Reheating. A thermal operation designed solely to heat stock for hot working; in general, no metallurgical changes are intended.

Rerolling quality. Rolled billets from which the surface defects have not been removed or completely removed.

Reset. Realign or adjust dies or tools during a production run; not to be confused with "setup," an operation performed prior to a production run.

Resink designation. Identification of a duplicate set of dies made to supplement or replace a die set.

Restrike on draw. Restriking a forging on the tempering heat of a heat treatment to produce closer alignment of sections.

Restrike. A salvage operation following a primary forging operation in which the parts involved are rehit in the same forging die in which the pieces were last forged.

Rib. A forged wall or brace projecting generally in a direction parallel to the ram stroke. See also *Web*.

Ring rolling. The process of shaping weldless rings from pierced disks or thick-walled, ring-shaped blanks between rolls that control wall thickness, ring diameter, height, and contour.

Rockwell hardness testing. A method of determining the relative hardness value of a material by measuring the depth of residual penetration by a steel ball or diamond point under controlled loading.

Roller (rolling impression). The portion of a forging die where cross sections are altered by hammering or pressing while the workpiece is being rotated.

Roll forging. The process of shaping stock between power driven rolls bearing contoured dies. The workpiece is introduced from the delivery side of the rolls, and is reinserted for each succeeding pass. Usually used for preforming, roll forging is often employed to reduce thickness and increase length of stock.

Rolling. The forging operation of working a bar between contoured dies while turning it between blows to produce a varying circular cross section.

Rolling edger. A combined edger and roller, employed for the distribution of metal in preparation for the finishing operation.

Rolling mandrel. In ring rolling, a vertical roll of sufficient diameter to accept various sizes of ring blanks and to exert rolling force on an axis parallel to the main roll.

Rolling table. Serves to carry the ring during the rolling process of the ring rolling mill. The table is horizontally displaceable with the rolling mandrel.

Rough machine. Remove excess or undesired metal from forgings or from forging stock in process by means of machine tools such as lathes and boring mills. The term includes most machining operations other than scalping.

Rope hammer. A gravity-powered forging hammer with ropes for raising the ram and upper die.

Rotary furnace. A circular furnace constructed so that the hearth and workpieces rotate around the axis of the furnace during heating.

Rub mark. A minor form of scratching consisting of areas made up of large numbers of very fine scratches or abrasions.

Ruptured metal. Forging stock, particularly on very thin sections, that has been hammered so severely as to cause broken fibers in the metal.

Saddling (mandrel forging). The process of rolling and forging a pierced disk of stock over a mandrel to produce a weldless ring.

Sand blasting. The process of cleaning forgings by propelling sand against them at high velocity. See also *Blast cleaning*.

Saw trim. The operation of removing flash from blocker or finished forgings by means of bandsaw equipment.

Scale. The oxide film that is formed on forgings, or other heated metal, by chemical action between the surface metal and oxygen in the air.

Scale pit. A surface depression formed on a forging due to scale in the dies during the forging operation.

Scalping. Machining operation in which the outside surface of rolled, pressed, or cast stock is removed to eliminate surface defects.

Scleroscope hardness testing. A method of measuring hardness of metal by the drop and rebound of a diamond-tipped hammer.

Scratch. A visible linear indentation caused by a sharp object passing over the surface.

Screw press. A high-speed press in which the ram is activated by a large screw assembly that is powered by a drive mechanism.

Seam. A longitudinal surface defect in the form of a seam that appears on a forging when opened by the forging action; a crack or inclusion on the surface of a forging. If very fine, termed a hair seam or hair crack.

Semifinisher (semifinishing impression). An impression in the forging die that only approximates the finish dimensions of the forging. Semifinishers are often used to extend die life of the finishing impression, to ensure proper control of grain flow during forging, and to assist in obtaining desired tolerances.

Set hammer. Forming tool used to make smooth, flat surfaces, especially in small areas. See also *Flatter*.

Setup. Preparing equipment or unit for operation; includes miscellaneous rearrangement of auxiliary facilities such as conveyors, skids, and hand tools.

Shank. The portion of the die or tool by which it is held in position in the forging unit or press.

Shear (defect). An indirect result of mismatch, a shearing action can occur: (1) by restriking mismatched forgings, or (2) by restriking in misaligned dies. The first way is more common, as it is generally employed as a remedial action for mismatched parts. The severity of the shear depends on the amount of mismatch on the parts; the acceptance or rejection of parts so treated depends on the resulting effect on forging dimensions. Because forging dies may wander "off match," restriking in misaligned dies can occur at die setup or at any time during the operation.

Shearing. A process of mechanically cutting metal bars to the proper stock length necessary for forging the desired product.

Shipping tolerance. See *Quantity tolerance*.

Shoe. A holder used as a support for the stationary portions of forging and trimming dies.

Shotblasting. A process of cleaning forgings by propelling metal shot at high velocity by air pressure or centrifugal force at the surface of the forgings. See also *Blast cleaning*.

Shrinkage. The contraction of metal during cooling after forging. Die impressions are made oversize according to precise shrinkage scales to allow forgings to shrink to design dimensions and tolerances.

Shrink scale. A measuring scale or rule, used in die layout, on which graduations are expanded to compensate for thermal contraction (shrinkage) of the forging during cooling.

Side thrust. Lateral force exerted between the dies by reaction of the forged piece on the die impressions.

Sinking. The operation of machining the impression of the desired forging into the forging dies.

Sizing. A process employed to control precisely a diameter of rings or tubular components.

Sizing (coining). The operation in a coining press performed in order to obtain closer tolerances on portions of a forging.

Slab. A flat-shaped semifinished rolled metal ingot with a width not less than 10 in. and a cross-sectional area not less than 16 in.^{2}.

Sliver. A slender fragment or splinter that is a part of the material, but that is incompletely attached. A torn fiber of metal forced into the surface of a forging.

Slot furnace. A common batch-type forge furnace where stock is charged and removed through a slot or opening.

Slug. (1) Metal removed when punching a hole in a forging. Also termed "punchout." (2) Forging stock for one workpiece cut to length. See also *Blank*.

Smith forging. See *Flat die forging*.

Smith hammer. Any power hammer where impression dies are not used for the reproduction of commercially exact forgings.

Snag grinding (snagging). The process of removing portions of forgings not desired in the finished product, by grinding.

Snip vents. An operation to remove metal projections resulting from vents in the die cavity and where such an operation is an independently scheduled operation and not performed in conjunction with the forging operation.

Soaking. A heating process during which metal is held at an elevated temperature for the length of time sufficient for the attainment of uniform temperature throughout the material, or for homogenization of elements.

Solution heat treatment. A process in which an alloy is heated to a suitable temperature, held at this temperature long enough to allow a certain constituent to enter into solid solution, and then cooled rapidly to hold the constituent in solution. The metal is left in a supersatur-

ated, unstable state and may subsequently exhibit age hardening.

Sonic testing. See also *Ultrasonic testing*.

Sow block. Metal die holder employed in a forging hammer to protect the hammer anvil from shock and wear. Also known as "anvil cap." Sow blocks are occasionally used to hold insert dies.

Spall. The cracking off or flaking of small particles of metal from the surface.

Special tolerance. Any tolerance that is closer or wider than "standard."

Sprue (gate). A small impression at one end of the finisher for forming a small projection that can be used to handle those forgings cut off from the forging stock before completion of the forging operations; permits connection between multiple impressions or the forging bar and impression.

Squeezing. Forming under pressure in closed dies.

Stamp (marking). An operation performed to identify the particular forgings as specified or requested by the customer.

Stains (black smut). A product of caustic action on aluminum; sometimes results from inefficient etching operations, hindering the visual inspection of parts. The condition is easily remedied by repeating the etching operations, taking care that the method of stacking and agitation is sufficient to result in complete removal of the etching products.

Standard tolerance. An established tolerance for a certain class of product; this term is preferred to "commercial" or "published" tolerance.

Spheroidizing. A form of annealing consisting of prolonged heating of iron-base alloys at a temperature in the neighborhood of, but generally slightly below, the critical range, usually followed by a relatively slow cooling. Spheroidizing causes the graphite to assume a spheroidal shape, hence the name.

Steam hammer. A type of drop hammer where the ram is raised for each stroke by a double-action steam cylinder and the energy delivered to the workpiece is supplied by the velocity and weight of the ram and attached upper die driven downward by steam pressure. Energy delivered during each stroke may be varied.

Stock. A general term used to refer to a supply of metal in any form or shape and also to an individual piece of metal used to produce a single forging.

Stock marks. In cutting forging stock to specified length for a die-forged part, the ends of the bar always contain surface imperfections caused by the cutting

tool; these are often retained on the surface of the finished part. If pronounced, such marks are removed by light grinding. On parts where repeated indications of stock marks are encountered, efforts are usually made to eliminate them by conditioning the stock ends prior to forging by polishing the cut ends and beveling the edge of the cut.

Stocks. Stocks are tong-like forging instruments that permit the operator to obtain a good hold on the hot metal and manipulate forgings at the hammer.

Straighten. Decrease misalignment between various sections of a forging.

Straighten, coin. A combination coining and straightening operation performed in special cavity dies designed to impart a specific amount of working in specified areas of the forging to relieve stresses set up during heat treatment.

Straighten, die. A straightening operation performed in either a hammer or a press using flat or cavity dies to remove undesired deformation and bring the forging within straightness tolerance.

Straighten, hand. A straightening operation performed on a surface plate to bring a forging within straightness tolerance. Frequently, a bottom die from a set of finish dies is used instead of a surface plate; hand tools used include mallets, sledges, blocks, jacks, and oil gear presses, in addition to regular inspection tools.

Straightening. A finishing operation for correcting misalignment in a forging or between various sections of a forging. Straightening may be done by hand, with simple tools, or in a die in forging equipment.

Stress relieving. A process of reducing residual stresses in a metal object by heating the object to a suitable temperature and holding for a sufficient time. This treatment may be applied to relieve stresses induced by quenching, normalizing, machining, cold working, or welding.

Striking surface. Those areas on the faces of a set of dies that are designed to meet when the upper die and lower die are brought together. Striking surface helps protect impressions from impact shock and aids in maintaining longer die life. Also termed "beating area."

Stripper. A lug or ring on the forging or an impression in the dies for a mechanical upsetter to ensure clamping the piece firmly in the gripper dies.

Structural streak. A streak revealed by etching or anodizing and resulting from structural heterogeneities within the product.

Structure. The size and arrangement of the metal grains in metal.

Sub-sow block (die holder). A block used as an adapter in order to permit the use of forging dies that otherwise would not have sufficient height to be used in the particular unit or to permit the use of dies in a unit where the shank sizes are different.

Suck-in. A defect caused by the "sucking in" of one face of a forging to fill a projection on the opposite side.

Surface peening. Shotblasting to increase the fatigue life of forgings.

Swage. Operation of reducing or changing the cross-sectional area by revolving the stock under fast impact of blows. Finishing tool with concave working surface; useful for rounding out work after its preliminary drawing to size.

Swaging. Reducing the size of the forging stock; to shape metal by causing it to flow in a swage by pressing, rolling, or hammering.

T Shape. Forgings generally in the approximate form of a "T."

Table mill. In ring rolling, a type of ring forging equipment employing multiple mandrels with a common main roll. Usually used in high volume production of small-diameter rolled rings.

Tempering. See *Drawing*.

Template (templet). A gage or pattern made in a die department, usually from sheet steel; used to check dimensions on forgings and as an aid in sinking die impressions in order to correct dimensions.

Tensile properties. The property data obtained from tensile tests on a specimen, including tensile strength, elongation, reduction of area, and yield strength.

Tensile strength. The maximum load per unit of initial cross-sectional area obtained before rupture in a tension test.

Tolerance. Allowable deviation from a nominal or specified dimension; the permissible deviation from the exact dimensions given on the drawing or model, or from a specification for any characteristic.

Tonghold. The portion of the stock by which the operator grips the stock with tongs. A small portion of metal projecting from the forging used to manipulate the piece during the forging operation, usually trimmed off.

Tongs. Metal holder used to handle metal pieces.

Tool steel. A superior grade of steel made primarily for use in tools and dies.

Tooling marks. Dies containing surface imperfections and dies on which some repair work has been done will impart indications on the surface of the forged part; these tooling marks are usually slight rises or depressions in the metal. Light grinding or polishing is used to remove the marks if they seriously affect the appearance of the product.

Tooling pad. See *Chucking lug*.

Tote box. Metal container used to convey forgings to the various processing operations.

Traffic marks. Abrasions that result from metal-to-metal contact and vibration during transit. These abrasions are usually dark in appearance because of the presence of a dark powder consisting of aluminum and aluminum oxide fines produced by the abrasive action of surfaces rubbing together.

Trim. A shearing operation to remove flash along the outer parting line of a blocked or finished forging, performed on a trim press.

Trim and punch. A shearing operation to remove both an inner and an outer section of metal from a blocked or finished forging. A combination of two operations whereby flash and punchout are removed simultaneously. The operation is generally performed on a trim press using a combination trim and punch die.

Trimmer. The dies used to remove the flash or excess stock from the forging.

Trimmer blade. The portion of the trimmers through which the forging is pushed to shear off the flash. The shearing edge may be in more than one plane in order to fit the parting line of the forging.

Trimmer die. The punch-press die used for trimming flash from a forging.

Trimmer punch. The upper portion of the trimmer that comes in contact with the forging and pushes it through the trimmer blades; the lower end of the trimmer punch is generally shaped to fit the surface of the forging against which it pushes.

Trimmers. The combination of trimmer punch, trimmer blades, and perhaps trimmer shoe used to remove flash from a forging.

Trimming. The mechanical shearing of flash or excess material from a forging by use of a trimmer in a trim press. This can be done either hot or cold.

Trimming press. A power press suitable for trimming flash from forgings.

Trimming shoe (trimming chair). The holder used to support the trimmer.

Trip hammer. A small power hammer that delivers blows in rapid succession.

Tryout. Preparatory run to check or test equipment, lubricant, stock, tools, or methods prior to a production run. Production tryout is run with tools previously approved; new dies tryout is run with new tools not previously approved.

Tumbling. The process for removing scale from forgings in a rotating container by means of impact with each other and abrasive particles and small bits of metal. A process for removing scale and roughness from forgings by impact with each other, together with abrasive material in a rotating container.

Twist. See *Bend*.

Type. A small, hardened block machined to the shape of a small portion of the impression and driven into this portion of the impression to determine the shape and dimensions accurately.

U Shape. Forgings generally in the approximate form of a "U."

Ultrasonic testing. A nondestructive test applied to sound-conductive materials having elastic properties for the purpose of locating inhomogeneities or structural discontinuities within a material by means of an ultrasonic beam.

Undercuts. Sections of a forging which, if driven into the impression while the metal is hot, would lock themselves into a die impression and prevent removal of the forging without distortion.

Underfill. A portion of a forging that has insufficient metal to give it the true shape of the impression.

Upend forging. A forging in which the metal is so placed in the die that the direction of the fiber structure is at right angles to the faces of the die.

Upset. Working metal in such a manner that the cross-sectional area of a portion or all of the stock is increased, and length is decreased.

Upset forging. A forging obtained by upset of a suitable length of bar, billet, or bloom; formed by heading or gathering the material by pressure upon hot or cold metal between dies operated in a horizontal plane.

Upsetter (forging machine). A machine, with horizontal action, used for making upset forgings.

Upsetter, mechanical. See *Mechanical upsetter*.

Vent. A small hole provided for escape of the air from a die cavity.

Vent mark. A small protrusion resulting from the entrance of metal into die vent holes.

Warpage. Term generally applied to distortion that results during quenching from the heat-treating temperature; hand straightening, press straightening, or cold restriking is employed, depending on the configuration of the part and the amount of warpage involved. The condition is governed by applicable straightness tolerances; beyond tolerances, warpage is a defect and cause for rejection. The term is not to be confused with "bend" or "twist."

Water stain. A superficial etching of the surface from prolonged contact with moisture in a restricted air space such as between layers of the product. The stain is generally white.

Ways. The fitted V-shaped grooves in the ram and columns of a hammer or press that guide the descent and ascent of the ram.

Web. A relatively flat, thin portion of a forging that effects an interconnection between ribs and bosses. A panel or wall that is generally parallel to the forging plane. See also *Rib*.

Web fire. See *Reducing atmosphere*.

Wide tolerance. Any special tolerance that is wider than "standard."

Y (double). A forging, such as a connecting rod, that is widened at each end. Abbreviation is "DY."

Y shape. Forgings shaped generally in a "Y," such as connecting rods or banjo-shaped parts. Piece where one end requires spreading into a U, V, disk, or similar shape, or a combination of one or more of the these.

Yield point. The load per unit of original cross section at which a marked increase in the deformation of the specimen occurs without increase of load. It is usually calculated from the load determined by the drop of the beam of the testing machine or by the use of dividers. The stress in a material at which there occurs a marked increase in strain without an increase in stress.

Yield strength. Stress corresponding to some fixed permanent deformation, such as 0.1 or 0.2% offset from the modulus slope.

Zyglo®. A method for nondestructive surface inspection of primarily nonmagnetic materials using fluorescent penetrants. Trade name of Magnaflux Corp.

Index

Symbols used after page numbers indicate where the entry is to be found: (F) figure, and (T) table.